John Milton Scudder

The Eclectic Practice in Diseases of Children

John Milton Scudder

The Eclectic Practice in Diseases of Children

ISBN/EAN: 9783337035211

Printed in Europe, USA, Canada, Australia, Japan

Cover: Foto ©berggeist007 / pixelio.de

More available books at **www.hansebooks.com**

THE

ECLECTIC PRACTICE

IN

DISEASES OF CHILDREN.

BY

JOHN M. SCUDDER, M. D.

PROFESSOR OF THE PRINCIPLES AND PRACTICE OF MEDICINE IN THE ECLECTIC MEDICAL
INSTITUTE OF CINCINNATI; AUTHOR OF "A TREATISE ON THE DISEASES OF
WOMEN," "THE ECLECTIC MATERIA MEDICA AND THERAPEUTICS," "THE
ECLECTIC PRACTICE OF MEDICINE," "THE PRINCIPLES OF MED-
ICINE," "SPECIFIC MEDICATION," "SPECIFIC DIAG-
NOSIS," "THE USE OF INHALATIONS," ETC.

REVISED EDITION.

CINCINNATI:
PUBLISHED BY THE AUTHOR.
1881.

PREFACE:

In presenting a new work for approval, the author hopes that it may be as favorably received as the former ones, and that its readers will find it a trustworthy guide in the treatment of the diseases of childhood.

The dedication should have been " to the children of this country," as the endeavor has been to free the practice of medicine from everything harsh and revolting, and to substitute those gentle means and appliances which, while successful in counteracting disease, entail no present or future suffering.

The practice of medicine in the past has been a chapter of horrors, which the truer civilization of the present will not tolerate. " We may endure the sufferings of disease, with some degree of equanimity, but we will not have those sufferings intensified by medicines." This is the feeling of the better class of people, and especially of parents with regard to their children.

If the author has aided to establish a better system of medicine, and the present work fulfills its mission of rendering the treatment of the diseases of childhood certain and pleasant, he will be well repaid for his labor.

It has been the object to make the descriptions succinct and explicit, stating what the author believes to be true, and avoiding theories. Thus, while the entire list of diseases has been studied, the work has been kept in that moderate compass that renders it of easy reference to the busy practitioner.

With regard to the treatment recommended, the author offers the voucher of an extended and successful practice. It varies greatly from the treatment of the standard works of the day, or the teachings of the schools, but upon it he proposes to risk such reputation as he may have made as a teacher or writer.

But few quotations have been made, and in these cases due credit has been given. While acknowledging no special obligations to any writer, the author begs to make that general acknowledgment to all who, writing on medicine, have revealed the germs of new truths, or have been instrumental in freeing old ones from the chaff in which they were enveloped, that they might be seen and applied.

In this revision the author has added the results of the last twelve years' experience. The first part, on infantile therapeurapeutics, has been entirely re-written, and forms a good materia medica. In the third part, the treatment has been mostly re-written, and the reader will find here the most successful treatment of the day. It is small doses of pleasant remedies for direct effect, which we believe will be the practice of the future.

CINCINNATI, June 15th, 1881.

TABLE OF CONTENTS.

PART I.

INFANTILE · THERAPEUTICS.

Gelseminum,   Quinine,   Bromide of Ammonium,
Belladonna,   Opium,   Carbonate of Ammonium,
Rhus.   Phosphorus,   Iodide of Ammonium.
Bryonia,   Camphor,   Ferro cyanuret of Potassium,
Lobelia,   Æther,   Ænothera,
Pulsatilla,   Chloroform,   Asafœtida.
Nux vomica,   Chloral,

Aconite,   Rhus,   Podophyllin,
Veratrum.   Lycopus,   Hamamelis,
Gelseminum,   Lobelia,   Cactus,
Eupatorium,   Digitalis,   Apocynum Canabinum,
Spiritus Ætheris Nitrosi.

PART II.

CARE AND MANAGEMENT OF INFANTS.

PART III.

DISEASES OF CHILDHOOD.

DISEASES OF CHILDREN.

PART I.

INFANTILE THERAPEUTICS.

CHAPTER I.

It is generally admitted that there are sufficient differences in the action of remedies upon the adult and child to demand a careful study of the subject. While all practicing physicians are impressed with this fact, every one will admit the difficulty of making the distinction, so that the student of medicine will readily appreciate it. This is especially true of the older treatment by indirect remedies, which, used as in the diseases of the adult, were very uncertain in their action; but in the special or specific medicines the difference becomes less and less, as we thoroughly understand their action.

These differences may be appropriately classified under the heads of: Their Action upon the Nervous System—upon the Organs of Circulation and the Blood—upon the Digestive Tract, and upon the Excretory Organs.

THE NERVOUS SYSTEM.—The nerve centers of the child are immature and more easily influenced for harm than in the adult. We notice, first, that they are very easily excited (*irritative*), and in consequence there is determination of blood. The immature structures do not resist this morbid influence well, hence we have excitation of function—restlessness, wakefulness, undue expenditure of nerve force, deranged circulation and nutrition, and convulsions. 2

Second, That there is a greater tendency to impairment of circulation (stasis), giving rise to congestion and consequent functional lesions. In this case structural changes take place rapidly, and prove a frequent cause of death.

Not only is there this tendency in disease to affect the nervous system of the child, but many remedies influence the nervous system in these ways. As we progress in the description of remedies, we will have occasion to study this action, and will here merely cite the well-known influence of opium as an example.

This fact, though it leads us to great caution in the use of remedies, and to a careful observance of those symptoms which experience has shown to foreshadow these nervous disturbances, points out the course of therapeutic study likely to prove most valuable. As all processes of life are, to a greater or less extent, under the influence of the nervous system, we may so guide them by medicinal action upon the nerve centers as to remove disease.

This subject should also be studied with reference to infantile *hygiene*. *Rest to the nervous system* is of the highest importance to the young child; especially is this the case, when from disease or other cause there is an imperfect performance of the functions of digestion, nutrition, or secretion.

What I mean by rest here is, an avoidance of all causes of irritation or excitement, whether physical or mental. In a case there will be some more or less easily removed physical irritation, which it is the duty of the physician to point out to the mother, and if necessary to give instructions for proper management. Frequently we find such irritation of the skin, from want of cleanliness, or more frequently from too frequent and hard rubbings with an irritant soap. The most common physical irritation is from confining the child with its clothing, and thus preventing a healthful change of position and the natural movements of its limbs and body. In the proper place I will describe what I deem a proper clothing for the child.

Most persons fail to appreciate that the child is a sentient being, capable of receiving mental impressions and being pleasurably or painfully impressed by them. If I should say that the child is more impressionable than the adult, receives more pleasure, suffers more pain, and that both influence the powers of life more, I would but state the case as it presents itself to those who have most thoroughly studied children. The child is helpless, wholly dependent upon the care of its mother, and grows into conscious life in an atmosphere of love. It responds pleasurably to care for its welfare, and suffers acutely from neglect.

We frequently meet with parents who seem to have no mental balance; now they are the most loving, then they are harsh, passionate, and wholly without reason. In sickness, and when occasion offers to give advice, the injury, both to the physical and mental welfare of the child should be pointed out, and the influence of the physician thrown in favor of a systematic effort for self-government.

In some diseases, rest to the nervous system is absolutely essential to life; in all it is of much importance. We will, therefore, carefully guard our own association with the child, that it may not be a source of irritation, calling the attention of parents to it, and prescribing remedies with reference to it. The good old times of physic, when a child could be thrown upon his back, his nose held, and the nauseous dose forced down his throat, as we would drench a horse, have passed by, for parents have learned quite as soon as physicians that it is not necessary to a cure.

The Circulatory System and Blood.—Derangement of the circulation forms a very important part of most diseases of children. Very slight and temporary diseases present often marked change of pulse, while in the more grave affections it forms the basis of both functional and structural lesions.

But while the circulation is thus easily deranged by causes of disease, it is readily reached and easily amenable to remedies. Possibly there is no part of the practice of medicine that will give more satisfaction than the use of remedies for this purpose.

The blood of the child seems less subject to zymotic influences than the adult. Typhoid and typhus fever are of very rare occurrence, and epidemic dysentery and like affections pass the little ones by. On the contrary, the susceptibility to the eruptive fevers is greatest in early life.

Lesions in blood-making are common in childhood, especially those that form the basis of that protean disease, scrofula; true, the lesion may be primarily of the lymphatic system, (*Principles of Medicine*, p. 236), but it manifests itself in imperfect elaboration of the blood.

Deficient blood-making, though not more common than in the adult, is less easily influenced by medicine. Possibly because the digestive processes can not be stimulated by the same active tonic medicines.

THE DIGESTIVE APPARATUS.—The digestive tract is very susceptible to the action of irritants, which unfavorably influence
or arrest the processes of digestion, giving rise to crude and imperfect products, and deterioration of the blood; not only this,
but the influence upon the sympathetic nervous system is very
marked, and is sometimes extended to the cerebro-spinal centers.
A very common as well as marked example of this is, convulsions from irritation of the stomach or intestinal canal. I have
seen the most severe and persistent convulsions produced by the
administration of some of the vermifuges, and even from castor oil.

There is a less marked influence upon the nervous system from
irritation of the intestinal canal, though frequently not less injurious. I speak of that which manifests itself in the form of
nervous irritability, restlessness, and sleeplessness, which, as we
have already seen, are to be carefully avoided.

As the vegetative functions are in a very active condition, the
child suffers more from deprivation of food than the adult. It
has always seemed to me that the most serious error in the treatment of diseases of children, was the use of such means as would
impair the power of the stomach to digest food, and destroy the
appetite.   Making no provision for a supply of nutritive material,
the waste of the body would be stimulated to excess, and the
child's life rapidly exhausted; using a homely simile, " it was
burning the candle at both ends."

THE EXCRETORY APPARATUS.—As there is less waste of tissue,
the excretory organs are not so active as in the adult, neither do
they possess the power of sustained activity under the influence
of medicines. While it will be found quite as easy to temporarily
increase excretion, it requires much care to sustain these processes.
Over stimulation frequently leads to such exhaustion that excretion
is almost wholly arrested, and the little patient dies, poisoned by
its own excreta.

I do not know of any one point in infantile therapeutics that
is more essential than this, in the more grave diseases. Given, a
disease that will require one or two weeks for its progress, and
we may see how essential it is that the daily waste of tissue
should be promptly removed. And we can never afford to hazard
the successful termination of such disease, for the uncertainty of
aborting it by over stimulation of the excretory organs.

DIRECT MEDICATION.

In the treatment of the diseases of childhood especially, I prefer direct or specific medication, as I do in the treatment of the adult; but I find more directness of action in the case of the child, possibly because it is more free from the effects of a false civilization.

The old methods of indirect medication may be employed when nothing better is known, but in the majority of cases the following rule will be found to yield the greatest success: "Never give a dose of medicine unless you see clearly the indication for that particular medicine, and have a reasonable certainty that its action will be beneficial." Of course, this cuts off hap-hazard medication, or treating the name of a disease, and if persistently followed, will lessen the amount of medicine administered several hundred per cent.

We have not yet specifics for every disease, possibly for no disease in its totality, but we have specifics for pathological conditions, and, properly employed in the order in which these pathological conditions take precedence, they offer a direct and very certain treatment for diseases.

In the employment of remedies we find greater success from doing *one thing at a time*, and, as a general rule, from the use of single remedies or very simple combinations. If we properly appreciate the different parts of a disease, we will find that they hold the relation to one another of cause and effect; that there are some primary, and others secondary and dependent upon the first; and in the use of remedies we give those first that meet the primary lesion, and follow with such as reach the indications of cure in the order named.

In the treatment of diseases of children, as elsewhere, there is the constant tendency to view the sum of pathological processes as a unit, and to meet this at once. Though a physician knows well, if he reflects, that recovery will occupy days, he gives remedies to accomplish it in hours. All the remedies that would seem to be indicated in the whole treatment are given the first and each succeeding day.

The true method is, to make a thorough analysis of the disease, and separate it into its component parts; determining also the order of these. Then remedies are selected, not with reference to the name, but with their application to correct a clearly understood pathological change of function or structure.

As an example, I may take inflammmation of the lungs. On analysis we find it to consist of an accelerated circulation, increased temperature, arrested excretion, difficult respiration, and cough, and more or less excitation of the nervous system. In addition to this, there is impairment of the digestive functions, and a gradually increasing lesion of the blood, and the processes of waste and nutrition. Such analysis tells us what the disease really is, and points out a specific or direct treatment.

The accelerated circulation is controlled by the use of veratrum, to which gelseminum is added if there is much nervous irritation. The action of these is aided by the general bath, hot foot-bath, and a mush-jacket to the chest, which also relieves the irritation of the respiratory organs. Following this, if necessary, gentle diaphoretics, diuretics, and laxatives are employed to stimulate secretion. If the inflammatory process does not yield, the use of inhalations of the vapor of water, or a specific, like small doses of ipecacuanha, completes the cure. In the meanwhile any gastro-intestinal irritation, or other lesion, passes away as the circulation becomes normal and secretion is established.

Very many times it requires but the remedies that control the circulation, for with this the entire series of morbid phenomena gradually disappears.

CHAPTER II.

FORM IN WHICH REMEDIES SHOULD BE ADMINISTERED

SUCCESS in the treatment of diseases of children will depend much upon the form in which the remedies are administered. The first object is to have the proper medicine; the second, is to introduce it into the circulation without exciting any morbid process. It matters little with what skill the disease is diagnosed, or the remedies selected, if they fail to gain entrance into the blood. I doubt not that any physician can call to mind cases which will verify this statement.

The form of a medicine, then, is very important, and deserves our careful consideration. The remedy should not be objectionable to the taste, if it is possible to avoid it; for the unpleasant taste excites disgust, followed by more or less nausea, and during this but little if any absorption takes place. If now such remedy is repeated every one, two, or three hours, the disgust and nausea may be rendered persistent, and the remedies fail to produce the desired effect, because they do not get into the circulation where they may act.

Many remedies undergo decomposition if retained in the stomach any considerable time, and either lose or change their medical properties; hence, another reason for proper attention to the form, as well as some other accessories to their proper administration.

Next to having the remedy in a pleasant form is, to have it in solution or readily soluble in the gastric fluids. To gain entrance into the blood it must be in perfect solution, and as time is an important element in treatment, it is well to see that the remedy has these desirable qualities given it before it is introduced into the stomach.

I prefer tinctures or solutions, and usually administer them with water. A very good method is to prepare the remedy in a glass of water, so that the proper quantity will be contained in a teaspoonful. I find that the slight taste is not objectionable, and to the child it does not look like medicine. If the child is old enough to observe, I prepare it in his presence, showing the bottles, the steps of the process, and then have him taste it and have his opinion of the result. In this way I find no difficulty in competing with homœopathy as to the pleasantness of medicine.

The mixing of medicines in compounds, using some syrup as a vehicle, and attempting to disguise the taste with peppermint, cinnamon, or something worse, is too disgusting now for the well-trained adult, and I do not wonder that it is objectionable to the child. I am very free to confess, that rather than have my children dosed with nauseous medicines, however good might be the physician, I should take the infinitesimal nothing of the homœopath, and trust to nature for a cure.

Powders are generally more objectionable than mixtures, not only on account of their taste, but the difficulty in swallowing them, and their unpleasant influence in the stomach. They are also frequently less soluble, and are absorbed with greater difficulty.

Syrups are not so objectionable in taste, yet frequently the sugar undergoes decomposition in the stomach, producing irritation. They are, however, very uncertain in their composition, and hence should be dispensed with.

Occasionally we may employ the more powerful medicines by the use of homœopathic pellets. They are easily prepared, but the dose is minute and not definite in quantity. I have used aconite, veratrum, gelseminum, strychnia, morphia, and others in this form.

Occasionally we may administer a remedy endermically when it can not be given by mouth. I have thus used quinia in intermittents and remittents, morphia, atropia, aconite, and some others. I do not recommend it however with any but the first two, and the use of these will hereafter be described.

THE DOSE OF MEDICINE.

The dose of medicine should be as small as will give the desired result. The harsh and immediate action of medicine is not usually desirable, but rather that gentle influence which is in the direction of healthy action, and which may appropriately be termed physiological. It is now generally admitted that the doses of medicine have been much too large, and that much injury has been occasioned by this over medication ; but there is no necessity of going to the opposite extreme of homœopathy.

In this, as in other things, there is a happy mean, which when found gives the best results. The doses recommended in this work are the maximum. The minimum, though not infinitesimal, would generally be regarded as too small to produce an effect. The human body, however, is a very delicate mechanism, and its processes most easily influenced, and we may readily believe that it may be acted upon by anything of positive quality and appreciable quantity.

A considerable experience in the treatment of diseases of children, has convinced me that the various processes of life are influenced with more certainty, and with far less liability of injury, through the sympathetic system of nerves, and that a very gradual influence is far more desirable than a speedy one.

Another very important point in infantile therapeutics is, that the medicinal influence be continued until the desired result is obtained. In other words, that, as the ultimate object is to be slowly obtained, and that by repetitions of the remedy, gradually increasing its influence from dose to dose, the doses should be so frequently repeated as not to lose the influence of one before another is given.

Medicine should never be left for the sick to be given in drops, especially with children, who would suffer most from deviation in this respect. Let the remedy be added to some vehicle—water is the best—so that the dose will be a teaspoonful. I have had one patient poisoned from neglect of this rule, and others injured, hence I place much stress upon it. There is, however, another reason for such dilution which may have some bearing : and that is, the medicine is placed in a form that it can be more readily absorbed.

CLASSIFICATION OF REMEDIES.

The classification of remedies is a work of no little difficulty, if we are to study them according to their specific action, which is the only method that will advance the practice of medicine. In the treatment of diseases of children, at least, we will abandon the old methods of large doses with their poisonous effects, and substitute the small dose of pleasant medicine, kindly in its influence upon the body, and directly antagonistic to the processes of disease.

The old practice never gave good success. It added additional pangs to the suffering, new wrongs to those existing, prolonging the duration of disease and increasing its mortality. A purely expectant practice has proven far more successful; good nursing with diet and rest has given much better success; the water-cure has yielded better results, and homœopathy has proven an incalculable blessing to the suffering little ones.

Whilst we administer remedies for the most part according to specific indications—special symptoms pointing out the remedies —some are given for their well-known physiological action, and because they influence special parts in a well-known way. A large number of our remedies may be classified according to their action upon special parts or functions, and I can not but think that such grouping will be of advantage.

We will, therefore, study our remedies as—1. Those which influence the nervous system ; 2. Those which influence the circulation ; 3. Those which influence the temperature ; 4. Those which influence the respiratory apparatus ; 5. Those which influence the digestive apparatus ; 6. Those which influence the urinary apparatus and its secretion; 7. Those which influence the skin and its function; 8. Those which oppose the malarial poison—antiperiodics; 9. Those which oppose the process of zymosis—anti-zymotics; 10. Those which oppose the process of sepsis—anti-septics; 11. Those which oppose rheumatism—anti-rheumatics; 12. Those which oppose the poison of erysipelous—anti-erysipelatous; 13. Those which oppose the syphilitic poison—anti-syphilitic, and 14. Agents that promote the processes of nutrition and increase life.

REMEDIES WHICH INFLUENCE THE NERVOUS SYSTEM.

Gelseminum,	Quinine,	Bromide of Ammonium,
Belladonna,	Opium,	Carbonate of Ammonia,
Rhus,	Phosphorus,	Iodide of Ammonium,
Bryonia,	Camphor,	Ferro-cyanuret of Potassium,
Lobelia,	Æther,	Œnothera,
Pulsatilla,	Chloroform,	Asafœtida
Nux vomica,	Chloral,	

As has been already remarked, " the nerve centers of the child are immature and more easily influenced for harm than in the adult." It might also be stated that the nervous system of the child suffers more from disease, and remedies influencing it assume a more important place in infantile than in adult therapeutics. It would sometimes seem as if the old doctrine—" that all diseases have their origin in wrongs of innevration "—was more than truth; at least, our acting upon it would be more nearly right than he who is always " acting upon the bowels."

The influence of the nervous system, even the brain, upon the processes of vegetative life, is much more marked than in the adult, and the relief of a wrong of innevration will many times occur, when remedies directed to the disease as classified, have no effect. Instances will continually come up in practice, when single remedies like Rhus, Gelseminum, Belladonna, Bryonia, etc., are sufficient for the cure of seemingly complex diseases.

The evidences of disease here will soon become plain to the careful observer, as he studies the position of the body, its movements and expression, the expression of the face, its color, the expression of the eyes, color, dryness and moisture, dilation or contraction of pupil, temperature, and the frontal, orbital and facial expression of pain. We do not claim that this is to be learned in a day or a year, but it will be learned in a satisfactory degree by him who gives it his attention.

If now, we take some common examples, we may see the truth of the above. For instance, here is a patient (a grave case of disease) who is restless, starts in sleep, cries out suddenly and shrilly; we look at the face, which is contracted, especially about the eyes or base of the brain; the eyes are bright, and evidently the little patient suffers pain, as shown by the contraction of the frontal and orbital muscles. We give Rhus in small doses, and all the unpleasant symptoms fade away. Here is another, suffering from fever or inflammation, which shows a flushed face, bright eyes and contracted pupils, with increased heat; there is restlessness, sleeplessness, it may be threatened convulsions, and thinking of the usual means, it is a bad case. We admisister Gelseminum with Aconite, and there is speedy and marked amendment. In still another, we find the patient dull, sleepy, comatose; the eyes dull, pupil dilated, and it may be a dull or dusky coloration of skin. We give Belladonna and these unpleasant symptoms soon pass away, and the patient convalesces. In another, the twitching of the facial muscles, the sudden movement of the head, and the involuntary movement of the extremities indicate the tendency to convulsions; the face is pale, brow contracted, circulation feeble, and we administer bromide of ammonium with a certainty of removing the unpleasant symptoms.

GELSEMINUM. *Specific Indications.*—The face is flushed, eyes bright, pupils contracted, temperature increased, nervous system excited, patient restless and wakeful.

Dose.—℞ Tinct. Gelseminum, gtt. v. to gtt. xx., water ℥iv.; a teaspoonful every hour. When the irritation is very great and the patient is threatened with convulsions the dose can be increased.

Homœopathic Indications, (3d to 6th decimal).—Bad effects from fright and fear, fright causes diarrhœa; vertigo with loss of sight, chilliness, accelerated pulse, double vision; great heaviness of the eyelids; paralysis of the eyelids and other parts; pupils dilated; dull expression of countenance; difficult articulation and deglutition; neuralgia, sharp, shooting pains through the face, eyes and head. Scarlet fever with great nervous excitement, tonsils swollen and very red, pulse rapid.—*Ehrmann.*

We use a saturated tincture of the green root; all other preparations are imperfect or wholly worthless for the purposes for which we employ it. The difficulty in describing this, as well as some other remedies, so that the practitioner can employ them with success is, the very great difference in the quality of the article, as sold by different druggists.

Gelseminum exerts an influence on the circulation like veratrum and aconite, but less marked, and not so certain, and it is not for this purpose that I employ it. It has a specific action upon the cerebro-spinal nerve centers, relieving irritation and determination of blood to these parts. This is a use that renders the Gelseminum a remedy of great value in treating the diseases of childhood. The action is so certain and definite, that we prescribe the remedy with great satisfaction. Given, a case of fever or inflammation, in which the child is uneasy, irritable and restless, the face is flushed, the eyes bright, the pupils contracted, and head hot, Gelseminum removes in a few hours all the unpleasant symptoms.

It is also one of our most valued antispasmodics. Its use is in those cases in which the convulsions are due to irritation of the cerebro-spinal centers. In a case of acute disease, the child presenting those involuntary twitchings of the muscles which indicates the approach of convulsions, there being the flushed face, bright eyes, and irritability, Gelseminum wards off the approaching spasms, and relieves the child of them when they have appeared.

Gelseminum also has an important action on the urinary organs, which it is well to notice here. Occasionally a case of dysuria will be met with, consequent upon irritation of the neck of the bladder or spasmodic stricture of the urethra. Gelseminum will overcome the obstruction and cause a flow of urine, in a majority of cases.

BELLADONNA. *Specific Indications.*—The patient is dull, drowsy, comatose, dull heavy headache, the face dull and expressionless, the eyes dull, pupils dilated or immobile, capillary circulation sluggish, as marked by deep color of skin or redness that is effaced by pressure of the finger, which leaves a somewhat persistent white line. Passage of large quantities of limpid urine, involuntary passages of urine.

Dose.—℞ Tinct. Belladonna, gtt. v. to gtt. x., water ℨiv.; a teaspoonful every one to three hours.

Homœopathic Indications, (3d to 6th decimal).—Throbbing headache, with congestion of blood to the head, vertigo, pulsation of the carotids, worse from motion, light and noise are intolerable; furious delirium, illusions, hallucinations, with flushed face and redness of the eyes; wishes to strike, bite or shriek, pupils dilated, eyes brilliant and staring; smooth, shining redness of the skin, hot, burning and itching; sore throat, with sensation of a lump on swallowing, and great dryness of the throat; involuntary discharges of feces and urine from paralysis of the sphincters; pains come on suddenly, and leave just as suddenly. — *Ehrmann.*

Belladonna causes contraction of capillary blood-vessels, doubtless by stimulation through the sympathetic nervous system. Hence it is the remedy for congestion. Stasis of blood, or congestion, occurs in dilated capillary blood-vessels; indeed this condition of the capillaries is the cause of congestion. This action of Belladonna was first observed by Brown-Sequard in some of his physiological investigations.

While this influence is exerted upon the entire circulation, it is most marked upon the circulation of the nerve centers. Thus we regard Belladonna as specific to congestion of the brain and spinal cord, and the dullness and coma which are symptomatic of it. In this respect Belladonna is the opposite of Gelseminum, which removes irritation and lessens determination of blood, the condition being an active one.

RHUS TOXICODENDRON. *Specific Indications.*—The patient is restless, starts suddenly in sleep, sudden sharp cry, contraction of muscles about eyes and at base of brain, sharp frontal pain, especially in left orbit, small pulse with sharp stroke, tongue shows small red papillæ at the tip, erythematous flushing of surface, small vesiculæ about mouth and sometimes upon the skin.

Dose.—℞ Tinct. Rhus, gtt. v., water ℥iv.; a teaspoonful every hour.

Homœopathic Indications, (3d to 6th decimal).—Great debility, feeling of soreness and lameness of the muscles, especially when at rest. Rheumatic pains, drawing and tearing in the muscles and joints. Stiffness of the joints, relieved by continuous motion. Paralysis of the limbs. Bad effects from getting wet. Putrid taste in the mouth, metallic taste. Very restless sleep before midnight. Violent stretching and yawning. Diarrhœa, stools watery or of bloody mucus, frothy or white. Involuntary stools at night while asleep. Dysentery with nausea and tenesmus. Involuntary discharge of urine. Typhoid fever, dull stupid expression of the face, muttering delirium. Tongue dry, red, smooth. Restlessness. Involuntary passage of stools and urine, great exhaustion. Erysipelas with intense itching and burning. Scarlet fever or small pox before the eruption itches a great deal, with restlessness. Pustulous eruptions.—*Ehrmann.*

This has a very extended use in the treatment of diseases of children. Commencing with the cases in which cerebral disease is a prominent complication, or indeed the principal element of disease, we find that this remedy relieves the excitement, promotes normal innervation through the sympathetic, and thus restores the circulation, waste and nutrition of the brain. It not only releives the brain and spinal cord, but it exerts a marked influence upon the circulation of blood, upon the temperature, upon waste and secretion, and upon nutrition. It presents us one of the most marked examples of the truth of the therapeutic maxim, " where a remedy is specially indicated, it will favorably influence every function of life, and sometimes do every thing that needs be done to effect a cure."

Thus a malarial fever having the pronounced indications for Rhus will be cured by it better than by Quinine. An inflammation of the eyes, of the lungs, of the bowels, of the urinary apparatus, or of any part, yields readily to Rhus, when it is specially indicated.

It is a remedy frequently demanded in erysipelas and erysipeloid diseases, in zymotic and typhoid fevers, and in many endemic and epidemic diseases of children. Its use in the eruptive fevers, and in diseases of the skin will be named hereafter.

BRYONIA. *Specific Indications.* Headache in right side, with flushed right cheek; pain in serous cavities, with tension of muscles, and moderate tenderness on pressure; sharp, lancinating pain, with tension of tissues; hacking cough with pain; dry skin, sensitive; hard pulse, with moderate fullness, sometimes corded; rheumatic pain.

Dose.—℞ Tinct. Bryonia gtt. v. to gtt. x., water ℥iv.; a teaspoonful every hour.

Homœopathic Indications (3d to 6th decimal).—In rheumatism and other complaints, marked aggravation on the slightest motion, feels best when quiet; typhoid fever, with dry mouth and lips, tongue is dry, rough, and cracked, and of a dark brown color;

delirium at night of the business of the day; patient wants to go home; urine dark with pinkish sediment; pneumonia and pleurisy, with sharp stitching pains on motion or breathing deeply; cough worse in warm room.—*Ehrmann.*

Bryonia will be found a prominent anti-rheumatic, and also a remedy for pain, especially when the pain is increased by pressure. . Physiologically it has been found to exert a special influence upon serous membranes, and when our little patient complains of pain in the articulations, abdominal pain with tenderness, pleuritic pain, headache, with tenderness of the eyes and temporal regions, tensive pain in the ears, we think of this remedy.

It is a prominent remedy in disease of the respiratory apparatus —the cough irritative, hacking or rasping, pleuritic pain, soreness as if the parts were bruised, or soreness and tenderness of larynx and supra-sternal notch. It does not matter whether the disease is pneumonia or bronchitis, the Bryonia acts well.

I wish especially to call attention to the abdominal pain and tenderness in typho-malarial fever, and in some zymotic diseases, as calling for this remedy. It may also be associated with Ipecac or Euphorbia in cholera infantum, when there is abdominal tension and tenderness, or swelling and pain of articulations.

BROMIDE OF AMMONIUM. *Specific Indications.*—Sudden movements of body or extremities, jerking of the tendons, twitching of the facial muscles, want of accommodatoin in the eyes, eyes turned upward; face unusually pale, and pulse small, though I should not be guided by this.

Dose.—For a child one to two years of age, I would prescribe, ʒj. to ʒij. to water ʒiv.; a teaspoonful sufficiently often to control the symptoms and give rest.

I employ bromide of ammonium to relieve the irritation of the brain and spinal cord, which gives rise to convulsions. It may be used to prevent the occurrence of convulsions when threatened, or to prevent their recurrence when arrested by other means. Regarding the remedy as a stimulant to the nerve centers, we would use it when there was evidence of an enfeebled circulation of blood to the brain.

There are a few cases of whooping cough in which bromide of ammonium will be found curative. In these there is an epileptiform movement of muscles of the extremities associated with the cough, and even the convulsive movements of the chest may be distinct.

When convulsive disease has assumed a chronic form, as in epilepsy, I regard the bromide of ammonium as one of our most certain remedies. Its influence in childhood is more marked ·than in the adult. And its curative action is of course greater at the commencement of the disease than when it has continued for several years. I have employed it in a large number of cases of epilepsy, with very good success, a permanent cure resulting in a fair proportion.

PULSATILLA. *Specific Indications.*—The child is restless, weary, cries frequently, sobbing even in sleep, face pale and expressionless, pulse small and feeble.

Dose.—℞ Tinct. Pulsatilla gtt. v., water ℥iv., a teaspoonful every three or four hours.

Homœopathic Indications, (3d to 6th decimal).—Mild, yielding disposition, with inclination to shed tears. Vertigo when stooping or lifting up the eyes. The pains of pulsatilla constantly change their position, flying from one part to another. Painful inflammation of the eyes, styes, lachrymation in the wind. Dryness, burning and itching of the eyes and lids, with inclination to wipe something away. Fistula lachrymalis. Obscurations of the cornea. Otalgia, darting, tearing pains. Discharge of mucus or pus from the ear. Measles with loose, rattling cough, photophobia, thick and yellowish coryza.--*Ehrmann.*

LOBELIA. *Specific Indications.*—The pulse is full and oppressed, breathing difficult, rattling of mucus in chest, dull expressionless face. Dull eyes, somewhat swollen, threatened convulsions, restlessness with uneasiness of bowels.

Dose.—The dose will vary from the usual gtt. v. to water ℥iv., a teaspoonful every hour, to that which will prove nauseant or emetic. To relieve restlessness and pain, and improve the circulation and respiration the old prescription—℞ Tinct. Lobelia Sem. ℥j., Comp. Spirits Lavender ℥iij., Simple Syrup ℥iss.—will be found very good.

Homœopathic Indications, (3d to 6th decimal).—Nausea and vomiting, with flow of saliva; sensation of a lump in the throat, impeding swallowing; asthma with burning feeling in the chest and tightness; urine deep red color.—*Ehrmann.*

Many of our readers will recall the old treatment of convulsions in which Lobelia played so important a part as an antispasmodic. Compound Tincture of Lobelia and Capsicum (King's antispasmodic) was advised in all cases, and physicians would often wonder why, when the little fellows were so thoroughly drenched with this nastiness, the convulsions would sometimes persist even to a fatal termination. It was the old error of prescribing for names instead of conditions. The Lobelia is a powerful antispasmodic, the conditions being as above named— an oppressed circulation. It is a stimulant to the brain and spinal cord, to the respiratory system, and to the ganglionic nerves. If the condition of the nervous system in convulsions is such as to require a stimulant, then Lobelia is a remedy.

To quiet irritation of the nerve centers and give rest, the remedy must be given in doses less than nauseant, and, as above remarked, it may be combined with an aromatic, or alternated with Aconite or Nux. It also relieves irritation of the stomach upon which the restlessness many times depends.

Nux Vomica. The child is restless, draws up its legs, contorts its body, screams violently, wakes suddenly from sleep with crying, face flushes—intestinal irritation, colic, increased nervous sensibility. The face is pallid, sallow, yellowness about mouth, tongue pallid, nausea and vomiting, child uncomfortable, restless, sleepless, pallid, expressionless face and tongue, food distresses it, urine free but high colored, colors the napkins.

Dose.—R Tinct. Nux gtt. j. to gtt. iij., water ℥iv.; from one-fourth to one teaspoonful every half-hour or hour.

Homœopathic Indications, (3d to 6th decimal).—Irritable; inclined to find fault, disinclined to talk; wants to be let alone; nausea and vomiting; constipation, large difficult stools; uneasy feeling as if the bowels ought to move; ineffectual efforts to pass the feces; painful hemorrhoids; loss of appetite; restless sleep, wakes at 3 A. M.; eructations and heart-burn; violent hiccough; colic of whisky and coffee drinkers; can not bear the clothing tight around the waist; incarcerated hernia; dysentery, stools small and frequent, with ineffectual urging; retention of urine; suitable to persons of sedentary habits.—Ehrmann.

Nux undoubtedly relieves irritation of the brain and spinal cord, when the circulation to these centers is enfeebled. One

3

might think, from its kindly action upon the intestinal canal and associated viscera, that the relief depends wholly upon this, but I am satisfied that it is not so. When the face is pallid and the pulse feeble, I have seen it relieve the patient from threatened convulsions, give rest and good sleep.

In infantile paralysis, imperfect movement, and want of development of the extremities, it is a remedy to be thought of if the circulation is feeble. When the respiratory movements are feeble, in malarial disease, I would suggest Nux. It is not so important in retention of urine as santonine, but it may occasionally be used with advantage.

It is the tonic of childhood, or rather the stomachic; for it is kindly received by the stomach and is not as objectionable to the taste as the most of bitters, while it increases the appetite and the power to digest food.

When there is great depression of the nervous system, without fever or inflammation, it may be used with advantage. As an adjuvant to quinia in periodic disease, especially in intermittent fevers, strychnia will serve a good purpose, and in some old cases, where quinia has been used to excess, we may arrest the periodic disease by its use.

QUINIA SULPHAS. *Specific Indications.*—There is distinct periodicity in the disease. When administered the pulse should be soft, the skin soft or inclined to moisture, the tongue moist and cleaning, and the nervous system moderately free from irritation.

Dose.—The antiperiodic dose of quinine may be stated at one grain for each year, a grain being the dose for a child six months old. This proportion will continue up to the sixth, eighth or tenth year, when it will stop until after puberty. In many cases the antiperiodic influence of quinine may be obtained from a minute dose, or, as we frequently use it, from its application to the skin by inunction.

Homœopathic Indications, (3d to 6th decimal).—Great debility resulting from loss of fluids, as in diarrhœa, hemorrhage, etc.; ailments which have a marked periodicity; metorrhagia, with paleness of the face and coldness of the skin; pulse imperceptible; ringing in the ears; diarrhœa, painless and debilitating, stools undigested, worse after eating or at night; intermittent fever, followed by profuse and debilitating perspiration; sensitive to the least draft of air.—*Erhmann.*

Many wrongs of innervation will be found to have distinct periodicity, whether they take the form of pain, restlessness, sleeplessness or tendency to convulsive action. In malarial regions the practitioner is continually on the watch for this, and even an obscure periodicity is thought to demand quinine, and its right use is followed by most satisfactory results.

Quinine has been regarded as a tonic and a nerve stimulant, but these actions are doubtful unless there is the specific indication. The inunction of quinine is frequently followed by increased appetite and digestion, a restful nervous system and better excretion, and patients recover rapidly from cholera infantum, slow forms of inflammation and obscure diseases, but I think in each of these cases a careful examination would determine the malarial influence and periodicity.

ERGOT. *Specific Indications.*—Dullness, stupor, dull eyes, expressionless face, pulse slow and oppressed, tendency to congestion, hemorrhages.

Dose.—℞ Tinct. Ergot gtt. v., water ʒiv.; a teaspoonful every hour to three hours.

Homœopathic Indications, (3d to 6th decimal).—Great anxiety, fear of death. Wild, staring look, distortion of the eyes, vision obscured. Sunken countenance, eyes sunken. Spasmodic distortion of the mouth and lips. Female complaints, especially in thin individuals. Vomiting of bile, with great weakness. Dry gangrene, beginning in the toes. Very debilitating diarrhœa. Cholera.—*Ehrmann.*

The influence of Ergot upon the system is very similar to Belladonna, and as there is no other place where it can be more properly described, we will place it here. I have employed it in cases where there was an oppressed condition of the nervous system, with a tendency to coma, a labored respiration, and full oppressed pulse. In those cases in which, with such symptoms, there are convulsions, it will be found particularly useful. I should prefer Belladonna in febrile diseases; Ergot is non-febrile.

IODIDE OF AMMONIUM. *Specific Indications.*—Headache with dizziness, unsteadiness of walk, feeble or sluggish circulation.

Dose.—℞ Iodide of Ammonium grs. x. to ʒss., water ʒiv.; a teaspoonful four times a day.

The indications for this remedy are occasionally met with in diseases of children. Sometimes the lesion of brain is purely functional, but at others it will be found strumous or syphilitic. In the latter cases, I employ this and the iodide of potassium, where the tongue is broad and pallid ; Donovan's solution when it is small and red.

CAMPHOR. *Specific Indications.*—Pallid, expressionless face, pain, sleeplessness, with tendency to exhaustive discharges.

Dose.—℞ Tinct. Camphor gtt. v., water ℥iv.; a teaspoonful frequently repeated.

Homœopathic Indications, (3d to 6th decimal).—Coldness of the skin, at the same time patient does not wish to be covered. Sudden prostration with diarrhœa. Retention of urine, with constant pressure on the bladder and desire to urinate. Pulse small, weak, and slow.—*Ehrmann.*

ÆTHER. *Specific Indications.*—Pain, or convulsive action, the face being pallid and cool, and circulation feeble. It is a very good remedy for the relief of headache from an enfeebled circulation.

Dose.—℞ Æther ℥j., simple syrup ℥j.; ten drops to ℥ss. may be given with a little water. As an anæsthetic it may be employed with a napkin in the usual way of administering chloroform.

With the indications named it will sometimes promptly relieve pain, give rest and sleep. It will also control convulsions, though the dose for internal administration is usually greater than that named. It is also sedative at least to the extent of lessening the frequency of the circulation.

It is now regarded as a safer anæsthetic than chloroform. and many physicians use it exclusively. Some employ it with an equal amount of chloroform, its action being quicker. It may be used by inhalation to relieve pain, arrest convulsive action, and prevent its recurrence, as well as for complete anæsthesia when surgical operations are to be performed. A spray of æther is used to produce local anæsthesia, in minor operations, also as a refrigerant in cases of inflammation.

CHLOROFORM. *Specific Indications.*— Pain. Convulsions. Gall stones. Surgical operations.

We prefer chloroform as an anæsthetic because of its speedier action, its more agreeable effect on the patient, and the greater case of its administration. We have used it for thirty years, without having a fatal case, and think that with care it is one of the safest remedies. Who can say as much for any of the agents in common use. Say Opium, Aconite, Veratrum, Gelseminum. The first I will guarantee has killed thousands, where one has been lost by chloroform.

Chloroform is administered as an anæsthetic by using a folded napkin or handkerchief, cupping it, and sprinkling the fluid in the cavity. This is held over the nose in such manner that the vapor may be inhaled with a sufficient quantity of air. Especial attention is given to the respiration, which, if free, assures us that the patient is in no danger. If it becomes slow, difficult, or ceases, the inhalation is suspended, the child is turned quickly, and two or three smart blows on the chest or buttocks restores the respiratory function.

We use chloroform by inhalation to relieve intense pain, and to arrest convulsions. Of course it is not often demanded for the first, and its use in the second case is frequently temporary—we control the convulsions for the time being, until the indicated remedy can be given, and has time to act. This is much better practice than to allow the convulsive action to continue whilst we are vainly striving to get a medicinal action through the stomach.

One of the most powerful antispasmodics at our command is found in chloroform. In the majority of cases it will be found safe as well as certain, but there is a class in which I think its use dangerous. There are cases in which there is congestion of the brain, and especially an impairment of innervation through the sympathetic nervous system, and a sluggish general circulation, with tendency to congestion of all important organs. It is somewhat difficult to diagnose such a condition ; but the beginner will do well to reject the remedy when the breathing is labored, the pulse not much increased in frequency, when the eyes are congested and the lips present a continued dusky or purplish appearance.

Chloroform has been administered in very large doses with safety, though I do not think they should be recommended. As much as a teaspoonful has been given at once to a child two years of age, and repeated. A half teaspoonful is quite frequently named as a dose. I think that ten drops will usually produce the effect desired, and order the vehicle in such proportion as to give this quantity in a teaspoonful. Simple syrup, glycerine, or mucilage, are good vehicles for its administration.

CHLORAL HYDRATE. *Specific Indications.*—To relieve pain, promote sleep, arrest convulsions, and relieve the cough of nervous irritation. It is contra-indicated when the circulation is sluggish, when there is atendency to congestion, pallid bluish lips, pale leaden tongue, dull eyes, and enfeebled respiration.

Dose.—The dose will vary from one-half to five grains, for a child two years of age, and it may be given with syrup and water, as—℞ Chloral grs. xx., simple syrup ℥ss. water ℥iss.; dose a teaspoonful.

The element of danger in the administration of Chloral must never be overlooked. Adults have died from a dose of but ten grains, whilst as much as three hundred grains have been taken with safety. If, however, the rule given above is observed, I think its use will be safe. It must be regarded as a temporary remedy only, and when the present emergency is passed it must be suspended for permanent remedies.

OPIUM. *Specific Indications.*—The pulse is small and open, waves short, the face pallid, eyes dull, pain, sleeplessness.

Homœopathic Indications, (3d to 6th decimal).—Mania; delirium tremens; frightful or pleasing visions, alternating with stupor; steady stertorous breathing, with half open eyes; bad effects of fright; congestion of blood to the brain, with strong pulsations; staring look, glassy immovable eyes; apoplexy with deep snoring breathing, mouth open; paralysis without pain; trembling and twitching of the limbs; constipation, stools in round, black, hard balls; all ailments accompanied by sopor.— *Ehrmann.*

Opium is a remedy that should be employed with very great care in diseases of children, though it need not be discarded. There are some cases, in which a judicious use of opium will calm irritation of the nervous system when all other means have failed, and in which this influence is essential to the preservation of life.

The danger from the use of opium is, that its influence on the brain be such as to impair its circulation and nutrition. It first stimulates the nerve tissue and increases the circulation to it; and, secondly, exhausts the nerve tissue and causes congestion.

The safest use of opium is in small doses to obtain its stimulant action, the remedy to be frequently repeated to continue the influence. Occasionally a case occurs in which the little patient is very much exhausted, and suffers pain from this cause, there being neither fever nor inflammation. In such cases opium may be used with advantage. The only trouble in such cases as it might be used with benefit is, that it tends to arrest secretion, and if arrested we have again a return of blood poisoning.

Occasionally opium is given in the form of some soothing syrup, cordial, or other preparation to quiet children, until we have a condition of chronic poisoning. The symptoms are very marked—general emaciation, languor, a withered, sallow countenance, red and swollen eyelids, derangement of the digestive organs with loss of appetite, and constipation of the bowels, with white stools.

For young children the safest preparation is the *compound powder of ipecac and opium* of our Dispensatory. I generally add five grains to an ounce of water, sweeten with sugar, and have it given in doses of half a teaspoonful frequently, until the patient is relieved.

AMMONIUM CARB. *Specific Indications.*—The pulse is feeble, circulation to the surface imperfect, skin pallid or dusky, respiration difficult, restless and sleepless.

Homœopathic Indications, (3d to 6th decimal).—Suitable for weak and nervous persons of sedentary habits. Scarlet fever when the rash is only faintly developed, with drowsiness, stupor, dry mouth, right parotid gland swollen, burning in the throat. —*Ehrmann.*

Ammonia is an excellent diffusible stimulant, and may be employed in any case in which this action is desirable. The sesquicarbonate of ammonia is a good form, and may be given in doses of one-half to three grains, in sweetened water. It is especially in the prostration from acute disease that it will be found serviceable.

PHOSPHORUS. *Specific Indications.*—In acute disease there is a low grade of inflammation, with feeble circulation, and cerebral anemia; urine contains mucus, sometimes pus, and is passed with difficulty. In chronic diseases, an enfeebled nutrition of the brain, imperfect retrograde metamorphosis and combustion, with cacoplastic or tubercular exudations.

Dose.—In acute diseases—℞ Tinct. Phosphorus gtt. j. to gtt. v., water ℥iv.; a teaspoonful every one to three hours. In chronic disease I prefer the hypophosphites : hypophosphite of lime gr. ss. to gr. j. three times a day ; or the compound syrup of the hypophosphites, one-fourth to one-half teaspoonful after meals.

Homœopathic Indications, (3d to 6th decimal).—Very weak, empty feeling in the abdomen ; heat up the back; constipation, long, hard, and dry stools, which are expelled with difficulty ; sour eructations and sour vomiting; desire for cold food and drink, which is vomited as soon as it becomes warm in the stomach ; vomiting of blood ; watery diarrhœa, coming away in a gush, followed by a sense of weakness ; sharp, shooting pains in abdomen ; hoarseness, loss of voice, croup, bronchitis ; hard, tight, dry cough, which is very exhaustive, worse from lying on the back or left side; expectoration salty, bloody yellow, purulent or of sour taste ; more expectoration in the morning; pneumonia, respiration oppressed, quick and anxious; circumscribed redness of the cheeks; phthisis pulmonalis, particularly in tall, slim persons; pulse rapid.—*Ehrmann.*

Ferrocyanide of Potassium. *Specific Indications.*—In chronic disease there is marked irritability of the nervous system, with restlessness and broken sleep; the pulse and respiration have their frequency increased; mucous membranes are pallid and lax with increased secretion, bowels irregular and tumid.

Dose.—℞ Ferrocyanide of Potassium ℨj., water ℥iv. ; a half teaspoonful to a teaspoonful three or four times a day.

This remedy may be used in catarrh, chronic pharyngitis, threatened tuberculosis, or intestinal diseases simulating tabes mesenterica. Though named in this class of diseases—imperfect nutrition with tendency to aplastic deposits—the remedy is indicated by the peculiar nervousness, with enfeebled circulation and nutrition of the brain.

Œnothera. *Specific Indications.*—A sallow, dirty skin, tissues full and expressionless, dull face and eyes, large sallow tongue, feeble innervation with broken sleep and tendency to cry on the slightest annoyance.

Dose.—℞ Tinct. Œnothera gtt. x., water ℥iv.; a teaspoonful every three or four hours.

Asafœtida.—This is one of the oldest antispasmodics in use, and is a very good remedy in some diseases of children. The objection to it is its offensive odor and taste, which render it difficult of administration. Its action seems particularly beneficial when the spasmodic or convulsive action arises from an irritation of the stomach, or parts associated with it. It is a gentle stimulant to the gastro-intestinal mucous membrane, and relieves irritation of it. The tincture of asafœtida with simple syrup offers the best means of administration. The dose of the tincture will be from ten to twenty drops.

REMEDIES WHICH INFLUENCE THE CIRCULATION.

Aconite,	Rhus,	Podophyllin,
Veratrum,	Lycopus,	Hamamelis,
Gelseminum,	Lobelia,	Cactus,
Eupatorium,	Digitalis,	Apocynum Canabinum,
Spiritus Ætheris Nitrosi.		

Lesions of the circulation are met with in almost all forms of disease, and in many are a principal element, so that if the circulation is brought to a normal standard and kept there, the disease passes away. Readers will recall the teaching that there is in a majority of cases a *first element of disease* (basic element), upon which the disease is based, and which if taken away, the other wrongs disappear of themselves. This is frequently the case with the circulation. Here is a fever or an inflammation with a pulse of 120 beats per minute, small, and upon this as a base, we have a temperature of 104°, arrested secretion, irritation of the nerve centers, and symptoms point to the development of a typhoid condition. Aconite is given in the usual small dose, and as the pulse comes down to the normal standard, the temperature comes down, secretion is established, the nervous irritation and pain pass away, the appetite returns, and the patient convalesces. But one remedy has been used, yet everything necessary has been accomplished.

The association between frequency of pulse and temperature will be recalled, for every ten beats of pulse an increase of one degree of heat. If the pulse increases in frequency the temperature increases, as the pulse comes down the temperature comes down. With a high temperature every process of life is impaired, the cause of disease is intensified, and molecular death is rapid. As the temperature comes back to a normal standard the various functions are better performed, nutrition commences, causes of disease are less active, and the body frees itself from devitalized material.

Increased frequency of pulse intensifies inflammatory action, and looks toward death of the part. In inflammation of the respiratory apparatus, increased frequency of pulse causes increased frequency of respiration, cough, determination of blood, pain and unrest. We control the inflammatory process as we control and obtain a normal circulation of blood.

ACONITE. *Specific Indications.*—The pulse is small and frequent. (The indication is short but explicit). The remedy influences special parts, as the throat, the larynx, bronchial tubes and intestinal mucous membranes, the indication being irritation with determination of blood.

Dose.—℞ Tinct. Aconite gtt. iij. to gtt. v., water ℥iv.; a teaspoonful every hour.

Homœopathic Indications, (3d to 6th decimal).—Chill, followed by hot, dry skin, quick, full pulse; great restlessness and thirst, fear and anxiety of mind, sudden sinking of strength; congestion to head, chest, and heart; anxious, labored breathing.—*Ehrmann.*

The pulse is *small and frequent*; it seems plain enough and yet it is sufficient for all our purposes. If it is a fever the temperature will be increased in proportion to the frequency of the pulse, and the remedy will then reduce the temperature as it brings the circulation down to the normal standard. If, however, it be the cold stage of a fever, or an Asiatic cholera, it will increase the warmth of the body as it gives a natural circulation. The frequent small pulse is the indication in inflammatory disease when there is marked irritation and determination of blood, (an active condition), but the remedy serves an equally good purpose if the circulation of a part is enfeebled—the pulse being small and frequent.

Looking at the action of remedies in the usual way, this seems very strange—a paradox—but there is much in nature that we have not unraveled. If we observe the action of Aconite in fever, we find that as the pulse diminishes in frequency it increases in freedom, and there is a better circulation. If we note its action in cholera we observe that as the pulse loses its frequency it gains volume and freedom, and there is a better circulation of blood. If we note its action in active inflammation we notice that it lessens determination of blood, quiets the irritation, checks the rapid circulation in the capillaries where it is too active, and increases the circulation where it is sluggish. If, as we think, it acts upon and through the ganglionic system of nerves, we can account for all of this by saying that it gives *right* innervation. I have been in the habit of saying that Aconite was a stimulant to the heart, arteries and capillaries, because whilst it lessened the frequency it increased the power of the apparatus engaged in the circulation.

In noting the special parts and tissues, influenced by Aconite, we may commence with the tonsils, the fauces and pharynx, where its influence may depend somewhat upon its topical action when swallowed. It has a direct influence upon the larynx, and is one of our best remedies in acute laryngitis and croup; indeed I would very much rather trust my patients with Aconite alone, than with all the old materia medica. Its influence upon the mucous membrane of the bronchia, even to the air cells, is very marked.

The topical action of Aconite upon the stomach relieves irritation, and frequently puts this organ in condition to receive remedies and food kindly. It acts directly in relieving irritation of the intestinal canal from stomach to rectum, and proves one of our most efficient remedies in diarrhœa and dysentery, as well as in acute diseases of the associate viscera.

It will be noted that Aconite has a wide range of use, especially in diseases of children. It may be called, par excellence, the child's remedy, for the indication—small and frequent pulse—is a common condition in the acute diseases of children. I can almost agree with a medical friend who remarked, that if he had only the choice between Aconite, and the remainder of the materia medica, in diseases of childhood, he would take the one remedy.

VERATRUM. *Specific Indications.*—The pulse is frequent and full, may be hard, but has the quality of strength. The circulation is active, the skin somewhat flushed. It is the remedy in inflammation when the circulation to the part, and in the part, is active, and where near the surface, when the surface is flushed red (the color of arterial blood).

Dose.—℞ Tinct. Veratrum viride, gtt. v. to gtt. x., water ℥iv.; a teaspoonful every hour.

Homœopathic Indications, (3d to 6th decimal).—Great arterial activity. Convulsions or mania. Meningitis, cold sweat on face, hands and feet. The skin looks shriveled. Opisthotonos. Chorea, twitchings and contortions of the body. Froth constantly on the lips. Chewing during sleep. Difficulty in swallowing. Neuralgia.—*Ehrmann.*

The use of Veratrum is clearly indicated above. It is the remedy for sthenia, when the heart and muscles have power, and the circulation is strong. It lessens the frequency of the pulse, gives a free and equal circulation, lessens the temperature, and thus promotes better innervation and a better performance of all functions. Thus in many cases it will do all that is necessary to cure a fever.

Its action upon any part engaged in inflammation is of the same character, whether it be taken internally or topically applied. It checks determination of blood by relieving irritation, slows the blood in the capillaries when it is running too rapidly, and at the same time it gives strength to the enfeebled tissues. This action may be witnessed in cutaneous or subcutaneous inflammation, or in erysipelas when the part has the flush of arterial blood. It seems to make but little difference where the inflammation is located, as it influences the entire circulation.

It is claimed by some that Veratrum is one of the best alteratives in our Materia Medica, and it is true that in its influence upon the ganglionic nerves and the circulation, it puts the excretory apparatus in the best condition for its work, and favors retrograde metamorphosis.

It does not depress the heart or the patient as has been taught. In medicinal doses it increases the strength, and favors normal functional activity, or it gives a better circulation of blood. Acting through the sympathetic system of nerves, it not only gives a slower and a better circulation of blood, but it influences every organ and part supplied by this system. It thus improves the condition of the stomach and intestinal canal, favors appetite, digestion and blood-making, and improves secretion.

Topically applied it relieves irritation, checks determination of blood, and arrests the inflammatory process. It has been used with the best results in this way, to control inflammatory action. It is employed as a local remedy in erysipelas when the inflammatory action is active, the part having the color of arterial blood.

GELSEMINUM.—A study of Gelseminum has been given under the head of "Remedies which Influence the Nervous System." Its influence upon the circulation is very direct, when the frequent pulse is based upon an excited condition of the nervous system. The flushed face, bright eyes, contracted pupils, with increased temperature, are the indications, but with this, it lessens the frequency of the pulse, lowers the temperature, controls the inflammatory process and favors secretion and excretion.

The Gelseminum case (the remedy being used in our dose) has a vigorous circulation, the pulse being frequent and free, usually full, the capillary circulation also being free. Locally there are evidences of determination of blood.

EUPATORIUM PERFOLIATUM. *Specific Indications.*—The pulse is full, free and strong, the surface flushed, temperature increased, tendency to perspiration, pain in the loins, fullness of chest with sense of oppression, sense of weight with pain in lumbar region, frequent desire to urinate, urine cloudy.

Dose.—℞ Tinct. Eupatorium gtt. x., water ℥iv.; a teaspoonful every hour.

Homœopathic Indications, (3d to 6th decimal).—Intermittent fever, paroxysm generally commences in the morning, thirst a long time before the chill, after the chill vomiting of bile, pain in the bones, as if broken, before the chill. The fever goes off by perspiration and sleep. During the apyrexia loose cough.—*Ehrmann.*

Our readers will recollect the old use of "boneset" following an emetic "to keep the fever down," and produce diaphoresis, or an infusion with the spirit vapor bath "to assist," or in the early stage of a malarial fever, when the patient "ached in every bone," to relieve pain—boneset it was called and bone-ache it relieved. Again, in rheumatic fever, when the patient would burn one hour and sweat the next, and also in acute rheumatism, when the pain would make him "sweat like a Turk," boneset proved a good remedy.

It has not been so extensively used in the small dose, but it will be found valuable in many cases if the indications are observed. I do not believe in substituting it for a better remedy, or using it when Lobelia or Veratrum would serve a better purpose, but when indicated it will give satisfaction.

RHUS.—The reader will find a study of Rhus under the head of "Remedies that Influence the Nervous System," and we only need to study it here as it influences the circulation.

With a frequent, small, *sharp* pulse, frontal headache (left orbit), and red papillæ at tip of tongue, this remedy will slow the pulse, lessen the temperature, relieve pain, and establish secretion. It is true, it is usually administered with Aconite, but if one will give it alone he will find it all-sufficient. Burning of the skin is another very good indication, as is bright eyes.

LYCOPUS. *Specific Indications.*—The circulation is active, pulse frequent and hard, determination of blood to a part—lungs, stomach, bowels, kidneys—with sanguineous discharge or hemorrhage, cough with sense of heat or burning in the chest, or sense of "rawness," with irritation.

Dose.—℞ Tinct. Lycopus gtt. v. to gtt. x., water ʒiv.; a teaspoonful every one, two or three hours.

This remedy is not in common use, and yet will be found of marked value, if the indications as above are followed. We employ it, more frequently, in diseases of the chest in the adult, and especially for active hemorrhage and an irritative cough. In the child, it will be especially a remedy for chronic cough, irritation of bronchial tubes and lungs, and determination of blood to the kidneys.

LOBELIA.—The *full, oppressed* pulse is the indication for Lobelia as a sedative. It is a stimulant to the circulation, giving power to the heart and arteries, and a better condition of capillaries for the blood to pass through. An enfeebled circulation in the lungs with a sense of fullness and oppression is relieved by Lobelia.

DIGITALIS. *Specific Indications.*—A small, frequent pulse with want of power in the heart, is the best indication. A feeble circulation in lungs, or elsewhere, with scant urine, may also be benefited by this remedy.

Dose.—℞ Tinct. Digitalis gtt. v., water ʒiv.; a teaspoonful every one or two hours.

Homœopathic Indications, (3d to 6th decimal).—Very slow pulse; harsh appearance of the face; stools gray or ash color; urine scanty; pulsations of the heart intermit, fainting; ascites; hydrocele; hydrothorax; jaundice with light colored stools, scanty brown urine; great debility; vertigo with trembling; dimness of vision, dark bodies like flies hover before the eyes.—*Ehrmann.*

Digitalis is not in common use, and has not been studied with the care given to others of this class. But in small doses it will be found an excellent remedy. Much of the Digitalis in market is of poor quality. Of a recent and well cured specimen of the herb, let a tincture be prepared in the proportion of four ounces to one pint of alcohol of 76 per cent.

The principal use that I will suggest is, as a stimulant and tonic to the heart in cases of anæmia and feeble circulation. In these cases it is associated with iron, cod-liver oil, and a nutritious diet.

SPIRITUS ÆTHERIS NITROSI. *Specific Indications.*—The pulse is frequent but free, the skin inclined to be moist, frequent but scanty passages of urine.

Dose.—The dose will vary from, gtt. j. to gtt. x. Usually we add a half teaspoonful to a half glass of water, and give it in teaspoonful doses every hour.

It is a little difficult for one who has been thinking of medicine in the ordinary way, and reading regular text-books, to think of Sweet Spirit of Nitre as a sedative. It is strong alcohol containing a portion of nitrous ether, and may be regarded as an alcohol, and as a stimulant, using this term as generally understood. But we have seen that the remedies classed as arterial sedatives are in reality stimulants, that is, they lower the frequency of the pulse by increasing the power of the heart and arteries, and by improving the condition of the capillaries.

The practitioner may occasionally use the Spirit of Nitre with marked benefit in simple fevers and inflammations, making it take the place of Veratrum or Aconite.

CACTUS GRANDIFLORA.—The pulse is irregular, sometimes frequent, occasionally slow; there is a sighing respiration, sobbing, uneasy dreams, from which the patient wakes crying or sobbing; face is pale, eyes expressionless; pain in top of the head; sense of oppression in region of the heart.

Dose.—℞ Tinct. Cactus gtt. v., water ℥iv.; a teaspoonful every one to four hours.

Homœopathic Indications, (3d to 6th decimal).—Heart troubles, where you find a great sense of constriction, as if the heart was firmly grasped by a hand or in a vise; difficulty of breathing, attacks of suffocation with fainting, cold perspiration and loss of pulse; hemoptysis, with convulsive cough; chronic bronchitis, with rattling of mucus; vomiting of blood, hemorrhages from nose, lungs, rectum, or stomach; œdema of the feet.—*Ehrmann.*

The indications as given above will be a sufficient guide to the use of the remedy. The child is in poor health, pale and inactive; very sympathetic, suffers greatly from slight unpleasantness; sleeps in the day, wakeful and fearful at night; appetite variable; abdomen full or doughy. It may be a remittent or intermittent fever, threatened phthisis, or they call it worms; we give Cactus and the patient is relieved.

PODOPHYLLIN. One would hardly think of Podophyllin as a remedy influencing the circulation, and yet it does so in a very marked manner. The *veins are full*, face full, abdomen tumid, tongue full, and the patient complains of giddiness and unsteadiness, with stupid headache. As a rule the pulse is full and oppressed. I have used it with most marked advantage in malarial fevers, in ague, and in inflammatory diseases, and hardly know what would take its place. The reader will notice, however, that nothing is said about " constipation of the bowels," and it makes no difference to us whether they are constipated or loose, if the indications for the remedy are present. It will also be noticed that stress is placed upon "fullness of tissue ;" when patients are pinched we never give it.

I use a first or second centessimal trituration (one to one hundred) in diseases of children, but even the old cathartic dose will prove curative when the symptoms are strongly marked.

HAMAMELIS. *Specific Indications.*—The veins are full and feeble, varicose, tissues soft, feeble, relaxed, swollen. There is fullness about the anus, prolapsus ani, difficult evacuation of feces, swelling of vulva or prepuce, œdema of legs, spongy throat, enlarged tonsils, abundant mucous discharge from nose, hemorrhage from atony.

Dose.—The distillate of Hamamelis may be given in doses of from one to ten drops, or in the usual way. ℞ Hamamelis gtt. x., water ℨiv.; a teaspoonful every one to four hours. As a local application it may be used of full strength, or diluted with two to ten parts of water.

Homœopathic Indications, (3d to 6th decimal).—Hemorrhoids, bleeding profusely with sensation of soreness, weight and burning in the rectum ; varices protrude and the anus feels sore and raw; stools covered with mucus; the back feels like breaking; restlessness at night; dryness of the mouth; varicose veins, hard, knotty, swollen and painful.—*Ehrmann.*

Hamamelis is an important remedy, and when the physician learns to use it he will value it highly. It is an admirable remedy in acute catarrh when secretion is established ; in chronic catarrh, chronic pharyngitis and tonsillitis, when the child's voice is husky or flat; in chronic bronchitis with free secretion; in mucous diarrhœa, abundant urine but painful micturition; in hemorrhoids, prolapsus ani, ottorrhœa, sprains, bruises, atonic inflammations, etc. Remembering that it strengthens and improves the venous circulation, freeing parts from congestion and giving them tone, we can hardly go astray in its use.

APOCYNUM CANABINUM. *Specific Indications.* — There is œdema of feet, cyclids, or of an inflamed part. The pulse may be frequent and full, or frequent and feeble, but capillary circulation is weak.

Dose.— ℞ Tinct. Apocynum gtt. v., water ℥iv.; a teaspoonful every one to four hours.

Homœopathic Indications, (3d to 6th decimal).—General dropsy, hydrothorax, urine high colored and scanty, considerable gastric disturbance, pulse weak and irregular, skin dry and husky, hoarse loose cough.—*Erhmann.*

This is one of our best remedies, and the indications are so clear that no one can go astray—œdema, or even the appearance which a tissue presents when it has been infiltrated with serum. In this as with other remedies it does not make any difference what name the disease has, if the indication is present. We use it with equal advantage in cholera infantum, or in scarlet fever. It lessens the frequency of the pulse, gives a better circulation of blood, stimulates all the secretions and improves nutrition.

If I notice in scarlet fever a fullness of the eyelids, or swelling of the feet, I give Apocynum. If in measles there is difficulty in breathing with a harassing cough, and the face or feet are puffy, I give Apocynum. So I would in croup (inspiration difficult), in infantile remittent fever, in inflammatory diseases, in rheumatism, or in any chronic disease.

REMEDIES THAT INFLUENCE THE TEMPERATURE.

Baths,	Veratrum,	Acids,
Food,	Rhus,	Alkalies,
Air,	Gelseminum,	Cod Oil,
Exercise,	Bryonia,	Phosphorus,
Aconite,	Baptisia,	Sulphur.

Among the means employed in the treatment and cure of disease, there are none of greater importance than those which rectify wrongs of temperature. The reader will recall the fact that a temperature of 98° is essential to health, and the performance of healthy function. Even with but a slight variation the person is ill, and the illness is in proportion to the amount of change. If the temperature falls below the normal standard, every function is impaired—the blood is not circulated well, respiration is feeble, appetite and digestion impaired, nutrition bad, and waste and excretion are imperfect. If the temperature is increased the pulse is more frequent, the nervous system excited, (it may be wrong in kind or oppressed), the appetite lost, digestion imperfect, .nutrition arrested, waste and excretion diminished, and changes of the blood go on more rapidly.

The body thermometer is one of our most certain means of diagnosis, and can not be dispensed with by any one who wishes certainty in practice. It measures the intensity of diseased action, and determines for us the gravity of disease, and the danger of a fatal termination. It is a little more difficult to use with children than with adults, but there is hardly a case that the temperature can not be taken at the axillæ.

In studying the means that regulate the temperature it may be well to recall something of the physiology of heat production, and the means that nature provides to maintain this constant temperature of 98°. The heat is furnished by the combustion of food, and there are two, possibly three, elements here to be taken note of: (a) the quality, quantity and preparation of the food ; (b) the introduction of oxygen by respiration for combustion ; (c) the presence in the blood of certain materials (phosphrus, sulphur, etc.,) which may be regarded as excitants of combustion. If a sufficient quantity of food is not taken the body is burned to supply the necessary heat. If the body is burned, the person suffers the excitation of burning.

The temperature is regulated, in so far as the production of heat by combustion is concerned, by the ganglionic or sympathetic system of nerves. It is also regulated by the skin which serves as a safety-valve for the body—opening to allow the escape of heat, when it is produced in too great quantity, and closing to retain it, when produced in too small quantity. Our remedies influence the temperature in both these ways; we control the processes of combustion through remedies influencing the ganglionic nerves, influencing the circulation, respiration and combustion ; we influence the skin so as to put it in better condition for its work, increasing secretion and respiration for the removal of heat, stimulating it and giving it tone so that heat may not escape.

BATHS.' The importance of baths in the treatment of disease has been clearly shown in the past.fifty years, so that but few will dispute their advantage, though in practice they may not use them. I think that the success of our school of medicine has depended as much upon our knowledge of bathing—the water cure—as upon the administration of drugs. From the days of Beach we have been using cold baths, hot baths, alkaline baths, acid baths, tonic baths, stimulant baths, vapor baths, general baths, local baths, baths to reduce the temperature, to stimulate the skin and increase the temperature, to promote elimination, to increase the respiratory function, etc.

The Cold Wet-Sheet Pack is not used as frequently as it might be because people do not like the first impression of cold, and are afraid of it. In young children the nervous excitement from the shock is sometimes injurious, but if well borne, and there is a vigorous circulation with increased temperature, its action is very kindly.

A sheet is wrung out of cold water and spread upon the bed, the little fellow is undressed, lain upon and wrapped up in it. Blankets and comforts are spread over and tucked in, and in a few minutes an agreeable warmth takes the place of the first impression of cold, the skin is moist, the pulse comes down, nervous excitement and pain pass away, and presently the child sleeps sweetly. It is allowed to remain in the pack for one hour, when the body is rubbed thoroughly dry, and dry clothing is put on.

The local cold pack may be employed in acute inflammation, with an active circulation, in pharyngitis, laryngitis, croup, inflammation of bronchia or lungs, inflammation of bowels, etc. A towel wrung out of cold water, and covered with dry flannel, is a very good way to use it.

The indication is an increased temperature with an active circulation.

The Hot Wet-Sheet or Blanket Pack is employed when the skin is enfeebled and inactive, especially in the eruptive fevers, and in local inflammations when the circulation is feeble. In measles, scarlet fever, or smallpox, the eruption fails to make its appearance at the proper time, the nervous system is oppressed, there is tendency to coma, and the skin shows a feeble capillary circulation. Here a hot blanket pack, using the water as hot as it can be borne, is attended by the best results.

The local hot pack is used in acute inflammations when the life of the part is impaired and the circulation is feeble. We thus use it in sore throat, in croup, in inflammations of the respiratory apparatus, in diseases of the abdominal viscera, local inflammations, etc.

To determine whether a bath, either a pack or hand-bath, shall be hot or cold, is important, and we are guided by this: If the circulation is active and the temperature increased, it will be cold; but if the circulation is feeble, it is to be hot. A hemorrhage, or an abundant discharge of mucus or pus, or an increased secretion, is best treated with the hot local or general bath.

Simply Sponging the surface with water lowers the temperature by evaporation, and puts the skin in better condition to perform its function. We thus use it several times in the day when persons are suffering with fever and carrying a high temperature. When necessary soap is used for the purpose of cleanliness.

The Alkaline Sponge Bath is one of the old Eclectic means, and a very important one. Originally it was *broke* water, wood ashes being used in sufficient quantity to make the water slightly slippery to the touch; after this pearlash or bicarbonate of potash was used in place of the ashes. In more modern days soda has taken the place of potash, and now the alkaline bath is a soda bath. The "broke water" or potash was decidedly the best in the majority of cases, its action upon the skin being kindlier, leaving it in better condition. When I speak of the alkaline bath, I wish to be understood as recommending water made alkaline with potash.

If one is in doubt whether the sponge bath should be alkaline or acid, he will do well to note the indications as presented by the tongue. If the tongue is broad and pallid the bath should be alkaline; if red, especially if deep red, it should be acid.

The Acid Bath (water acidulated with vinegar) will be found to exert a very pleasant influence upon the skin, when indicated, leaving it soft and in better condition to do its work. The alkaline bath may have been used, leaving the skin dryer and harsher than it was before, whilst the acid bath gives relief. The child's face and head are hot and dry, and the ordinary use of water does no good; it is sponged with vinegar and water and is relieved, and presently sleeps.

In some rare cases we use water acidulated with muriatic acid; in a larger number, especially where the symptoms are typhoid, and it is almost impossible to keep the child free from unpleasant odors, we use sulphurous acid.

Fatty Inunction is a most important means of putting the skin in better condition and rectifying the wrongs of temperature. Sometimes we use lard alone, the child being thoroughly rubbed with it, and then rubbed clean with soft flannel. In some cases we add quinine to lard, ʒss. to ʒj. to ʒij., especially when there is a malarial influence, or when we wish to stimulate the brain or spinal cord. In other cases when stimulation is wanted, oil of cinnamon, cloves, or eucalyptol, is added.

In scarlet fever we sometimes use a " bacon rind," or a prescription of creosote and salt with lard, as a stimulant to the congested skin. It lowers the temperature, improves the circulation, gives better functional activity, and favors the appearance of the eruption.

In malarial or typho-malarial fevers I have seen the inunction of quinine lower the temperature two degrees in as many hours, lessen the frequency of the pulse, and in a short time arrest the disease, when quinine internally had not only failed to do good, but had proven harmful. In cholera infantum, especially if attended by fever, the quinine inunction is one of our most important remedies.

We use inunction in chronic disease when innervation is feeble, the appetite and digestion poor, nutrition imperfect, and skin atonic. In some cases it exerts a very salutary influence, and patients improve from the commencement of the rubbing.

The Stimulant Bath is occasionally useful when the circulation is feeble. It may be a mustard-water pack, when there is sudden and great prostration ; quinine with proof-spirit when there is a malarial influence ; or an infusion of Xanthoxylum, Polygonum, or remedies of like character.

Tonic or Astringent Baths are employed in some cases where the skin is atonic and the circulation is feeble. An infusion of Hydrastis may represent the first, and of Quercus rubra the second.

Food.—It might be thought that a study of foods would come more appropriately elsewhere, and we will have occasion to notice them again under the heading, Restoratives. But heat is set free by the oxygenation of food, or of tissue, its only sources, so far as we know. Foods are calorifacient or heat-producing, and histogenetic or tissue-making. The first contains no nitrogen, the second contains nitrogen. The first may be represented by sugar and starch; the second by the albumen and gluten of vegetable substances, and the albuminoid and muscular tissues of animals.

The majority of foods are in part calorifacient and in part histogenetic, and we determine their value in tissue-building by the per centage of nitrogen in them. When this is low, the food is principally heat-producing, where it is large the food is eminently tissue-making. For the child, milk contains these elements in the best proportion, though we use a starchy food, or fat, when heat is especially wanted, or beef tea when a stimulant, tissue-making food is wanted.

If a child is carrying a high temperature in fever and inflammation, and is taking no food, the body is burned to supply the heat. The body can not be burned, even slowly, without irritation of the nervous system and suffering. I am sure that this will account, many times, for the restlessness, sleeplessness, and final wearing out of the nervous system in febrile and inflammatory disease.

One of the maxims of a rational practice of medicine is—to keep the stomach in such reasonably good condition that a portion of food may be taken and digested, and we look as carefully after the food of the sick as their medicine.

In many cases we find that if we relieve a temporary irritation or atony of the stomach, so that the child can take a moderate amount of milk or other food, the temperature comes down, and the irritation of the nerve centers passes away, and the child rests and sleeps. By giving food we supply the materials for combustion, and save the tissues of the body.

In some cases of chronic disease, we will observe a deficient temperature, or the heat is not properly distributed, the feet and hands being cold. In such cases a supply of such calorifacient food as may be digested, not only furnishes the 98° of heat which is an essential condition of life, but such other force as may aid nutrition and give the force necessary for exercise.

AIR. Air is usually studied as a condition of healthy life, and as a cause of disease, but if a condition of life, then it must prove a remedy in some cases of disease. A certain amount of air taken through the respiratory apparatus is necessary for combustion and heat production, and if patients are not supplied with it they must suffer. We place stress on the necessity for good ventilation—a sufficiently free admission of air to the apartment, and a current of air from it, usually by means of an open fire.

The open air is necessary for a cure in some cases. of chronic disease in children, as it is for the adult. If the little patients can have that moderate exercise which calls into action the respiratory apparatus, it is so much the better.

A dry air, from stove heat, is sometimes so irritant that it excites the nervous system, increases the frequency of the pulse, and will increase the temperature. A very moist atmosphere may be so depressing that the functions of life are feebly performed, and we will be obliged to rectify this if we cure the patient. Again, the air may be so loaded with dirt, or unpleasant gases, that it will not sustain life, and in severe cases is absolutely poisonous. A cure comes in such cases from free ventilation, cleanliness, and the use of agents which change or destroy the unpleasant material, as chloride of lime, chlorinated soda, sulphurous acid, etc.

EXERCISE. As we have just seen, where the respiratory movement is feeble, the processes of combustion go on slowly and a sufficient amount of heat is not produced. In addition to this, the burning of waste material and old tissues is not properly performed, and the body and blood are loaded with effete material. We find some chronic diseases of children where the exercise that a child will get, if a rug is thrown upon the floor and it is allowed to tumble around upon it, is better than medicine.

In the neighborhood of the sea, children are sent to the sea-shore and allowed to roll on and dig in the sand. In cities nurses take children to the parks and allow them to make sand and gravel houses or fortifications of the gravel walks. What we want to know in these cases is, that every effort should be made to induce the child to play and amuse itself, and thus get the necessary exercise, and that nursing in the arms is the worst possible way to obtain good health.

But the knowledge of when *rest* is necessary is just as essential. Rest to body, rest to nervous system and sleep, are essential in the treatment of diseases which have a high temperature. Restlessness and want of sleep are almost certain to increase the temperature; even the continuous fretting of a sick child for something it wants, or for its mother, will send the thermometer up one or two degrees.

A part suffering from inflammation needs rest, sometimes the absolute rest that we get by the use of a splint. We arrest the progress of a morbus coxarius, or a disease of the knee joint by a plaster of Paris or other dressing which will give absolute rest. A flannel bandage so nicely adjusted that it will support the abdomen well, is sometimes one of our best prescriptions for cholera infantum.

ACONITE. This remedy has been twice studied, but it will do no harm to look at it as a remedy influencing the temperature. We have already noted the relation between the pulse and the temperature, an increase of one degree of heat for each ten beats of pulse. If the pulse is lessened in frequency by remedies the temperature comes down, and *vice versa*. All arterial sedatives are therefore thought to lessen the temperature, when above the normal standard, and this is especially true of Aconite and Veratrum.

But this remedy influences the functions of calorification directly through the ganglionic and respiratory nerves. We have been in the habit of thinking that it always lessens heat production, and it very certainly does so in many cases when the temperature is too high. But if one will administer the remedy in a case of congestive chill or cholera, when the pulse is small and running over one hundred beats per minute, he will see it increase the temperature, and parts that were cold regain their heat.

VERATRUM. This, like the preceding remedy, has a direct influence upon the function of calorification, both as it influences the circulation, respiration and combustion. All these are too active and characterized by strength, and Veratrum lessens the excitement and the activity. When indicated it will frequently lower the temperature from one to three degrees in twelve hours, and in minor diseases, like the febricula, or at the commencement of inflammations, it may bring it down to the normal standard in a very short time.

Its influence in superficial inflammation, when topically applied, is sometimes so marked as to excite surprise. The irritation is relieved, determination of blood is arrested, and the temperature falls in a short time.

RHUS. We have studied Rhus as a remedy influencing the nervous system, but its influence in controlling the temperature is not to be neglected. It allays irritation of the nerve centers, especially of the ganglionic system, and slows the process of combustion. In the larger number of cases, when indicated, it markedly reduces the temperature; indeed its action in this direction is quite as marked as any remedy of the materia medica.

BRYONIA. This is another remedy that lessens the function of calorification, (when the temperature is high,) by relieving irritation of the ganglionic nerves, and possibly by its action upon the blood. It will be remembered that pain is one of the principal indications for Bryonia, and pain is the evidence of unrest.

GELSEMINUM. Gelseminum lessens the temperature by allaying cerebro-spinal irritation which is back of ganglionic excitement. When indicated by flushed face, bright eyes and contracted pupils, its influence is very marked in this direction.

BAPTISIA. This remedy is fully studied under the head of Antiseptics, but it also exerts a marked influence in lowering the temperature when above the normal standard. Like other remedies it requires the special indications—the full purplish face, like one who has been exposed to severe cold, and a bluish-red tongue. In this case the high temperature is dependent, in part, upon a peculiar sepsis, and this being antidoted the temperature falls rapidly. I have seen it come down, under the influence of Baptisia alone, from 106° to 100° in twenty-four hours.

ACIDS. *Specific Indications.*—The tongue is dusky red or deep red, and frequently small; mucous membranes and sometimes the skin show the same color. The coatings of tongue and sordes upon the teeth brown, growing darker as the disease progresses.

Dose.—Usually we order muriatic acid ʒi., water, syrup, aa. ℥i.; add to water so as to make it pleasantly acid, and give as patient will take it. In some cases a sharp sparkling cider is used and given a teaspoonful to a tablespoonful in water every three or four hours. In other cases an acid whey (lactic acid) will be taken by the patient, whilst the others will be rejected.

The truth of specific medication is well illustrated by the action of acids and alkaline agents in the cure of disease. That

they are curatives no one will dispute; that they act equally well
in the same cases, or that they can be taken by chance and get
uniform good results, no one will claim. If they are to be used
at all we must have some means of determining when we shall
use the one, and when the other. I determine this by the
color of tongue, and where blood shows freely as in mucous membranes and some portions of the surface. If the color is deep
or dusky red, an acid is wanted; if it is pale, pallid, the patient
requires an alkaline salt; if neither the patient requires neither
the one nor the other. If any one has a better means of determining these points, then the better light should be followed; if
not we will continue to be guided by the color.

When indicated the acid exerts a marked influence in lessening,
an exalted temperaturé, and will sometimes reduce it when sedatives and baths have wholly failed. This is especially true when
typhoid symptoms are present, as the acid antidotes the process
of sepsis. Thus in England and on the Continent, the continued
fevers have been treated with acids alone, (diet, rest and good
nursing added,) in thousands of cases with most marked success,
the mortality being reduced in some cases to two or even one
per cent. If we can give our little patients a pleasant acid drink
in place of the nauseous drugging of the olden time we should be
thankful.

Not only is the acid of advantage when given internally, but
the acidulated bath is a means not to be neglected. There are
cases when the ordinary bath or the alkaline bath seems to make
no impression upon the skin or the temperature, but when a little
vinegar added to water is refreshing and cooling, the hot face and
head sponged with vinegar and water have given relief, the irritation is quieted and the patient sleeps. The tense, hot abdomen,
with skin like parchment, is softened, cooled and relieved of
irritation by an acid bath or pack. It will not do to neglect the
"small things" in the practice of medicine.

ALKALINE SALTS. *Specific Indications.*—The tongue is broad
and pallid, and its coatings are white and pasty.

Dose.—If there is no special indication for another, we will
employ a salt of soda, for soda is the salt of the blood. We add
it, usually the bicarbonate, to water in such quantity as will make
a pleasant drink, and let the patient take it freely.

If there is marked muscular debility we use a salt of potash in
place of the soda.

If there is a tendency to subcutaneous inflammation, or inflammation and suppuration of cellular tissue, lime-water or sulphite of lime is to be given.

The indications for the use of the alkaline salts are so clear that no one can mistake them. Why they should lessen the temperature (when in excess) as do the acids, is more than we can say with our present knowledge; only this, that they correct a wrong of the blood and the fluids of the body, and righting this wrong, they remove the nervous and vascular excitement which was caused by it.

The one thing in therapeutics that we can not have impressed upon us too forcibly is, that the prominent lesion, as indicated by those symptoms that point us to remedies, is very frequently the basis of the disease, and if taken away the entirety of the disease passes away of itself. ·

In our Western country, and where malarial diseases prevail, the alkalies are much more frequently indicated than the acids; indeed, in some seasons diseases are cured by these alone. I recall a year of malarial fever in which sulphite of soda or chloride of sodium was more certain than quinine; indeed they would cure ague when quinine had failed. In rheumatic fever this is also the case in some seasons, and the indications for the alkaline salts being marked, they cure rheumatism.

PHOSPHORUS. *Specific Indications.*—Phosphorus is indicated by a low temperature, cold extremities, doughy or waxy skin, enfeebled nutrition of the nerve centres with want of innervation, sensation of weight and pressure in peritoneum and pelvis, irritation of urinary passages and burning on passing urine.

Dose.—The tincture of phosphorus is used in low grades of inflammation of the respiratory apparatus, in disease of the prostate, bladder and urethra; the hypophosphites to improve waste and excretion, and to aid blood-making and nutrition. The phosphites are used to aid digestion (laxative) and to favor nutrition.—℞ Tinct. Phosphorus gtt. v. to gtt. x., water ℥iv.; a teaspoonful every one to four hours. Hypophosphite of lime gr. ½ to gr. j., every four hours. Phosphate of soda gr. j. to grs. ij., twice or three times a day.

It will be noticed from what has been said above, that phosphorus has an extended use in medicine, but it must be employed with care. I employ the tincture in diseases of children but rarely, selecting those cases where there is an atonic condition of the respiratory apparatus, or where there is disease of the bladder and urethra.

To stimulate calorification, and increase the production of heat, and also to burn waste material, I prefer the hypophosphites. The hypophosphite of lime is preferred where there is tendency to deposit of tubercle or disease of cellular tissue; the hypophosphite of soda when the tongue is broad and pallid, and a salt of soda would be given. The compound syrup of the hypophosphites is sometimes useful both to increase calorification and to aid nutrition.

COD LIVER OIL. *Specific Indications.*—The extremities are cool; the skin relaxed, doughy or dirty; tongue dirty at base; bowels irregular; pulse lacks strength; small boils; inflammation of cellular tissue; ulceration of the skin; bad blood.

Dose.—The dose of cod oil for a child will vary from one-half to two teaspoonfuls three times a day. A pure inodorous oil will frequently be taken without difficulty.

The indications given above will point out the cases which will be benefited by cod oil. The processes of combustion do not go on well, the waste of tissue is not burned as it should be, and fitted for excretion, and the blood is loaded with it. This impairs the new blood that is making, the new tissues which are building, and enfeebles all the functions of the body. In this way the patient grows the conditions which give cacoplastic and aplastic deposits.

In some cases all the functions seem to be stimulated by the administration of the oil. The skin becomes clean and active, the appetite improves and digestion is better, and there is an increase of tissue.

QUININE. We have already made a full study of quinine under the head of Remedies which Influence the Nervous System, and we will only consider it here as it influences the temperature. It was shown by Prof. I. G. Jones, that in the purely malarial fevers—uncomplicated—this remedy could be administered with the effect of lowering the temperature, diminishing the frequency of the pulse, and establishing a complete intermission of the fever, if not arresting it. In the remittent fevers of the Scioto valley the administration of quinine would commence as soon as the highest point of fever was passed, and then repeated every three hours, there would be a marked abatement of the disease.

In diseases of children in malarial regions, the action of quinine is frequently desirable. The temperature is high, the pulse frequent, the secretions checked, and we are sure of the periodic character of the disease—the proper quantity of the remedy as an antiperiodic is given in divided doses, and the fever is arrested.

In the very minute dose, or when used by inunction, it will sometimes exert a very marked influence in controlling the temperature, even when the disease is not malarial. We do not wish to recommend it, however, when other remedies are indicated, for it is used far too frequently now, and to the injury of many patients.

NUX VOMICA. This remedy has been fully studied elsewhere, and we have only to notice it here as a stimulant to the respiratory system, and an excitant of calorification. In cases where there is enfeebled spinal innervation and consequent respiratory function, nux or strychnia will be thought of as remedies. Sense of oppression about the præcordia, increased difficulty of respiration when asleep, tendency to retention of urine, slow movement of the bowels, and nausea, are indications.

STIMULANTS. There are a few stimulants which can be used with advantage to stimulate heat production, but their use is limited. Alcoholic stimulants are very rarely used in diseases of children, and possibly the only cases in which we would think of them is in the slow convalescence from acute disease. I am sure that their use during the progress of fever or inflammation is injurious. A teaspoonful of brandy or good whisky in four to six tablespoonfuls of water (hot or cold, as suits the patient best) sweetened, sometimes is a good form.

Tinct. Xanthoxylum may be used occasionally, as may a Tinct. of Asarum, or once in a long while the Comp. Tinct. of Cajeput. Probably the best child's stimulant will be found in the old mixture: ℞ Comp. Spirits of Lavender ʒiij., Tinct. Lobelia ʒi., water and syrup ʒiss. Mix. This stimulates the respiratory nerves, improves the circulation, and gives an agreeable sense of warmth.

SALICYLIC ACID. Salicylic acid will be studied under the head of Anti-rheumatics. Here we wish only to note its effect in diminishing the temperature of the body. As yet we know simply the fact that it has this influence in some cases, but in what particular cases, and what the specific indications are, we do

not know. I should say that the best effects are to be expected
when the patient suffers considerable pain in the extremities, and
when, though the temperature is increased, the skin is inclined
to be moist.

SALICIN. This remedy stands between quinine as an antiperiodic and salicylic acid as an anti-rheumatic, and may be
thought of when with periodicity there is pain of a rheumatic
character. In such cases it will lessen the heat of the body.

NITRIC ACID. This remedy will be fully studied with the next
class, though it is difficult to properly classify it. When indicated—violet colored tongue—it will lessen the temperature in
a very marked manner in cases of fever or inflammation, and in
malarial fever will prevent the recurrence of the exacerbations.
In slow infantile remittents, when the pulse has a range of 100°
to 102°, I have occasionly seen the most marked benefit from its
administration.

REMEDIES WHICH INFLUENCE THE RESPIRATORY APPARATUS.

The function of respiration is one of the most important in
the economy, and the organs engaged in it may be regarded as
almost the center of life—in the olden times classified as "noble
organs." Through this apparatus the blood receives its supply
of oxygen, and is freed from carbonic acid gas. It is intimately associated with the circulation, both in its structure and
its innervation, and has sympathies with the entire body.

Remedies influence the respiratory apparatus and its function
directly, as they influence other parts of the body, and a study
of them in this relation can not but be profitable. The principal
of these are :—

Aconite,	Stillingia,	Senega,
Veratrum,	Lycopus,	Scillœ,
Bryonia,	Drosera,	Sticta,
Ipecacuanha,	Sanguinaria,	Rumex,
Lobelia,	Euphorbia,	Grindelia,
Phytolacca,	Phosphorus,	Nitric Acid.
Eupatorium Perfoliatum,		

ACONITE. We have made a study of this remedy two or three
times, and need but note here its direct influence in relieving
irritation and controlling inflammation of this apparatus. It has
a direct action upon the tonsils, and will sometimes arrest a

quinsy, if used in the early stage. It is our best remedy in laryngitis or croup, giving results that can not be obtained with the old nauseant or emetic treatment. In mucous croup I rarely think of using any other internal remedy, and in the pseudo-membranous form it will be one of the most important. In acute bronchitis and in pneumonia it is the sedative usually indi-cated in childhood, and forms a part of a good treatment.

VERATRUM. Veratrum is the remedy selected when the pulse is full and frequent, and its action is direct in arresting inflammation of any part of the respiratory apparatus. It controls irritation of the pneumogastric, allays cough, and improves the respiratory function.

BRYONIA. This remedy influences the pleura and parenchyma of the lungs, lessening irritation and arresting the inflammatory process. It is indicated by pain in the chest, sense of soreness, with catching pain on inspiration, pleuritic pain, and a short harassing cough. The flushed right cheek is a good indication.

It is one of our most important remedies in this relation, and will be in frequent demand in bronchitis, pleuro-pneumonia, and in pleurisy.

IPECACUANHA. Ipecacuanha exerts a specific action in relieving irritation of mucous membranes, and we employ it in the first stages of bronchitis and pneumonia with the most marked benefit. In years past I have treated infantile pneumonia with Ipecac alone with much success. It was rubbed up with sugar, and given in doses of one-fourth to one grain, sometimes producing slight nausea. Now we use the tincture, combining it with Aconite, the proportions being—℞ Tinct. Aconite gtt. iij., Tinct. Ipecac gtt. v. to gtt. x., water ℥iv. ; a teaspoonful every hour.

LOBELIA. This remedy exerts a very marked influence upon the respiratory apparatus, improving the innervation and circulation. The indication is—an oppressed respiration, congestion, increased mucous secretion, moist blowing sounds, mucous rattling in the chest. When patients are old enough to complain it is of a sense of weight and oppression, especially about the præcordia.

The most marked benefit is obtained when there is a tendency to congestion, and when the respiratory tubes are filled with mucus. In asthenic bronchitis, when the child breathes with great difficulty, and the rattling of mucous can be heard all through the chest, I know of no remedy so certain to give relief. It is also one of our best remedies in infantile asthma.

In ordinary practice we use it with the sedatives, gtt. v. or gtt. x. of a tincture of the seed being added to water ℥iv. In asthenic bronchitis and in asthma, I frequently make the old prescription. ℞ Tinct. Lobelia (seed) ℨj., Comp. Spts. Lavender ℨiij., Syrup ℥iss. Mix. Give in small portions, frequently repeated, just short of nausea. This will be found an admirable form for the remedy, as it is readily taken, is kindly received by the stomach, and it relieves nervous irritation and gives rest.

A description of Lobelia would not be complete without reference to its old use as a nauseant. For this use in diseases of children, I prefer the acetous tincture to other preparations. It is prepared in the proportion of four ounces of the herb lobelia to the pint of dilute acetic acid, or vinegar; or the acetous tincture may be formed into a syrup by the addition of two pounds of sugar to the pint.

Given in nauseant doses, lobelia relaxes the respiratory passages, and thus gives temporary ease to the breathing. In croup, in asthma, in bronchitis, and in whooping-cough, this action is very important. Continuing this nauseant influence there is increased secretion from the mucous membrane of the respiratory passages, and it is thinner and less tenacious than the secretion during the inflammatory process. The engorgement of the vessels is somewhat relieved by the secretion, and the mucus is removed with greater ease.

In administering the nauseants for this purpose, they should be repeated so frequently as to keep up a continued action; for if given at long intervals, the alternation of relaxation and determination of blood proves injurious; and to obtain this action, they should never be given to produce speedy emesis; indeed emesis is not desirable in any case, unless to remove accumulations of mucus already secreted.

EUPATORIUM. When the pulse is frequent, full and free, and the skin inclined to be moist, the Eupatorium will be found a good remedy in bronchitis or pneumonia.

PHYTOLACCA. This remedy exerts a special influence upon the fauces and pharynx—the throat—and when this is inflamed or irritated, and is a source of cough or respiratory difficulty, we think of Phytolacca. A pallid tongue with red spots, sore mouth, sore throat, enlarged lymphatic glands or soreness or pain of mammary glands, associated with disease of the respiratory apparatus, call for Phytolacca.

STILLINGIA. *Specific Indications.*—Sense of rawness and tickling in the throat; sense of irritation behind the fauces or the velum pendulum.palati; burning, itching of the larynx, which causes a short cough and inclination to hawk and free the throat; croupal cough and voice.

Dose.—℞ Tinct. Stillingia gtt. x., Simple Syrup ℨij., one-fourth to one-half teaspoonful. I like the Linamentum Stillingia (℞ Oil of Stillingia ℨij., Oil of Cajeput ℨi., Oil of Lobelia ℨss., Alcohol ℨj. Mix) in doses of one-half to one drop on sugar.

Stillingia is one of our best remedies for the relief of cough, when it is caused by a sense of irritation of the throat. We use it in chronic bronchitis, in acute bronchitis when secretion is established, and especially in laryngeal disease, and croup. Even the external application over the larynx will cure croup, and is much safer than the old treatment. Usually I administer Aconite internally and apply the Stillingia liniment to the throat.

LYCOPUS. The pulse is frequent and somewhat full or hard, the cough paroxysmal, expectoration of muco-pus, difficulty in urination, hemorrhage from lungs or kidneys, deposit of tubercle.

Dose.—As a cough medicine I administer it in drop doses on sugar; for other purposes—℞ Tinct. Lycopus gtt. x. to ℨij., Alcohol ℨss., water ℨiijss.; a teaspoonful every one to four hours.

I like the action of Lycopus very much in chronic bronchitis, pneumonia, or tendency to phthisis pulmonalis. It relieves the cough, quiets pain, gives rest, diminishes the temperature, and brings the pulse down to a normal standard. It will also give good results in chronic inflammation of the kidneys, bladder and urethra.

SANGUINARIA. *Specific Indications.*—A sense of burning and constriction in the fauces or pharynx, with irritative cough and difficult respiration. The patient is nervous and restless, redness of nose with burning and thin acrid discharge, spots of bright redness on face or on chest, redness and burning of the ears.

Dose.—The dose will depend upon the action wanted. Many

5

times I use it in very minute doses—℞ Nitrate of Sanguinaria gr. ss, water ℥iv.℥; a teaspoonful every two or three hours, (or Tinct. Sanguinaria gtt. ij. to gtt. v., water ℥iv.) In other cases I would use the acetous tincture in full doses to nausea for a temporary effect.

Homœopathic Indications, (3d to 6th decimal).—Congestion of blood to the head, with ringing in the ears and flushes of heat. Sick headache, beginning in the morning with vomiting of bile, worse from motion, stooping, noise or light. Periodical headache. Coryza, with loss of smell. Nasal catarrh. Nasal polypus. Ulcerated sore throat. Tongue feels sore as if burned, coated white. Sensation of emptiness in the stomach soon after eating. Croup. Asthma. Pneumonia with very difficult respiration, cheeks and hands livid, pulse soft and easily compressed.— *Ehrmann.*

Like the Lobelia, for the old use, I prefer the acetous tincture of sanguinaria to any other preparation. It is prepared with two ounces of the ground root to one pint of dilute acetic acid, or vinegar. A syrup may be prepared in the same manner as named for the Lobelia.

Sanguinaria is rarely used alone; but in combination with Lobelia it gives us our most efficient nauseant expectorant, and is the remedy we prefer in cases of mucous and pseudo-membranous croup. In small doses, so that it does not produce nausea, it becomes a stimulant expectorant, and will check secretion from the bronchial mucous membrane.

NITRATE OF SANGUINARIA. The nitrate of sanguinaria is one of the few really good concentrated preparations. It is rarely used in the form of powder, being too acrid; but, combined with simple syrup, in the proportion of one grain to four ounces, it furnishes a very desirable remedy. Its action is rather that of a stimulant to the respiratory apparatus, and it should not be used in nauseant doses. The dose of the syrup will be ten drops for a child two years of age.

EUPHORBIA. This remedy, which is fully studied with those which influence the digestive apparatus, has an action very much like ipecac, and may be used to quiet irritation of the bronchial tubes and to check profuse secretion.

Dose.—℞ Euphorbia Hypericifolia gtt. v. to gtt. x., water ℥iv.; a teaspoonful every hour.

PHOSPHORUS. This remedy, which has been fully studied in other relations, influences the respiratory apparatus in a direct manner, stimulating a better innervation and circulation. It may be prescribed in low grades of inflammation of the lungs and bronchia in minute doses as—℞ Tinct. Phosphorus gtt. j. to gtt. iij., water ℥iv.; a teaspoonful every hour.

SENEGA. *Specific Indications.*—The cough is deep, succussive; much rattling in the chest; free expectoration of mucus, or muco-pus; skin is harsh and dry and the epidermis desquamates, or it is relaxed and the surface looks dull and dead.

Dose.—The Syrup Senega may be used in doses of one-fourth of a teaspoonful, as a stimulant to the respiratory apparatus. Or the tincture may be employed, as—℞ Tinct. Senega gtt. v. to gtt. xx., water ℥iv.; a teaspoonful every one to three hours.

The stimulant influence of Senega upon the throat and bronchial mucous membrane is well known, and is probably its most important use. For this purpose I prefer to use it in the form of tincture to that of syrup as commonly employed. In chronic bronchitis with profuse secretion, it may be combined with small doses of Ipecac and Veratrum.

Its influence upon the kidneys and reproductive organs needs to be studied, and I have no doubt some important uses will be found for it. I have employed it in squamous disease of the skin, and like its action very much; it is one of a very few remedies that influence these diseases.

SCILLÆ. *Specific Indications.*—Cough with secretion of a yellowish muco-pus, mucus rattling in the chest, scanty urine, feeble circulation.

Dose.—℞ Acetum Scillæ ℥ss., Syrup ℥iss.; from one-fourth to one-half teaspoonful every one to three hours.

Homœopathic Indications, (3d to 6th decimal).—Whooping cough, sounding loose, with sneezing and watering of the eyes and nose. Catarrhal affections, with loose sounding cough; more expectoration in the morning. Wheezing breathing. Pneumonia. Pleurisy. Asthma. General anasarca. Hydrothorax. Frequent desire to urinate, with profuse discharge of pale urine.—*Ehrmann.*

STICTA. *Specific Indications.*—The patient complains of pain in the shoulders extending to neck and back of the head; the child will be observed to draw its shoulders upward, throw the head backward and move it uneasily. There is a harsh, dry cough, evidently from irritation and not to remove secretion.

Dose.—℞ Tinct. Sticta gtt. v., water ℥iv.; a teaspoonful every one to three hours.

Homœopathic Indications (3d to 6th decimal.) Dry coryza, with constant desire to blow the nose. Chorea, constant involuntary motion of the feet. Excessive dryness of the nasal passages and soft palate; deglutition painful on this account. Dry, hacking cough.—*Ehrmann.*

Sticta is a very fine remedy when the indications are clear as above. Why a pain in the shoulders, neck and to occiput, call for sticta, is more than I can tell, any more than I can tell why a sick person should have a pain in his shoulder and back of neck and head. But all we want to know in medicine is, this relation between a definite symptom or symptoms, and the curative action of a remedy. Sticta is an excellent remedy for cough, relieves irritation, and improves respiration, acts kindly upon the stomach, and is one of our best anti-rheumatics.

RUMEX. *Specific Indications.*—Cough with sensation of fullness in the chest, sighing, yawning, efforts to take a full inspiration.

Dose.—℞ Tinct. Rumex gtt. v., water ℥iv.; a teaspoonful every one to three hours.

We employ Rumex in cases of *bad blood*, with disease of the skin; in such cases it is certainly one of the most valuable alteratives we have. In these cases we not only use it internally, but as a local application. In scrofulous disease, with deposit in glands and cellular tissue, with tendency to break down and feeble repair, I think the Rumex unequaled. Here, also, we use it internally and locally.

GRINDELIA. *Specific Indications.*—The breathing is labored and asthmatic; the cough hard with rattling of mucus; sense of soreness and rawness of chest; chronic ulceration, with feeble venous circulation.

Dose.—℞ Tinct. of Grindelia ℥j., Glycerin, Syrup aa. ℥j.; one-fourth to one-half teaspoonful every two or three hours.

Grindelia may be employed as a cough remedy in fleshy children with feeble circulation, and in asthma with secretion but want of power to expectorate. It is a stimulant to the respiratory apparatus and to the respiratory function. Locally (in the proportion of ℨj. to water Oj.) it may be used in the cure of old ulcers, scrofulous ulcers, and as a means of discussing scrofulous enlargements.

HYPOPHOSPHITES. The compound syrup of the hypophosphites will prove an admirable remedy to relieve irritation of the lungs with atony, checking cough, and giving increased respiratory freedom. At the same time it improves digestion, blood-making and nutrition. The hypophosphite of lime is one of the most certain remedies I have ever employed in the early stages of pulmonary tuberculosis.

REMEDIES WHICH INFLUENCE THE DIGESTIVE APPARATUS.

It cannot be too often impressed upon the physican, that a good condition of the digestive apparatus is of first importance in the treatment of any form of disease. If stomach and bowels are in fair condition we are careful not to disturb them. If there is anything wrong with them, the first object of treatment is to right this wrong.

That system of medicine which irritates the stomach and keeps it in a state of unrest is intrinsically bad, increasing the suffering of the sick, prolonging disease, and greatly increasing the mortality. That system of medicine which disturbs the bowels, causing irritation or atony, wrongs the sick, intensifies suffering, and increases the death rate.

If any one will call up his past experience in this direction he will realize the truth of these statements. How have you felt when suffering from nausea and vomiting, or even from gastric irritation short of this? Did you find it conducive to comfort, to rest, to appetite, digestion and normal functional activity? How would you like a teaspoonful of Lobelia or Sanguinaria before each meal—as a steady diet? How have you felt when suffering from a good old fashioned diarrhœa of six to ten evacuations a day? Is this conducive to comfort, to rest, to appetite, to normal functional activity? When you are on the outside of two or three grains of Podophyllin, one or two drachms of compound powder of Jalap, or a teaspoonful of Cascara, was it comforting to the inner man, and did it give strength to the legs and ability to work? If a well man does not take kindly to these sensations, and finds that they impair his life, and all his functional activities, what must it be with the sick man, woman or child? These are questions which one should put to himself, and then see that they are answered.

If we think of the function and relations of this apparatus we can see additional reasons for the rule I have named. It is the inlet for all the fluid and the foods required by the body, and these are required for sustaining the life. In the olden time it was thought that the sick person required no nourishment, but we now know that the sick may be starved, and are frequently starved to death. A certain digestive power is necessary, even though we are careful to furnish food that requires but little digestion, and this requires a reasonably good condition of stomach and intestinal canal.

If one only thinks of the administration and absorption of medicines he will see the necessity of the rule. Unless the stomach is in fair condition medicines are not kindly received and absorbed. The right remedy may be selected, which, if it gained entrance to the blood, would do that which is needed for the cure; it is given, but wholly fails because absorption can not take place.

The sympathies of the gastro-intestinal canal are very numerous and very sensitive, indeed it seems to be the center of morbid sympathies. If a distant part is involved in disease, the stomach speedily suffers; if the body at large is diseased, the stomach suffers. It is abundantly supplied with ganglionic nerves, and the solar plexus, the center of this system, lies immediately behind it, and is almost directly influenced from it. It is thus related to the circulatory, respiratory, and excretory apparatus.

EMETICS.

The act of emesis seems natural to the nursing child, relieving the stomach of repletion and nourishment that fails to digest. This would point out the first indication for emesis, to relieve the stomach of food that can not be digested, or that is undergoing decomposition.

For this purpose there is nothing better than warm water given freely, and its action in some cases of emergency assisted by tickling the fauces with the finger. A solution of common salt is also very good in such cases, and leaves no bad influence.

The second indication for the use of an emetic is, when there are morbid accumulations in the stomach, from undigested and decomposing food, or from an increased secretion of gastric mucus. It is met with occasionally in the first stages of severe disease and is an unfavorable complication, for such condition of the stomach precludes the taking and digesting of food, and the proper appropriation of remedies. Frequently, in such cases, the medicines given will be ejected from the stomach two or three times a day, or may not be tolerated at all.

The third indication for the use of an emetic is, for the removal of material from the respiratory organs. This is generally mucus, occasionally· mucus and pus. The emetic is only employed in this case, when we have such evidence of the loosening of the mucus as will lead us to believe that it may be removed in this way. Emetics are employed in nauseant doses to aid in softening, diminishing the plasticity, and loosening such accumulations.

The fourth indication for the use of an emetic is, to rouse the nervous system from severe depression, and restore a uniform circulation of blood. For this purpose it is employed in scarlatina maligna, in the severer forms of rubeola and variola, and occasionally in other diseases.

IPECACUANHA. I administer ipecac to fulfill the second indication of an emetic. It is given in the form of powder, mixed with warm water, and assisted in its action by warm water or some warm tea. The dose for a child two years old, will be from three to five grains; for an infant, half to one grain.

ACETOUS TINCTURE OF LOBELIA AND SAGUINARIA.—℞ Lobelia, Sanguinaria, Ictodes, aa., ℥ij.; distilled vinegar, Oij.; alcohol, ℥ij.; make Oij. of tincture by percolation; dose from five to thirty drops.

To fulfill the third and fourth indications for an emetic, I prefer this preparation to any other that I have employed. It is repeated every five, ten, or fifteen minutes, until thorough emesis is produced, and its action is aided by warm drinks.

After the use of any emetic, the child should have warm drinks for some hours, but it is not necessary that they should be objectionable to the taste; a thin corn meal gruel or common tea, or hot water, does very well.

CATHARTICS.

The employment of cathartic medicines for every ailment, and in all conditions, was not only an absurd, but a very injurious practice. It arose from a misconception of the use of the intestinal canal. Instead of being a *cloaca* or drain for the effete materials of the body, it performs the most important part of the process of digestion; as an excretory organ its function is less than the skin, and two-thirds less than the kidneys.

Cathartics act upon the entire digestive tract, and in a manner subversive of natural processes. As the function of digestion is so important to health, furnishing the material for the nutrition of all structures, we should be very careful how we interrupt it and set up unnatural actions.

The frequent use of cathartics depends somewhat on the empirical benefit that follows their use in slight diseases. A person has a headache, a cathartic is taken, and the next day he is well; or he has an indigestion, and a cathartic preventing his eating for the time being, makes him feel better in a day or two; or he has caught cold, and an active cathartic, acting as a derivative, gives relief. But in these cases the influence is more apparent than real, time and abstinence from food being the important requirements.

It may be asserted that a person in the habitual use of cathartics can not enjoy good health, and that the occasional use is injurious in the same proportion. I hold that they should never be employed unless there is a special indication for their use, which indication we will now consider.

The first indication for the use of a cathartic is, to remove accumulations from the bowels that are proving irritant. These are more frequently of undigested food with the natural secretions. The symptoms are, an uneasiness of the child, manifestly from the abdomen; impaired or arrested appetite, and digestion; a uniformly coated tongue, usually with a yellowish shade; and occasionally a peculiar puffy, expressionless appearance of the face.

A cathartic is never indicated when the child is well, though the bowels have not moved for days.

The second indication for the use of a cathartic is, to produce revulsion or counter-irritation in case of serious disease of important organs. Thus, in the past, it was the principal means of reaching a determination of blood to the brain, inflammation of the brain, congestion of the brain, and occasionally inflammation of other parts. Our means of cure, and especially of reaching these diseases, have so increased that this use of cathartics will be rare.

The third indication for the use of a cathartic is, to promote the absorption of dropsical effusions. Even this use is becoming obsolete by the discovery of specific medicines for this purpose, without disturbance of the digestive tract.

Cathartic medicines are employed in small doses to stimulate the digestive tract, and increase its innervation and circulation. They increase the secretion of the digestive fluids and thus improve digestion, and at the same time increase the activity of its excretory glandulæ. For such purpose the remedy is thoroughly triturated with sugar, or sugar of milk, and the dose is below that which would prove laxative. Used in this way, some of this class furnish the most certain and efficient alteratives.

AMYGDALUS. *Specific Indications.*—There is irritation of stomach with nausea, vomiting, sense of heat and burning. The tongue is elongated and pointed, with reddened tip and edges.

Dose.—We prefer an infusion of the fresh bark of the green twigs in half teaspoonful doses, but the tincture may be used in the proportion of gtt. x. to gtt. xx., water ℥iv., (a little ice may be added), in doses of half to one teaspoonful every fifteen to thirty minutes.

The peach-tree bark will be found an excellent remedy to relieve irritation and determination of blood, with its attendant nausea and vomiting. As it relieves gastric irritation it will be found to give rest to the nervous system, and a better circulation of blood.

RHEUM. *Specific Indications.*—There is nausea, vomiting, uneasy sensation in stomach, irregularity of bowels, diarrhœa, with light colored discharges.

Dose.—℞ Tinct. Rheum gtt. v. to gtt. xx., water ℥iv.; half to one teaspoonful every half hour or hour, or compound powder of Rhubarb ℥j., boiling water ℥iv.; make an infusion and strain, and give in half teaspoonful to teaspoonful doses.

Homœopathic Indications (3d to 6th decimal.) Sour smelling diarrhœa in children. Colicky pains before or during stool. Difficult dentition. Longing for various things, but the first morsel satisfies.—*Ehrmann.*

Rhubarb is an old and favorite remedy with our school. The old compound powder in infusion was used to relieve irritation of the stomach, check nausea and vomiting, and cure diarrhœa. For 'the first purpose it was used in small doses frequently repeated; for the second it was continued in teaspoonful doses until the discharges had the color of the medicine, then less frequently. Diarrhœa from cold was readily relieved by it, and it would cure the simpler cases of cholera infantum or summer complaint. The tincture will sometimes be found an excellent stomachic, improving digestion as well as relieving irritation.

LOBELIA. In very minute doses Lobelia allays irritation of the stomach and checks nausea and vomiting. When the tongue is broad and full—atonic—the remedy will sometimes (minute doses) act as a tonic and improve digestion. Its action as an emetic has been referred to.

IPECACUANHA. *Specific Indications.*—There is irritation of stomach, small or large intestine, with determination of blood. Nausea, vomiting, diarrhœa, dysentery; the discharges in each case being somewhat violent and painful. Violent and expulsive cough, with sense of irritation and burning; mucus or muco-purulent expectoration; globular sputa, rusty; hemorrhage.

Dose.—℞ Tinct. Ipecac gtt. v. to gtt. x., water ℥iv.; a teaspoonful every hour.

Homœopathic Indications (3d to 6th decimal.) Constant sensation of nausea, with vomiting of mucus, bile or blood; the nausea continues after vomiting. Colic and diarrhœa, stools look like yeast, smelling sour. Suffocative attacks of breathing, respiration oppressed, anxious, quick; cough, with rattling of mucus in the bronchial tubes. Intermittent fever, nausea and vomiting predominate; chill with thirst, followed by fever; cases in which quinine has failed.—*Ehrmann.*

Ipecac is a favorite remedy in disease of the stomach and intestinal canal. It relieves irritation and quiets nausea, checks vomiting, and improves the functional activity of the stomach. It is a very certain remedy in diarrhœa from irritation, with determination of blood, or muco-enteritis, from the simplest form to the severer cases of cholera infantum. It is also a prominent remedy in dysentery, quieting irritation, relieving pain, and lessening the frequency of the discharges. In all of these cases we generally use it in combination with Aconite.

EUPHORBIA HYPERICIFOLIA. *Specific Indications.*—Diarrhœa, sense of heat in stomach and abdomen; abdomen is hot to the hand, some tenesmus with discharges, which are at times acrid. Bronchitis, with thin, acrid secretion.

Dose.—℞ Tinct. Euphorbia gtt. v. to gtt. xx.; water ʒiv.; a teaspoonful every one to three hours.

The Euphorbia is a most excellent remedy in the treatment of cholera infantum and in diarrhœa, filling a similar place to Ipecac. In some years it will be found preferable to this remedy, in other years the Ipecac will prove the best.

COLOCYNTH. *Specific Indications.*—Wandering pains in the abdomen, seemingly in the course and from contraction of the intestine; the intestines are felt to change their position; noise from the movement of the intestinal contents; tormina and tenesmus; dragging from the umbilicus; frequent desire to stool from pressure in the rectum, with burning sensations.

Dose.—℞ Tinct. Colocynth gtt. ij. to gtt. v. water ʒiv.; a teaspoonful every one to three hours.

Homœopathic Indications, (3d to 6th decimal).—Violent pain in the abdomen, causing the patient to bend double; diarrhœa, worse after eating or drinking, stools frothy, smelling acid or putrid; dysentery, discharges of mucus and blood with tenesmus, morbus coxalgia; when there is a sensasion as of being encircled with an iron band; urine viscid; general shortening of the tendons.—*Ehrmann.*

With the symptoms as above named, Colocynth is an excellent remedy for diarrhœa and dysentery, and for colic. It relieves pain, checks the discharges, and promotes normal functional activity. In infantile colic, with free, acrid discharges, it will be found an admirable remedy if the dose is very small, gtt. j. to water ʒiv., or a Homœopathic dilution is used.

HAMAMELIS. This remedy has been fully studied, and we notice it here as a remedy which gives tone to stomach and intestinal canal, allaying irritation and promoting functional activity. It may be used when the child persistently throws up its food mixed with mucus; in diarrhœa with large, light colored discharges, and when there is prolapsus ani. If the abdomen is full and doughy, the remedy (distillate or Pond's Extract) may be used as a local application, and if there is relaxation of the perineum with prolapse of the bowel, it may be locally applied to the parts.

CHIONATHUS. *Specific Indications.*—Fullness in right hypochondrium; pain in hypochondria, extending to umbilicus; pain in right shoulder; yellow (jaundice) coloration of eyes and skin; colic with excessively green discharges from the bowels; high colored urine, coloring the clothing yellow.

Dose.—The dose will vary from gtt. v. in water ʒiv., a teaspoonful every one or two hours, to gtt. j. to gtt. v. at a dose.

The Chionanthus will be found a most valuable remedy in the treatment of jaundice, and those painful affections of the bowels, associated with irregular action of the liver. When the child suffers with infantile dyspepsia, and there is fullness in the region of the liver, it may also be employed.

UVEDALIA. *Specific Indications.*—Enlargement of the spleen; full abdomen, doughy; enlargement of the liver (liver-grown); enlargement of any part, the circulation being feeble and the tissues atonic.

Dose.—For children I only recommend it as a local application, the affected part being thoroughly rubbed with the ointment of Uyedalia, or with one part of the tincture to two or three parts of cod-liver or sweet oil.

With the indications named there is no remedy equal to it. I have seen the enlarged spleen in the malarial fevers of infancy as early as the third month, and it will be found more frequently than physicians suspect. With this disease of the spleen the patient can not easily be cured of the fever, or if the fever were stopped he would still suffer from impaired blood-making '(lukœmia). It is also valuable in chronic inflammations of any part, the circulation being feeble, and tissues atonic.

NUX. This remedy, fully described in the first class, has a direct action upon the gastro-intestinal canal. It is indicated by evidences of atony, and an enfeebled circulation. Nausea and vomiting with a pallid, expressionless face, is speedly relieved by minute doses, as gtt. j. to water ʒiv. We use it in the cure of diarrhœa, when the discharges are large, the abdomen full and relaxed, and when there is pain simulating colic.

It is one of our best remedies for infantile colic, if there are no evidences of irritation and determination of blood. The pulse is feeble, the extremities cool, and abdomen full.

It is an excellent stomachic, and if an indigestion depends upon an enfeebled innervation and circulation, the patient will be benefited by it. A sallow, expressionless face, yellowness about the mouth, slight yellowness of the eyes, and clay-colored discharges are indications for Nux.

In cholera infantum we say that Nux is the remedy when there is atony of the bowels, with feeble circulation. Aconite where there is irritation with determination of blood; Ipecac or Euphorbia being associated with either.

CHELIDONIUM. *Specific Indications.*—Fullness in hypochondrium, tongue much enlarged and somewhat pale; mucous membranes full and pale; skin full and sallow, sometimes greenish; tumid abdomen; light colored feces; no abdominal pain; urine pale, but cloudy and of high specific gravity.

Dose.—℞ Tinct. Chelidonium gtt. v. to gtt. xv., water ℥iv., a teaspoonful every two or three hours.

Homœopathic Indications, (3d to 6th decimal).—A fixed pain under the inner and lower angle of the right shoulder blade, in chest or liver affections; orbital neuralgia of the right side, with profuse lachrymation; great sense of tightness around the neck above the larynx, hindering deglutition; constipation, stools like sheep's dung; gallstones with jaundiced complexion; reddish or greenish urine.

PODOPHYLLIN. *Specific Indications.*—The tongue is full, face full, abdomen full, véins full; enfeebled innervation through the sympathetic; dull pain, dull headache, dizziness.

Dose.—In diseases of children I prefer a second decimal trituration, which may be given in doses of from one-eighth of a grain to one grain. In older children granules containing 1-40 of a grain of Phodophyllin with 1-8 grain of Hydrastia will sometimes prove beneficial as a stimulant to the stomach and intestine.

Homœopathic Indications (3d to 6th decimal.) Depression of spirits. Giddiness, with sensation of fullness over the eyes. Difficult dentition, very offensive stools, moaning during sleep, half closed eyes and rolling the head from side to side. Diarrhœa, especially in the morning or soon after eating. Prolapsus ani. Prolapsus uteri. Pain in the ovarian region. Suppression of the menses, with bearing down sensation, better when lying down.

Pain in the sacrum, with uterine troubles. Whooping cough, with constipation and loss of appetite.—*Ehrmann*.

We employ Podophyllin and Podophyllum as a stimulant to the sympathetic nervous system, improving innervation to all parts supplied with ganglionic nerves. The indications for its use in wrongs of the digestive apparatus are those of atony—full expressionless tongue, full abdomen, impaired functional activity. In gastric and intestinal dyspepsia, with these evidences of atony, a trituration of Podophyllin will be found an excellent remedy. The liver may be stimulated by it, the portal circulation improved, and the spleen relieved of its overflow of blood.

We employ it as a remedy for diarrhœa when the abdomen is full and doughy, the discharges light in color, mucous, or containing undigested food. Of course the dose is small, much smaller than most physicians use. Some cases of cholera infantum are cured with a trituration of Podophyllin, when the ordinary treatment has wholly failed.

HYDRASTIS. *Specific Indications.*—The mucous membranes are flushed; papillæ of tongue prominent and red; uneasiness in stomach; loss of appetite; impaired digestion; mucoid matter with stools; circulation to surface and extremities feeble. Sore mouth, with increased mucus secretion, thick and tenacious saliva; sore throat, with muco-purulent secretion; sore eyes, with muco-purulent secretion.

Dose.—I prefer for use the yellow alkaloid Hydrastis or Berberin, which is very soluble; one grain to four ounces, make a very good tonic, and is about the strength we would use as a wash for the mouth and throat or as a collyrium.

Homœopathic Indications (3d to 6th decimal.)—Sensation of sinking at the epigastrium with palpitation of the heart; loss of appetite and fainting paroxysms; yellowish leucorrhœa of a very tenacious character, sometimes offensive; malignant and cancerous forms of ulceration; cancer of the breast; constipation with gastric disturbance, flatulence; small-pox when the pustules are dark colored and there is great prostration, the face is very red and facial œdema quite marked; throat very sore.—*Ehrmann*.

DIOSCOREA. *Specific Indications.*—Abdominal pain, shifting, paroxysmal, relieved by pressure or by supporting the abdomen and keeping the patient still. Skin soft, and feels as if perspiration were about to start; extremities cold; uneasy sensation in

lower part of the chest with sense of constriction in epigastrium.
Dose.—℞ Tinct. Dioscorea gtt. v. to gtt. x., water ℥iv.; a teaspoonful every fifteen minutes to an hour.

Dioscorea is an excellent remedy in infantile colic, if it presents the symptoms named, but it will not do to use it when Nux
or Colocynth are indicated. It is also an excellent diaphoretic,
but when used for this purpose I administer it with hot water.
Children will frequently show the evidences of cold, with tendency
to cough, and the respiratory movement will be observed to be
short, as if there were some obstruction in the lower part of the
lungs. The Dioscorea is a good remedy in this case.

APOCYNUM. This remedy has been studied in the second class,
as it influences the circulation. We employ it here as a stimulant to the entire gastro-intestinal tract, the indications being a
full, tense abdomen, the skin glistening, with œdema of some
part. In very small doses it increases the activity of the bowels,
and overcomes constipation, and if the dose is a little too large
will cause diarrhœa. It will be found an admirable remedy in
acute hydrocephalus, as it is in the chronic form of the disease.
Irritation of the nervous system, with prominent eyes, fullness of
the fontanells, or opening of the sutures. A very common indication for Apocynum is œdema of the eyelids, or that wrinkled
appearance that the eyelids present when they have been swollen.

CHAMOMILLA. *Specific Indications.*—Infantile dyspepsia with
irregularity of the bowels; diarrhœa with flatulence and colic,
discharges contain curdled milk or other undigested food; the
person is irritable and restless and the surface alternately flushed
and pale.
Dose.—℞ Tinct. Chamomilla gtt. v., water ℥iv.; a teaspoonful
every one to three hours.
Homœopathic Indications, (3d to 6th decimal).—Great sensibility to pain, making the patient cross and uncivil; children
crying and fretting, must be carried about in order to be appeased;
diarrhœa which smells like rotten eggs, and looks like chopped
eggs and greens, especially during dentition; one red cheek while
the other is pale, with great irritability and thirst; flatulent colic
of infants.—*Erhmann.*

IRIS. *Specific Indications.*—Fullness of thyroid gland, enlargement of lymphatic glands, fullness of spleen, grayish or coppery coloration of skin, chylous discharges from the bowels.

Dose.—℞ Tinct. Iris gtt. v. to gtt. x., water ʒiv.; a teaspoonful every one to three hours.

Homœopathic Indications, (3d to 6th decimal).—Sick headache, with vomiting of mucus, tasting sweet; fullness and heaviness of the head, head and face cold; colic relieved by bending forward; brown and very offensive diarrhœa, with cutting pains, nausea and vomiting, emmission of fœtid flatus.—*Ehrmann.*

LEPTANDRIA. *Specific Indications.*—Fullness of abdomen, doughy sensation to touch; tongue full, pallid, and covered with pasty fur; stools papescent and light colored; tawny, dirty skin, dirty eyes.

Dose.—Leptandria triturated one to ten may be given in half grain doses, or the tincture may be used with glycerine or syrup so that the patient will get from one-fourth to five drops at a dose.

Homœopathic Indications, (3d to 6th decimal).—Black fluid stools, great urging, with difficulty in retaining the stool; cutting pains about the umbilicus; stools like tar; chronic diarrhœa, worse in the afternoon.—*Ehrmann.*

SULPHATE OF MANGANESE. *Specific Indications.*—A pale, leaden tongue, dirty, with pendulous abdomen and sluggish bowels; jaundice with enlarged liver, fullness and weight in right hypogastrium, dropsy.

Dose.—For administration to children I prefer a second decimal trituration, of which one-half to one grain may be given every three hours.

MALT. *Specific Indications.*—The digestion of calorifacient food is imperfect, nutrition is impaired, and there is a tendency to scrofulous or tubercular deposits.

Dose.—Any of the good extracts of malt may be used in doses of one-fourth to one teaspoonful three times a day. Or instead of this an infusion of malt may be prepared and given to the child with its food, or immediately after eating.

It is believed that malt contains a similar ferment to the saliva (ptyalin), and to a less extent to the pancreatic fluid (pancreatin). The first has an especial influence upon starch, changing it into grape sugar; the second will peptonize proteids, emulsify fats, and convert starch into sugar. When these processes are imperfect and children suffer from impaired nutrition, and cacoplastic or aplastic deposits, malt has been found to give good results.

PEPSIN. *Specific Indications.*—Gastric digestion is impaired, and nutrition is imperfect; there are eructations of food and gas, chylous or lienteric diarrhœa, abdomen full, urine cloudy.

Dose.—Of a good pepsin one-fourth to one half grain may be given with or after the taking of food.

OXIDE OF ZINC. *Specific Indications.*—The tongue has a pasty coat, breath bad, eructations of food, waterbrash, gastrodynia; secretions of mouth become so acrid that they excoriate the mucous membrane and lips, and produce soreness of the nipple.

Dose.—It may be used in pill form (granules), the dose being one-tenth to one-fourth grain, or it may be employed in the second decimal trituration in doses of one grain.

Homœopathic Indications (3d to 6th decimal.)—Weakness of memory. Indisposed to converse. Great sensitiveness to noise. Hydrocephalus. Soreness of the eyes, lids and inner angle of the eyes, great itching. Paralysis of the upper eyelid. Otalgia, discharge of fetid pus from the ear. Great burning in stomach after taking sweet things. Great greediness when eating, can't eat fast enough. Nausea and vomiting. Metallic taste in the mouth. Patient can't keep still, must be in motion all the time. Varicose veins which give rise to fidgetiness of the limbs. Puerperal convulsions. Scarlet fever when there is retrocession of the eruption. Child unconscious and motionless. Involuntary jerking and twitching of the muscles. Grinding of the teeth. Screaming spells. Occiput very hot and forehead covered with cold perspiration. Pulse thread-like and difficult to count. —*Ehrmann.*

SANTONINE. *Specific Indications.*—Fullness of the upper lip, white line around the mouth, picking at the nose, foul tongue, fetid breath, full pendulous abdomen, tendency to retention of urine.

Dose.—As a remedy for worms I usually combine it with Podophyllin as in the following—℞ Podophyllin gr. j., Santonine gr. x., Sugar, or Sugar of Milk ʒj.; triturate thoroughly and make twenty powders, of which one may be given night and

6

morning. In retention of urine I have it triturated with sugar, so that the child may have a dose of one-eighth to one half grain every one to three hours.

Santonine is one of the best remedies we have to expel the ascaris lumbricoides, and to so influence the mucous membrane that the intestine will not be a habitation for these vermin. The combination with Podophyllin is very good when there is atony with increased secretion of mucus, but if there is irritation of the intestine, the remedy may be triturated with sugar and a minute portion of Ipecac.

In retention of urine in childhood it is *par excellence* the remedy, and I have not known it to fail in an experience of twenty years. It is not a question of how or why it influences the bladder, but the fact that when there is retention of urine from one to three doses of Santonine will cause its passage.

ALOES. *Specific Indications.*—There is atony of large intestine and rectum, mucoid discharges, prolapsus ani, pruritus ani, ascaris vermicularis.

Dose.—The remedy is a very nauseous one, and it is difficult to use it on this account. A second decimal trituration may be given in doses of one grain, or even a trituration of one to ten may sometimes be used. To remove the ascaris vermicularis, and break up the conditions under which it propagates itself, I have found that the following prescription does well: ℞ Tinct. Aloes, Comp. Tinct. Cardamom, aa. ℥ss., Syrup ℥j.; dose, one-half to one teaspoonful until it acts upon the bowels.

Homœopathic Indications (3d to 6th decimal.)—Sensation of weight or heaviness in the rectum; morning diarrhœa, very urgent, must go at once; rumbling and rolling in the bowels before stool; hemorrhoids protruding, feeling hot and sore; when urinating sensation as though something had passed from the bowels; stools consisting of jelly-like mucus.

REMEDIES THAT INFLUENCE THE URINARY APPARATUS.

We find in practice that, in a majority of acute diseases, the function of the kidneys will be re-established so soon as the circulation is controlled. Hence, in the common diseases of childhood diuretics are not required. It is principally in the malarial fevers, and in zymotic diseases that they will be demanded. In chronic disease, with deficient waste and excretion, they become our most important remedies.

Here, as in the adult, we recognize the two actions, *hydragogue* and *depurant*. In the first, the water of the urine is increased; in the second, the solid constituents are increased. The first may be used to lessen the volume of the circulating fluid, to remove irritations of the urinary passages by diluting the urine, and to promote the absorption and removal of dropsical deposits. The second, increasing the solids of the urine, are used to depurate the blood of worn-out and imperfectly-formed material. They also stimulate the processes of retrograde metamorphosis, and thus facilitate the breaking down and removal of tissue, and indirectly its renewal.

We might give a long list of remedies influencing the urinary apparatus, but a few will serve our purpose, as in diseases of children especially, the kidneys will do their work if the general conditions are right. The agents we will study are :—

Sweet Spirits of Nitre,	Mentha Viridis,	Cucurbita Citrullus,
Acetate of Potash,	Gelseminum,	Rhus Aromatica,
Eupatorium Pur.,	Santonine,	Agrimonia,
Hydrangea,	Apis,	Belladonna,
Benzoate of Lithia,	Local applications to the Loins.	

SWEET SPIRITS OF NITRE. I name this remedy first, because it has been employed so extensively, and with good advantage, in diseases of childhood. In this case it is not only diuretic, but, to a slight extent, sedative—lessening the force and frequency of the pulse. It is only in the simpler forms of disease that I would recommend it. To two ounces of water, in a glass, add a teaspoonful of spirits of nitre, keep covered, and give a teaspoonful every two hours. This is the dose for a child from one to two years of age.

MENTHA VIRIDIS. Though not usually regarded as of much importance by the profession, and as having but feeble diuretic properties, I think in diseases of children it will be found one of the most certain of remedies. It has this advantage, that, prepared in strong infusion and sweetened, it is quite palatable, and it may be given freely without danger.

I give it alone and in combination. With sweet spirits of nitre a very powerful hydragogue action may be obtained. If there is irritation of the urinary apparatus, I generally add tincture of

Gelseminum in small doses.   If the pulse is excited I give Veratrum at the same time.   Or, if there is congestion with tendency to coma, Belladonna.

CUCURBITA CITRULLUS.   An infusion of watermelon seeds is a very mild unirritating diuretic in diseases of children.   The only difficulty we find in using it is, that children will not drink unless they are thirsty, and in sickness they are very particular what they drink, and may not like watermelon-seed tea as a substitute for water.

ACETATE OF POTASH. *Specific Indications.*—The tongue is full, pallid, slightly leaden, and coated with a pasty fur; the abdomen is full and doughy, and the skin dirty.

Dose.—℞ Acetate of Potash ℨj., water ℥iv.; a teaspoonful every two or three hours, with as much fluid as the patient can be persuaded to take.

Of the class *renal depurants*, Acetate of Potash is the most efficient.   It is employed to increase the solids of the urine, and remove the waste from the blood.   When given in large doses, and for some time, it breaks down feeble tissues, and thus hastens the removal of worn-out material.   But having such action, it may be carried too far, to the weakening and breaking down of sound tissue, or to the removal of material faster than it can be replaced.   It is a good remedy in its place, but used to excess it is injurious.

In many cases it may be added to the child's drink, so that the necessary quantity will be taken in the course of the day, without its knowledge.   For a child two years of age the medium quantity will be half a drachm, each twenty-four hours.

Citrate of potash, bitartrate of potash, nitrate of potash, and acetate of soda may be used for the same purposes.

SANTONINE.   We think of Santonine as a vermifuge only, yet it has other desirable properties.   One of them is, its influence over the bladder in retention of urine.   In some diseases there is sometimes a tendency to retention which ordinary remedies will not reach, and which at last proves fatal.   Santonine thoroughly triturated with sugar, in doses of from half to one grain every two hours, affords very certain relief.   It is also very effectual in relieving burning, scalding, tenesmus, and other unpleasant sensations of the urinary passages.

EUPATORIUM. *Specific Indications.*—There is a sensation of weight and dragging in the loins, retraction of testicles, frequent desire to urinate, scanty secretion of urine.

Dose.—℞ Tinct. Eupatorium gtt. x. water ℥iv.; a teaspoonful every one or two hours.

When there is a continued scanty secretion of urine, though the general wrongs have been looked after, we may think of the Eupatorium. The indications as given above, will be found a good guide to its administration.

GELSEMINUM. Gelseminum has been studied in our first class, and we notice it here as a remedy directly influencing the kidneys, relieving irritation and determination of blood. The indications for it are, sharp pains in the loins and back, frequent desire to pass urine, which is passed in very small quantities with tenesmus.

It is not only a remedy for irritation and determination of blood to the kidneys, but for irritation of any part of the urinary passages as well. In cystitis it is a favorite remedy, as it is in irritable (spasmodic) stricture of the urethra.

ERYNGIUM. *Specific Indications.*—Frequent desire to pass water, with vesical tenesmus, contraction of abdominal muscles, drawing up of the thighs, scanty urine, acrid, irritating the external parts.

Dose.—℞ Tinct. Eryngium gtt. v., water ℥iv.; a teaspoonful every one to three hours.

Eryngium is an admirable remedy to relieve irritation of the bladder, and I have no doubt that it exerts quite as strong an influence upon the pelvis of the kidneys and ureters. It also increases the secretion of urine, which is less acrid and irritant.

BELLADONNA. This remedy has been fully studied, and we notice it here as a remedy for congestion or enfeebled circulation of the kidneys, and for diabetes and incontinence of urine. In cases where the patient complains of fullness, weight and dragging in the loins we prescribe Belladonna. When the urinary secretion is too free, the urine of low specific gravity, the administration of Belladonna internally, or the application of a Belladonna plaster to the loins is good treatment.

In incontinence of urine this remedy has given most excellent results, especially in those cases where there was inability to hold the urine during the day, and a dribbling away which soiled the clothing.

RHUS AROMATICA. This agent is highly recommended for incontinence of urine, especially for nocturnal incontinence. It is also used in irritable bladder; in chronic cystitis and urethritis; in phosphuria, and when there are mucoid discharges in the urine.

Dose.—℞ Tinct. Rhus Aromatica gtt. v. to ℨj., water ℨij.; a teaspoonful every two or three hours.

AGRIMONIA. *Specific Indications.*—Renal colic, pains extending from the hypochondria to the loins, and thence down to the bladder. Irritable kidneys; irritation of kidneys or the urinary passages, with cough; cough attended with expulsion of urine.

Dose.—℞ Tinct. Agrimonia gtt. v. to gtt xxx., water ℨiv.; a teaspoonful every one to three hours.

HYDRANGEA. *Specific Indications.*—Irritable bladder and urinary passages, urine passed with difficulty and with tenesmus, blood in the urine, strangury, irritation of urinary passages with cough.

Dose.—℞ Tinct. Hydrangea gtt. v. to gtt. x., water ℨiv.; a teaspoonful every one to three hours.

APIS. *Specific Indications.*—Itching of the genitals; itching along the urethra; burning in the bladder; itching of the anus; itching and burning of the skin; eyes itch and burn; child wants to rub them frequently.

Dose.—℞ Tinct. Apis gtt. ij. to gtt v., water ℨiv.; a teaspoonful every hour to three hours.

Homœopathic Indications (3d to 6th decimal).—Burning and stinging pains with scanty secretion of urine; breathing labored, with fever without thirst; dropsy without thirst, and waxy appearance of the skin; œdematous swelling of the face, especially about the eyes; hydrocephalus, with sudden shrill cries; boring of the head in the pillow, squinting, grinding of the teeth, urine scanty, twitching on one side while the other is paralyzed.— *Ehrmann.*

With the indications as above named the Apis will be found an admirable remedy. I have seen it start a free secretion of urine in a few hours, when it had been nearly arrested. It relieves irritation of the nerve centers, gives rest, improves the circulation, and especially removes hyperæsthesia of the cutaneous nerves.

BENZOATE OF LITHIA. *Specific Indications.*—There is difficult urination, discharge of mucus, muco-pus or phosphates with the urine; the skin is doughy or waxy; the tongue dirty; the breath fetid, and general evidence of imperfect waste and nutrition.

Dose.—It may be given to a child two years old to the amount of one grain each day in the water the patient takes.

Homœopathic Indications (3d to 6th decimal).—Urine scanty, of dark brown color, and strong urinous odor; nocturnal enuresis; rheumatism, with strong smelling urine; hypochondriasis, sore throat and diarrhœa, with the above characteristic urine.—*Ehrmann.*

REMEDIES THAT INFLUENCE THE SKIN.

It is hardly necessary to call attention to the importance of a healthy skin, as associated with good health, and to its impairment as a frequent cause and constituent of disease. This organ not only removes a large amount of nitrogenized waste, but it also regulates the temperature, and assists in the respiratory function. If its function as a regulator of the temperature (safety valve) is impaired, we will have a wrong of this condition of life which will work a wrong of every function of life. If it fails to do its work of excretion, the blood must suffer from the retained material, unless the kidneys and bowels do vicarious work. Impairment of the skin thus imposes additional labor upon the lungs.

In studying the pathology and therapeutics of the skin we must not forget its intimate relationship with mucous membranes. The skin and mucous membrane may be regarded as one piece—folded in, forming the internal lining or envelope, covering the body as an external envelope. The sympathy between the two is so intimate that wrongs of the skin work wrongs of mucous membrane, and wrongs of mucous membrane work wrongs of skin. An impairment of the functions of the skin may work such wrong of mucous membrane as to imperil life.

We see this illustrated in some of the simpler eruptions upon the skin. A child suffers from *heat*, or some of the simpler forms of erythema or urticaria, but as long as it is out freely upon the skin, the functions of the body are well performed. From some cause the eruption disappears (goes in), and at once the child is very sick, has bronchial irritation, croup, or intestinal irritation with diarrhœa. A cure in such cases will come from bringing the eruption to the surface, prevention of disease will come by keeping the eruption on the outside.

The remedies we will study in this connection are the following:

Baths,	Sedatives,	Rhus,
Diluents,	Stimulants,	Apis,
Asclepias,	Inunctions,	Senega,
Saffron,	Nepeta,	Sulphur,
Belladonna,	Serpentaria,	Arsenic.
Eupatorium,		

SEDATIVES. In proportion to the frequency of the pulse and the increase of temperature is the arrest of secretion. If, therefore, there is a frequent pulse, the first step toward obtaining a normal action of the skin is, to bring it down to the normal standard. The arterial sedatives thus become the most important diaphoretic means. With small and frequent pulse Aconite is our best diaphoretic ; with a full, frequent pulse it is Veratrum ; with an oppressed pulse Lobelia, etc.

In a large number of cases of fever and inflammation, if proper means are used to lessen the frequency of the pulse and bring the temperature to a normal standard, secretion from the skin is established in proportion as these effects are obtained. If secretion does not start as the pulse comes down, very mild diaphoretic remedies will be sufficient to accomplish the purpose. With a frequent pulse and high temperature no diaphoretics will act.

STIMULANTS. There are cases where, from a feeble general circulation, and an enfeebled skin, secretion can not go on. This will be known by the dull color, want of elasticity, coldness, and want of sensibility. In such cases anything that will stimulate the general circulation, and will stimulate the skin and its circulation, will improve functional activity. I have used in such cases something like the following, with good results : ℞ Tinct. Lobelia ʒj., Comp. Tinct. Lavender ʒiij., Syrup ʒiss.; give in teaspoonful doses, mixed with hot water every hour. ℞ Tinct. Ipecac gtt. x., Comp. Tinct. Cardamom ʒss., Syrup ʒiss. ; teaspoonful every half hour or hour, with hot water. An infusion of ginger or of Asarum canadense may sometimes be given.

BATHS. We have already studied the use of baths, as a means of rectifying wrongs of the temperature, but we can not learn too much about these means, and a little repetition will not be injurious, and we now want to know how they influence secretion.

The success of Eclectic practice has, I think, depended quite as much upon the skilled use of the bath, as upon the administration of internal remedies, and from the days of Beach great stress has been put upon "bathing the sick." It came in at a time when water was thought to be injurious, whether applied to the surface or given as drink, for water and mercury were antagonistic elements, and mercury the sick must have. In simple opposition, therefore, if for no other reason, we washed our patients frequently, and gave them as much water as they could drink.

The Alkaline Sponge Bath.—" The alkaline bath," was the one most generally recommended and used, and this was a potash, not a soda bath. In early days it was made by adding wood-ashes to water (broke water) so as to render it slightly slippery ; now we use bicarbonate of potash. It has a very kindly influence upon the skin, softening it, lessening the temperature, promoting a better flow of blood and increased secretion. In some cases soda may be substituted for the potash with advantage, but as a rule the old alkaline bath is the best. There are cases in which the skin is influenced better by simple soap and water.

The Acid Sponge Bath.—When the tongue and mucous membranes show a deep-red or dusky color, and the skin also shows dusky redness, an acid bath will sometimes be preferable. If the ordinary bath is used it seems to leave the skin drier and harsher than it was. If we now add a portion of vinegar to the water, it softens the skin, lessens the temperature, and aids secretion.

Cold Wet-Sheet Pack.—In recent disease where the attack is acute and the patient strong, the cold wet pack may be used to advantage. The temperature is increased, the pulse frequent and strong, and the skin dry. The only objection to it is, the fear of the parents that some harm may follow, and the indisposition of the child to have such liberties taken with it.

Resolving on a cold wet pack, we have the little fellow stripped, a sheet of proper size wrung out of cold water and wrapped around him, and then tuck him in his crib or bed, with plenty of covering. Half an hour to one hour is about the time we should allow him to be in it, when he may be thoroughly dried with rubbing, and have dry warm clothing put on.

Local packs may be used to any part which is suffering from increased heat and arrested secretion. The point we place stress upon is, that there shall be a vigorous circulation. It will not do to trust to the *reaction* to obtain a good circulation and temperature, though sometimes the reaction from the cold is sufficient for this.

Hot Wet-Sheet or Blanket Pack.—When the skin is feeble, its circulation impaired, and especially when we want to bring an eruption upon the surface, we use the hot blanket pack. Sometimes an internal congestion will demand speedy means of relief. In the eruptive fevers, if the eruption fails to make its appearance the patient becomes dull, sleepy or comatose, and the skin has a dull pallor or is dusky—here the hot pack is a good thing.

In using this, a blanket is wrung out of hot water, and applied around the child, which is then warmly covered in its crib; in some cases mustard may be added to the water to make it more stimulant. As the child does not like the hot application any better than it does the cold, and will not bear it hot enough for benefit, we sometimes apply the hot blanket over a thin undershirt and drawers; it can then be very much hotter.

Local hot packs are employed when parts are feeble, and have a sluggish circulation.

The Hot Foot Bath.—Every person thinks that he or she knows all about the use of this simple means, but yet we find in practice that not one in ten can use it so as to benefit a child. The adult will have a *full* bucket of *hot* water, and the limbs covered; the child will have one or two inches in a shallow basin, not very hot, and the limbs exposed up to the body. In the majority of cases the nurse dashes the water upon the child's legs with the hands, thus producing rapid evaporation, and of course refrigeration. Necessarily, the desired influence is not produced, and very frequently it has the contrary effect to that desired. So confident am I that the foot bath will be used in this way, that I never order it without giving specific directions, and many then misuse as above.

Hot Sponge Baths. Sponging the surface with water as hot as it can be borne is sometimes a most admirable stimulant to the surface, and very much better than the hot bath. If the skin is very inactive, with a feeble circulation, I recommend the use of hot water with a sponge lightly and briskly applied, the patient being covered with a blanket. In the eruptive fevers, when there is a tardy appearance of the eruption, no better means can be found than this; in some cases the eruption seeming to follow the hot sponge. In these cases I am in the habit of directing the nurse to apply the hot sponge to any part that suffers pain or uneasiness, or wherever the eruption may not come freely to the surface.

A hot iron wrapped in flannel and wetted with vinegar so as to raise a vapor is a still more active application, and one of the most powerful stimulants known. It is claimed that a current of electricity is thus generated, and we faradize the surface as well as stimulate it with heat.

The Hot Bath.—A general hot bath may be given in any vessel large enough to hold the child. The water should have a temperature of 125 degrees at least, and should cover the entire body. The child should not remain too long in it, and when removed should be wrapped in a warm blanket until thoroughly dry. Common salt may be added to the water, when there is much debility, or a tendency to congestion. With care, such a hot bath may be used with advantage in the early stage of acute diseases.

Inunctions.—There are cases where fatty inunctions answer a better purpose than baths. The bath leaves the skin dry and harsh and inactive; the inunction leaves it soft and pliable, and in better condition for functional activity.

Sometimes we use lard alone, or, if there is a tendency to aplastic deposits, cod-liver oil. When there is periodicity, or when we wish to get the action of quinine it is added to the lard. If we wish the inunction stimulant, we add one of the essential oils in small quantity (they are very active when combined with fatty matter).

In scarlet fever we sometimes have the patient rubbed thoroughly with a bacon-rind or piece of fat bacon. Or, we may order it from the drug store as follows:—℞ Creasote gtt. xx., Common Salt ℨss., Lard ℥ij. Ammonia added to the lard makes a very stimulant application, and I sometimes order it when the skin is very feeble, or when it is necessary to bring an eruption to the surface, as—℞ Bicarbonate of Ammonia or Hydrochlorate of Ammonia ℨss., Lard ℥ij. Mix.

DILUENTS. A certain amount of fluid is necessary to the proper performance of function, and I am satisfied that many times the sick suffer severely from the want of it, either because the physician does not think it worth his while to look after such small matters, or the condition of disease is such that it can not be introduced. Here was one of the evils of the old practice—they used remedies which were incompatible with water, or which kept the stomach in such condition that water could not be absorbed by the stomach.

Warm Drinks.—An infusion of catnip, sage, chamomile, asclepias, and many other simples, or even hot water, will prove diaphoretic, and is sometimes useful in slight ailments. The difficulty in their use is the repugnance the child has to anything unpleasant, and to the quantity necessary to be given to get a diaphoretic influence.

If warm drinks can not be taken, or sometimes if they can, we think it better to administer cold. I make it a rule to ask if my little patient has expressed a desire for water, and advise that it be given frequently if the child has any thirst. Sometimes we will find that the little one takes it eagerly, and following its use the irritation and restlessness subside, and it goes to sleep. Again, we will find that the frequent giving of cold water in moderate quantities lessens the temperature and the frequency of the pulse, and the child is better in every respect.

In some cases the child may have lemonade to good advantage, either hot or cold, as it will take it best. In others, whey answers a most excellent purpose, and is taken with pleasure.

When the temperature is high, and fluids can not be taken by mouth, I think it well, in some cases, to use it by enema. Four ounces of cold water thrown up the rectum will lower the temperature of the entire body two or three degrees. If there is continued nausea a very small portion of common salt may be added to the fluid used in this way.

ASCLEPIAS. *Specific Indications.*—The pulse is strong, the patient has *sharp* pain in some part, inflammation of serous membranes, skin hot and dry but not constricted, urine scanty, painful cough.

Dose.—℞ Tinct. Asclepias gtt x., water ℥iv.; a teaspoonful every hour.

Asclepias may be taken as the type of a diaphoretic, acting kindly, and promoting that normal activity, " insensible perspiration." The skin softens, the temperature comes down, the pulse is less frequent, and the nervous system is relieved. It is especially a good remedy in pleuritis or pleuro-pneumonia, and in rheumatism involving synovial membranes.

EUPATORIUM PERF. The Boneset has a direct influence upon the skin, and may be employed to good advantage when the pulse is full and strong, and the skin gives the sensation that perspiration is almost ready to break out. It may be employed alone or in combination with the sedatives.

CROCUS SATIVUS. *Specific Indications.*—The skin is dull, mottled and inactive, afterwards becomes exceedingly dry and harsh with irritation of mucous membranes. An eruption shows itself beneath the skin, or there has been a retrocession.

Dose.—Saffron may be used in infusion, or the tincture may be given with warm water.

True saffron is one of the most certain, as it is one of the kindliest of diaphoretics, and when it can be obtained it should be employed, especially in the eruptive fevers. It will be found very much better than the old remedies, which irritate the stomach and bowels.

NEPETA CATARIA. *Specific Indications.*—The skin is dry and harsh, mucous membranes of the air passages irritable, urine scanty, more or less pain in abdomen.

Dose.—Catnip may be given in sweetened infusion in irritation of the respiratory and intestinal mucous membranes, or the tincture may be added to water in the proportion of gtt. x. to gtt. xx. in water ℥iv., and given in teaspoonful doses.

It must not be supposed that because catnip is such a common domestic remedy, it is worthless, for it will sometimes answer a better purpose than the stronger and harsher medicines. Used with the sedatives it gives excellent results in fevers complicated with irritation of mucous membranes.

SERPENTARIA. *Specific Indications.*—The skin is inactive, harsh, desquamation of epidermis, epidermis feels rough and dead, weight and dragging in the loins, scanty urine, fullness of chest, difficult respiration.

Dose.—℞ Tinct. Serpentaria gtt. v. to gtt. x., water ℥iv.; a teaspoonful every one to three hours.

BELLADONNA. Though a remedy which is studied with especial reference to the nervous system, it also influences the skin directly, and is thus one of our most important remedies. The reader will remember the statement that " it causes contraction of capillary blood-vessels, and thus overcomes congestion." This is its most important action, and this it does in the skin and kidneys as well as in the brain.

Congestion of the skin is known by the deep coloration, dusky red, or when bright red, and the redness is effaced by pressure, a persistent white line is left. In the eruptive fevers its influence is to bring the eruption to the surface. This is shown in the simpler eruptions, erythema and urticaria, and in the severer eruptive fever—scarlatina.

Dullness of mind and disposition to sleep are almost always associated with congestion of the skin, but in some cases there will be marked impairment of the capillary circulation of the skin before the brain suffers.

RHUS. This remedy exerts a direct influence upon the skin, being indicated by bright redness and burning pain. It is a remedy in hives, with burning and itching; in scarlet fever when the skin is hot, dry and burning; in erysipelas where there is vivid redness with burning pain.

Of course we are not altogether guided by symptoms from the skin, though these would be sufficient. The *sharp* stroke of pulse, frontal headache, and prominent red papillæ on tip of tongue will be additional indications.

APIS. Apis is indicated by itching of the skin or outlets of the body. It is a most valuable remedy when thus indicated, as in "heat," hives, scarlet fever, prurigo, erysipelas and some cases of eczema.

SENEGA. Senega is a stimulant to the skin, and may be employed where there is a feeble circulation, coldness of the extremities, and dry, harsh, dead epidermis.

SULPHUR. This remedy, which is fully studied under the head of restoratives, exerts a direct influence upon the skin, improving its nutrition and functional activity. Dryness and separation of the epidermis, vesicular or pustular disease, or change of color will indicate it. It has been used both locally and as an internal remedy, and in both ways proves beneficial.

ARSENIC. *Specific Indications.*—Inactive skin, wants elasticity; dull, sallow, or pallid color; eczematous eruption; feeble circulation; periodicity not cured by quinine.

Dose.—For children, in ordinary cases, I prefer a second or third decimal dilution, though I sometimes use the pellets wetted with Fowler's Solution, or with a first Homœopathic dilution. In skin disease where a stronger action is wanted we prescribe—℞ Fowler's Solution gtt. v. to gtt. ·x., water ℥iv.; a teaspoonful every·four hours, or three times a day.

Homœopathic Indications (3d to 6th decimal).—Great restlessness and anxiety with burning pains, violent thirst with frequent drinking of but little water at a time; vomiting and diarrhœa, stools watery, offensive, and undigested, dysenteric stools; patient wants to be in a warm room, or warmly covered; attacks of anguish, with fear of death; face pale and haggard, great exhaustion; suppression of the urine or bloody urine; oppressed breathing, œdema of the eyelids, urticaria, eczema, pityriasis, falling out of the hair, trembling, stiffness and contraction of the joints; aggravation of symptoms after midnight, on lying down with the head low, and in the cold air.—*Ehrmann.*

ANTI-ZYMOTICS.

We have been in the habit of classifying such remedies as oppose the putrefactive process in blood as anti-septics, but it is better perhaps to study them as antizymotics. Zymosis is more ╷ like a process of fermentation; as Dr. Dunglison defines it, it is "any epidemic, endemic, contagious, or sporadic affection, which is produced by some morbific principle acting on the organism similar to a ferment—as the major exanthemata."

The symptoms of zymosis are prostration and impairment of life, and those that are usually designated as typhoid. The tongue is *dirty,* or the coatings are *brown* or deep colored; in the first case the tongue is moist, and in the second it is inclined to be dry. In both cases sordes accumulate around teeth and gums, and the discharges from the body become peculiarly offensive.

The causes of zymosis it is claimed, act similarly to if not as ferments, setting up in any fluids of the body septic processes which produce material like the original. Thus the poison of typhoid fever introduced into the body will grow a condition of disease like the original, and will generate a poison similar to that introduced. We notice the same in erysipelas, in zymotic dysentery, in diphtheria, and in other diseases.

An anti-zymotic is a remedy, therefore, which, introduced into the body (blood), will antagonize such ferment, and arrest the process that is going on to the destruction of life. In the case of some of these remedies, we can see how they act, for they will stop the process of fermentation outside of the body as well as within. But in others, we can not tell how this is accomplished, because the remedy has no chemical action, and is used in very minute doses. But whether we can account for this action or not, if experience proves that definite results follow the administration of a remedy, we should be satisfied to give it when indicated.

It will also be noticed that we have a group of these remedies, all antidoting the process of zymosis, yet each having an individual action, and not being interchangeable one for the other. Indeed, they are so different in their construction that we would hardly expect them to act in a similar manner. As we can not interchange them, and must have the right remedy, if we wish good results, we will be especially careful in examining the symptoms that point to the remedy.

The remedies we will study under this head are :—

Acids,	Sulphurous Acid,	Hydrochloric Acid,
Baptisia,	Sulphate of Soda,	Chlorate of Potash.
Phytolacca.		

Acids. We have studied these under the head of remedies that influence the respiratory function, to which the reader is referred, but it is well to examine them in this relation. The indications for their use is, a *deep-red* color of tongue and mucous membranes, the tongue being usually dry and contracted, coated brown, or slick and like raw beef; deep redness of parts, when blood comes freely to the surface, may also be noted.

As muriatic acid is studied separately we will notice here that acetic and lactic acids will sometimes fulfill the indication. We give the first in the form of a good sharp cider, which is also a mild stimulant; and the second as whey, formed by curdling milk. Both of these will be taken kindly by children, when muriatic acid would be refused.

Hydrochloric Acid. *Specific Indications.*—The tongue is dry and contracted, and the color *deep-red* or dusky-red. Coatings of tongue are inclined to be brown, growing darker as the disease advances, and there is dark sordes about the teeth.

Dose.—It is added to water so as to make a pleasant drink, a little syrup being added, and given *ad libitum.*

Homœopathic Indications (3d to 6th decimal).—In typhoid fever, when the patient has a tendency to slide down towards the foot of the bed; great debility; diarrhœa with great itching and soreness of the anus; hemorrhoids excessively painful to the touch, bleeding profusely; involving discharges of feces while urinating; sad and silent during the menses; early and profuse menstruation; scarlet fever, rash intensely red with great dizziness; sore throat, foul breath, acrid discharge from the nose.— *Ehrmann.*

With the indications named no remedy will give better satisfaction than this acid. In some cases it seems to do everything that is necessary in the case. It lessens the frequency of the pulse, brings the temperature down, relieves irritation of the nerve centers, places the stomach in better condition for the reception and digestion of food, softens the skin, and aids secretion. It makes no difference where or what the disease is, if the remedy is indicated it proves beneficial.

SULPHUROUS ACID. *Specific Indications.*—The tongue is of normal redness and moist, but *dirty*, the coating is pasty and sometimes looks like fecal matter, the tongue is dull and atonic and has no papillæ, the fauces and phaynx full and relaxed, and the mucoid secretion brownish and dirty, moist sordes about the teeth, fetid breath.

Dose.—This is a feeble acid and may be given in doses of gtt. v. to gtt. xx., and the following prescription will be very good for children from two to five years: ℞ Sulphurous Acid ℨj., water ℨiij.; a teaspoonful every three hours.

Sulphurous acid is a valuable remedy in diphtheria, cynanche maligna, malignant scarlet fever, in some cases of small pox, and indeed in any contagious or zymotic disease where the above indications are found. It is not only a good remedy when administered internally, but it is also an excellent local application in diphtheria and cynanche, and where a part shows a tendency to brake down.

7

SULPHITE OF SODA. *Specific Indications.*—The tongue is broad and pallid, the coating pasty, white and dirty, frequently the breath is fetid.

Dose.—The dose for a child will be from one to five grains, according to age, repeated every three hours.

If one will test the sulphite of soda when indicated as above, he will become a convert to specific medication. It acts kindly and directly, not only in controlling zymosis, but on the circulation, temperature, nervous system, and. in establishing excretion. It is a fair example of the therapeutic maxim, "when a remedy is indicated by a special symptom or symptoms, it proves curative in any form of disease."

CHLORATE OF POTASH. *Specific Indications.*—Fetid odor, as from cynanche maligna, fetid lochial discharge, as from decomposing animal matter. As a local application it is indicated by bright redness of parts.

Dose.—For internal administration 'to children' the dose will vary from one-eighth of a grain to one grain; as a local application a solution of grs. x. to grs. xx., in water ℥iv.

Chlorate of Potash has been so extensively used that there is less need of recommending it to readers, than of caution against its abuse. When indicated it is a most excellent remedy, none better; but when not indicated, especially when tissues are dull and dusky in color, and the patient has a feeble circulation, it may do serious damage.

It should be used with great care in the eruptive fevers, or when there is irritation of the kidneys, for it will sometimes impair the functions of these organs to such an extent that the patient will suffer, or may die from uræmia.

BAPTISIA. *Specific Indications.*—The face is full and purplish, like one who has been exposed to severe cold; the tongue and mucous membrane of fauces and throat are full, and bear the same purplish or leaden discoloration.

Dose.—℞ Tinct. Baptisia gtt. x., water ℥iv.; a teaspoonful every hour.

Homœopathic Indications (3d to 6th decimal).—Typhoid symptoms, dull headache, soreness as if in the brain, sleeplessness, can not go to sleep because "she can not get herself together;" great dryness of the mouth and tongue, tongue coated, dry and brown;

putrid, offensive breath, very fetid stools, preceded by colicky pains in the hypogastrium.—*Ehrmann.*

The symptoms indicating this remedy are so clear that one can hardly go astray. The first sight of the person is frequently sufficient—the full, discolored face, like one that has been long exposed to cold—if not the bluish or purplish coloration of tongue and mucous membranes, is sufficient. It exerts an admirable influence in typhoid disease, whether fever or inflammation, in some cases of diphtheria and cynanche, and in zymotic dysentery. When the indications are strong it will sometimes do everything that one could expect of medicine—lessening the frequency of the pulse, reducing the temperature, relieving the nervous system, putting the stomach in good condition for the reception of food, and establishing secretion and excretion.

PHYTOLACCA. *Specific Indications.*—There is sore mouth; small blisters raise on the tongue, mucous membrane of the mouth and fauces; apthæ; sore throat; diptheritic exudation; soreness and rawness of the mucous membrane of the nose; sore lips, they look blanched, and the epidermis separates; enlargement of parotid and submaxillary glands; the cervial lymphatics are enlarged; disease of the breasts; orchitis.

Dose.—℞ Tinct. Phytolacca (fresh root) gtt. v. to gtt. x., water ℥iv.; a teaspoonful every one to three hours.

Homœopathic Indications, (3d to 6th decimal).—Syphilitic and mercurial rheumatism; nightly pains in the tibia, with nodes and irritable ulcers on the leg; sensation of roughness and rawness in the throat; feeling of a lump in the throat, causing a continuous desire to swallow; deep ulcers on the tonsils or base of the tongue; very offensive odor, and great pain on swallowing; habitual constipation, feeling of fullness in abdomen before stool; capacious flow of urine; albuminous urine; inflammation, swelling and suppuration of the mammæ.—*Ehrmann.*

Phytolacca is one of the most certain remedies in the materia medica, meeting the most grave forms of disease. It is indicated in a large number of cases of diphtheria, and seems to antidote the zymotic cause, and in some seasons will cure nearly every case. It is also the remedy for parotitis, or mumps. We prescribe it for infantile sore mouth with most excellent results, as we do for the sore nipples of the mother which are associated with it. It has a very favorable action in scrofula when it manifests itself in the cervical lymphatics.

ANTISEPTICS.

We will embrace in this group those agents which destroy septic poisons, and arrest the processes of sepsis outside of the body. The classification may not be the best, but I hope it will not create confusion. The anti-zymotics are really antiseptics, though as we are in the habit of studying a group of diseases as produced by zymotic poisons, and their influence upon the body as zymosis, it is quite as well to think of the internal remedies which meet this as anti-zymotics.

In medicine, cleanliness takes precedence of godliness, and many times the need of antiseptics may be prevented by strict attention to cleanliness. There is an intimate relation between " dirt and disease," especially if the dirt is the waste of the human body, or that which it subsists upon. Soap and water will stand first in the list of antiseptics.

Good air is a requisite of good life, and many germs of disease overtake people and gain entrance to their bodies through bad air. The air is defiled in many ways, and all these should be thought of and avoided. The sick should have their supply from such sources as will give it as pure as possible. Whilst air is admitted to the room, from without, provision should be made for the egress of the air that has been used. An open fire-place with a fire is the best means of ventilation in ordinary houses; if stoves must be used, they should have the doors open so as to permit a draught.

Fecal matter is especially poisonous in zymotic disease, and should be removed from the room at once, and the vessels thoroughly cleansed and possibly disinfected. When removed, care should be taken that it is not emptied where it will defile the water supply, either by running into the well, or into some stream which supplies other persons.

We will study the following antiseptics :—

Thymol,	Sulphate of Iron,	Chloride of Lime,
Carbolic Acid,	Sulphurous Acid,	Chloride of Lead,
Permanganate of Potash,	Liquor Sodæ Chlorinatæ.	

THYMOL. Thymol, in weak solution in alcohol and water, is used as an antiseptic dressing to wounds and in surgical operations, or to abscesses when they are discharging. Thymol may also be used to disinfect the chamber utensils, and dressings for wounds. In some cases the object is to prevent infection by septic germs, in others it is to arrest putrescence in a feeble part. It may be used of the strength of—℞ Thymol gr. ij. to gr. iv., alcohol ℥ss., water ℥iijss.

CARBOLIC ACID. Carbolic acid may sometimes be used with advantage to arrest the process of sepsis in injuries and surgical operations, and as an excitant to feeble parts. Usually we prescribe—℞ Carbolic Acid ℨj., Glycerine ℨj., water ℥v. Mix. In this strength Carbolic Acid is an excellent dressing for burns and scalds.

We employ this acid in full strength to destroy warts, working it down to the base of the growth with a pine pencil. It causes suppuration, and in this the growth is permanently removed.

SULPHATE OF IRON. Sulphate of Iron in solution is a fair disinfectant for privies, drains, water closets, chamber utensils, etc., though it is not as good as the other named.

SULPHUROUS ACID. This remedy, studied with the antizymotics, is also a most excellent antiseptic; bowels may be cleared with it, a diseased surface sponged with it, dressing thoroughly disinfected with it, and finally, if used with a spray apparatus, the air may be disinfected.

The old method of burning sulphur in a room answers an excellent purpose. It is easily done by holding a pan, containing the sulphur, over the chimney of a coal oil lamp. In the olden time they sprinkled the sublimed sulphur over a pan of live coals. In cases of very severe small-pox I have seen the air so freshened by burning sulphur that the patient was improved within an hour.

CHLORIDE OF LIME. I like the old antiseptic, Chloride of Lime, quite as well as any for the disinfection of water closets, drains, sewers and privies, and sometimes to purify the air of rooms. This agent should be fresh, and as they now put it up in pound and half pound cans, it can be procured and used of full strength. For use in the house it may be placed in a saucer and sprinkled with vinegar.

LIQUOR SODÆ CHLORINATÆ. The officinal solution of chlorinated soda is a valuable disinfectant and antiseptic. It is especially useful to cleanse vessels, surgical dressings and bandages, water closets, drains and sewers; used with an atomizer or spray apparatus it will purify the air of a room and destroy disease germs. In severe diphtheria, in scarlatina maligna, and small-pox, its frequent use is of advantage to the patient, and prevents the spread of the contagion.

CHLORIDE OF LEAD. Take of Nitrate of Lead half a drachm, dissolve it in a pint of boiling water; disolve two drachms of common salt in a bucket of water, and pour the two solutions together. The clear, supernatent fluid will be a saturated solution of chloride of lead. "A cloth dipped in this solution and hung up in a room will sweeten a fetid atmosphere instantaneously; or the solution thrown down a sink or water closet or drain, or over a heap of refuse, will produce a like result." It is a cheap disinfectant and antiseptic, and may be used freely at but little expense.

PERMANGANATE OF POTASH. We rarely employ the permanganates internally on account of their irritant action upon the stomach. But as a local application to arrest the septic process, and to prevent rapid destruction of tissue, it has no superior. It is used in superficial and phlegmonous erysipelas, suppurative skin diseases, in the treatment of superficial inflammations when parts manifest debility, in the treatment of abscesses, and other suppurations the result of inflammation, and the treatment of wounds and injuries. The solution for external use will be made in the proportion of from one-half to one drachm of the salt to one pint of water. When the tissues are greatly debilitated it may be used as strong as ten grains to the ounce of water.

ANTI-RHEUMATICS.

We have a peculiar inflammatory process which we call rheumatic, and a variety of pain which we also call rheumatic, and both of these are due to a special cause. Not only have we rheumatic inflammation and rheumatic pain, but we frequently find the rheumatic cause influencing other diseases. What this cause is, is not accurately known. It has been thought to be *lactic acid* generated during retrograde metamorphosis, or by decomposition or mal-digestion of food. Whether lactic acid or not, it is certainly a product of retrograde metamorphosis or food decomposition, and may be acid in one case, an ammonia product in another, and neutral in another. Whatever it is, however, it is specific in its nature, and is met by specific remedies.

We might have studied this group of remedies under the head of "rheumatism," but I think it will serve our purpose better to examine them here, especially because we employ them in diseases that would not be called rheumatic. Then again, these remedies have other actions that are important, and we can see them all to better advantage.

The group of anti-rheumatics will be as follows :—

Macrotys,	Sticta,	Salicylic Acid,
Bryonia,	Apocynum,	Alkalies.
Phytolacca,	Acids,	

MACROTYS. Specific Indications.—Muscular pain, pain is intermittent, as if due to muscular contraction, parts swollen tense, pain somewhat throbbing (no evidence of suppuration), pains in inguinal region and groins, lumbar pains.

Dose.—℞ Tinct. Macrotys gtt. v., water ℥iv.; a teaspoonful every one to three hours.

With the indications as above named Macrotys is a very certain anti-rheumatic. As there is fever and inflammation, we usually associate it with the Veratrum. In some seasons it seems to be indicated in almost every case of rheumatism or rheumatic pain, so that we are inclined to think that it will cure every case, and possibly the next year the disease has so changed that it will relieve none.

BRYONIA. This remedy has been fully studied, and we notice it here only as it influences rheumatism. The indications are sharp pain, the serous membranes are involved, the pain is tensive, the pulse is full and hard, and the patient inclined to cough.

APOCYNUM. Apocynum was studied under the head of remedies which influence the circulation. The indications for it in rheumatism and rheumatic pains are—œdema of feet, eyelids or other part, and puffiness or œdema of part affected.

SALICYLIC ACID. Specific Indications.— The tongue is full, leaden-colored or purplish, the temperature high, but without dryness of the skin; the breath is sometimes fetid; the tongue is broad with leaden pallor.

Dose.—The best form to administer Salicylic Acid to children is as a Salicylate of Potash, say—℞ Salicylic Acid gr. x. to grs. xx., Acetate of Potash ℥ss., water ℥iv.; dose half teaspoonful

to one teaspoonful every two or three hours. As a local applica-
tian it may be solved with Borax or Chlorate of Potash, the
first being the best antiseptic, as—℞ Salicylic Acid grs. v. to
grs. x., Borax or Chlorate of Potash grs. x., water ℥iv.

Where the case is properly selected Salicylic Acid lowers the
temperature, controls the circulation, and the specific gravity of
the urine is markedly increased. But when not indicated it
irritates the nervous system. As a local application we employ
it in the treatment of nasal catarrh, and in diseases of the throat
with muco-purulent secretion. It is a good local remedy in some
cases of cynanche maligna, malignant scarlatina and diphtheria.
In these cases it is used with a spray apparatus. With Borax it
may be used as a dressing to wounds and superficial inflammations, if the parts are relaxed and atonic.

PHYTOLACCA. This remedy, studied under the head of anti-
zymotics is also an anti-rheumatic, if the indications for it are
present: Pale tongue, sore mouth, vesicular eruption on tongue
and about the mouth, enlargement of lymphatic glands.

STICTA. Sticta Pulmonaria, studied under the head of remedies
that influence the respiratory apparatus, is a most excellent anti-
rheumatic if the following symptoms present: The neck is
stiff and painful, the head drawn back or to one side, pain in
shoulders extending to neck and occiput.

ACIDS. If one has read the medical journals of the past
twenty-five years, he will be surprised at what seems to be a
fashion in the selection of remedies. For months, the medical
journals will teem with articles in favor of acids in the treatment
of rheumatism, and reports of cures ; and then it all seems to
change, and alkaline salts are just as strongly recommended, and
quite as many cases reported. It is all rheumatism, and nothing
is said about a difference of cases, and we might suppose it was
all chance. Yet the cases cured by acids could not be reached
by alkalies, and *vice versa.*

We have no doubt of the cures in both cases, for it is a fact
that acids will relieve rheumatism, as it is equally the fact that
the cause of rheumatism may be antagonized by alkalies. But
these are not the same conditions of disease.

Lemon juice is anti-rheumatic when the tongue is markedly red, and the papillæ prominent; the coating of the tongue is usually thin and white.

ALKALINE SALTS. If the tongue is broad and *pallid*, we use an alkaline salt, and sometimes it is all that we require to cure any rheumatism. It is usually bicarbonate of soda to the extent of one-half to one drachm daily for children two years old. If the tongue has a leaden pallor, nitrate or acetate of potash are the better remedies.

ANTIPERIODICS.

If physicians will not concede anything else, they will concede that the group of remedies called antiperiodic, and quinine especially, has specific action. They will go further, when pressed, and admit a well defined and positive indication for this group of remedies. Pushed a little further, and they will admit the certainty of its action when thus clearly indicated.

This is what we claim with reference to all remedies which have been carefully studied, and we believe that any one may satisfy himself that we live in a world of law, medically as well as physically, the same causes always producing the same effects.

The indication for the antiperiodics—periodicity of disease—is so clearly defined in many cases, that he who runs may read. There are exacerbations of the disease, whatever it may be, and there are remissions or intermissions in it. The most common type of periodic disease is intermittent fever or ague, a chill or cold stage, a hot stage, a sweating stage, and period of intermission. There is usually regularity in the recurrence of them, so that the disease having made one " revolution," the time of each may be predicted.

It would be interesting and profitable to know the cause of periodic disease—the malarial poison—not that it would enable us to treat it better, but that we might the better avoid it, or possibly remove it. But in so far as the administration of remedies is concerned, the essential thing is to determine periodicity as the basis of diseased action.

Whilst physicians will admit the positive action of antiperiodics in malarial disease, and claim that they are as nearly " specifics" as anything we have in the materia medica, they will yet confess to a great many failures. These failures prove to them that " there is no certainty in medicine;" the quinine cures the ague to-day, it fails to-morrow; it cures Smith, Jones continues to shake (?).

Here the diagnosis has been half made. In the one case the periodic cause, whatever it may be, was the basis of the disease; in the other case it was but a part, and there was something else at the bottom. There are many cases in which the antiperiodic can not act, because of some morbid conditions of the body; correct these wrongs, and at once it acts kindly. Proper preparation for the action of the antiperiodic is many times essential to its success.

The remedies we will study in this connection are :—

Quinia Sulph.	Arsenic,	Alstonia Constricta,
Cinchonidia,	Uvedalia,	Nitric Acid.

QUINIA SULPHAS. *Specific Indications.*—There is distinct periodicity in the disease, whatever else may be its character, the periodicity being marked by exacerbations and remissions, or intermissions.

Dose.—In the ordinary use of the remedy the antiperiodic quantity will be one grain for the first year, two grains for the second, and so increase one grain each year up to the fifth or sixth, when the dose will vary according to the strength and development of the person. In some cases very minute doses prove antiperiodic; in others its limited absorption by the skin when used as an inunction, or applied to the skin with alcohol, is sufficient.

Periodicity being the indication, we want to know if the patient is in such condition that the remedy will act, and will act kindly. In this case, as in many others, a remedy must gain entrance to the blood before it can antagonize the disease. We put it in the stomach in an insoluble form—the stomach must first furnish an acid menstruum for its solution, and then reverse the process of osmosis to cause its absorption. If either of these fail, the agent will not prove curative ; indeed, we might as well put it in the patient's shoes.

Again, the remedy acts kindly in a certain condition of body, unkindly in other conditions, and a knowledge of these is essential to its right use. We say *that if the pulse is soft, the skin soft, inclined to moisture, the tongue moist and cleaning, and the nervous system free from irritation,* quinine will act kindly, and if there is periodicity, will arrest the disease. On the contrary, if the pulse is frequent and hard, the skin dry and constricted, the tongue dry, and the nervous system excited, it will do the patient harm. It is true that in some severe periodic disease, quinine will lessen the frequency of the pulse, bring down the temperature, control irritation of the nervous system, and start secretion, but this is an exception to the general rule.

In many cases of marked periodic disease, we first prepare the patient for the kindly action of the remedy, and then administer it. The child has its Aconite or Veratrum to control the circulation, baths to influence the temperature, Gelseminum, Rhus, Belladonna, or whatever may be needed to relieve the nervous system, and sometimes means to promote secretion.

The use of quinine by inunction is an important means in the treatment of children. In many cases of distinct malarial disease the remedy seems to be quite as efficient an antiperiodic, as if given by mouth. In nervous persons, and when the remedy unpleasantly influences the nervous system it does much better. In obscure periodic disease, and when the nervous system requires stimulation, it will be found an excellent method of treatment.

Whilst I regard quinine as a most valuable remedy, in its place, I insist on the indication, *periodicity*, for its administration. It will not cure continued or typhoid fever; it never has arrested a case, and never will, but it has done an immense amount of harm, intensifying the disease, increasing the suffering, prolonging the duration, and increasing the death-rate. Neither should it be used as a stomachic or tonic, for it is neither. It is a stimulant to the nerve centers in some cases, but the indications for this use are as yet unknown.

CINCHONIDIA. Cinchonidia possesses the same properties as quinine, and has the same indications. It is claimed that it is better tolerated by the stomach, and is more easily taken by children. The dose will be the same.

ALSTONIA CONSTRICTA. *Specific Indications.*—The disease is periodic, having exacerbations and remissions, or intermissions; the skin is dirty or tawny, the tongue dirty, and the urine is turbid and deposits a cloudy sediment.

Dose.—The dose of Alstonia will be somewhat less than of quinine, say, one grain the first and second year, and then an increase of half a grain a year.

Whilst the Alstonia is not so certain an antiperiodic as quinine and the other alkaloids of the Cinchona barks, it stands next to these, and will meet cases where they have failed. If the reader will notice the indications—a dirty, tawny skin, a dirty tongue and mouth, atonic fullness of spleen and liver, he will see that it reaches very unpleasant cases. When it fits the case, it is much more likely to effect a radical cure, in ague, than quinine.

Alstonia is also an excellent tonic and restorative in the cases named, stimulating retrograde metamorphosis and excretion, as well as nutrition. It is a new remedy, and needs further study to determine its full value.

ARSENIC. *Specific Indications.*—The pulse is soft and feeble; skin inelastic, pallid or dirty; tongue pallid and expressionless; eyes dull; tendency to hemorrhage; periodicity in disease.

Dose.—We generally employ Fowler's solution, adding from one to five drops to four ounces of water, and giving a teaspoonful four times a day. Sometimes pellets wetted with Fowler's solution answer a good purpose, and we occasionally use the third to the sixth decimal trituration.

We have made a study of this remedy under the head of remedies that influence the skin, and need only notice it here as it is antiperiodic and restorative. As a rule, we do not employ it in acute ague, or when quinine will cure; it is the old and hard cases, and when other remedies have failed. There is impairment of the vital forces, and marked by the indications given above, we use it as a vital stimulant. The reader will notice that it is the minute dose that cures, and we do not use it in the old doses for any purpose.

NITRIC ACID. *Specific Indications.*—The tongue has a *violet* color, showing the ordinary redness beneath; sometimes the lips and finger nails will also show this violet coloration.

Dose.—℞ Nitric Acid gtt. x., Water and Syrup aa. ℥j.; one-half to one teaspoonful every three hours.

Homœopathic Indications (3d to 6th decimal).—Irritable disposition; headache, relieved on lying down or riding in a carriage; pain in the bones of the skull, worse at night; hardness of hearing, relieved by riding in a carriage or in the cars; fetid, yellow discharge from the nose, bad odor from the nose, complete obstruction of the nose; sore throat on swallowing, with sensation of a sharp splinter in the throat; putrid smell from the mouth;

the urine has an offensive, strong odor, resembling horses' urine; bad effects of frost-bite; caries.—*Ehrmann.*

Nitric Acid is principally used in chronic ague, though we find cases where it will break the paroxysms and cure the acute. The ordinary remedies have been used without success. If now we find the violet coloration of tongue, we may be pretty sure of a cure with this.

It is a remedy for any form of disease when this indication is present. It will influence the appetite, digestion and blood-making, give a better circulation of blood, and improved waste and excretion. It is one of the remedies recommended for whooping cough, and if the indication is present it is very prompt in its action. It is also a very good remedy in chronic bronchitis and some cases of phthisis.

UVEDALIA. *Specific Indications.*—Enlargement of the spleen (ague cake), abdomen full and doughy, fullness in hypochondria, enlargement or hypertrophy of any part, if the tissue is atonic, doughy and inelastic.

Dose.—For internal administration the dose will vary from the fraction of a drop to five drops. The common use is as a local application, an ointment of Uvedalia being employed. In ague cake, and asthenic inflammations, this ointment is rubbed on and toasted in with a hot iron.

This remedy is not properly an antiperiodic except as it influences the spleen, and by restoring it to a normal condition, removes the cause of the continuance of the ague. For this purpose it has no equal. In hypertrophy and inflammatory engorgements of parts, when there is marked atony, it has proven a useful remedy.

ANTI-ERYSIPELATOUS.

This may seem like a singular classification, and the reader may think that this group should have been included with the anti-zymotics. But the disease is not always zymotic, and neither can it be classed with diseases of the skin, in so far as its therapeutics are concerned. The erysipelatous poison is destructive, involves the blood, mucous membranes, and secreting organs, and deep-seated structures, as well as on the skin.

If it were one poison, and the body were one, then we would need but one remedy. The fact is, however, that there are a number of remedies, and one cannot take the place of the other. We use them by special indications; if indicated they are anti-erysipelatous, if not indicated they are of no advantage and may do harm.

One will need no better illustration of the truth of specific medication than is shown in the study of this group. If a remedy is indicated, and this erysipelatous poison is the basis of the disease, the single remedy does all for the patient that one can ask. If for no other reason than to see this clearly, we should make this study.

The remedies we will study under this head are :—

Tinct. Muriate of Iron,	Rhus,	Veratrum,
Sulphite of Soda,	Apis.	

TINCT. MURIATE OF IRON. *Specific Indications.*—The tongue is deep red, mucous membranes deep red, the local disease of the skin shows the same deep or dusky redness.

Dose.—℞ Tinct. Muriate of Iron, ʒij, syrup ʒij. I give a half teaspoonful to one teaspoonful every two or three hours. As a local application, ℞ Tinct. Muriate of Iron ʒss, Glycerine and Water aa. ʒiij. Mix.

When iron is indicated, and it is in a large number of cases, its curative action is wonderful. Here is a patient with a most intense inflammation of the skin, with a frequent, hard pulse, a high temperature, an excited nervous system, a dry, harsh skin, scanty urine, and constipated bowels, and the intensity of the disease increasing from hour to hour. Muriate of iron is not regarded as a sedative, yet it brings the circulation and temperature down to the normal standard; it is not thought of as a remedy to relieve irritation of the nervous system, and yet it does it promptly; it is not classed with diaphoretics, diuretics, and cathartics, but yet it restores these excretions. At the same time it controls the local disease, relieving irritation and determination of blood, and restores the circulation when it has been impaired.

It has been remarked above that the erysipelatous poison may influence any part, mucous membranes as well as skin, glandular structures, and even nerve substance. If we find its evidences, say in a sore throat or disease of the stomach or intestines, or elsewhere, with an erysipelatous tongue, or erysipelatous spots on the surface, we will think of tinct. muriate of iron as a remedy.

SULPHITE OF SODA. This remedy has been thoroughly studied, and we notice it here to show the difference in disease and in remedies, even when cured by one name. Here is a case of erysipelas, frequently deep seated, with a broad, pallid tongue, dirty, and we give sulphite of soda with as prompt relief as we had from the muriate of iron in the preceding case.

RHUS. If the local disease shows a vivid redness, with burning pain, the pulse having a sharp stroke, the tongue showing bright red papillæ at tip,with frontal headache, rhus will be the remedy. In this case it controls the frequent pulse, gives freedom to the circulation, relieves irritation of the nervous system, and cures the local inflammatory disease of the skin.

APIS. The indications for Apis are, itching and burning of the skin, itching and burning of the urinary passages, and pinkish coloration of the part involved. If these indications are found, this remedy gives prompt relief.

VERATRUM. If the pulse is full and strong, and the part shows the redness of a common acute inflammation, no remedy will prove more promptly curative than this. It is given internally in the usual doses, and the part is wetted with it, of full strength, or down to one part to seven of water.

RESTORATIVES.

Under this heading might be included all those agencies that improve the life, the power to make a body, and that increase its functional activity. Good functional activity comes from good nutrition, and right conditions for the performance of function. We have already seen that a right temperature, a right circulation, right innervation, and right excretion, are necessary to healthy life, as is the right quantity and quality of the food, good digestion and blood-making, and right appropriation of the material in the building of tissue. The last of these may be profitably studied under this head.

Some of these have been already studied, as food, air and exercise, and some of the remedies that aid digestion by mouth, stomach and intestine, and the reader is referred to the chapter on remedies that influence the digestive apparatus, and those which influence the temperature.

The agencies that we will study here are:—

Air,	Iron,	Cuprum,
Light,	Phosphorus,	Silica,
Exercise,	Sulphur,	Soda,
Food,	Cod Liver Oil,	Lime.
Tonics,		

AIR. As a restorative, air is of prime importance, as well as a vehicle for the supply of oxygen to the body. Not that the body is able to appropriate its nitrogen, or withdraw from it carbon, as do the plants, but it is a stimulant to normal functional activity.

We find persons suffering from impaired nutrition when the air is *dead*, *i. e.*, where its motion is impeded, and when it fails to receive its supply of sunshine. These are relieved by a storm, accompanied by lightning, which revivifies the air. They are also relieved by mountain air, or sea-air, for similar reasons.

The air of a house, if it is continuously closed, becomes dead air, even though it is not defiled by human excretion, or other forms of dirt. The air is rejuvenated by being set in active motion, and becomes a prominent restorative. Dead air looks towards impaired functional activity and death, air in motion towards life.

LIGHT. All life is from the sun, and it comes to this world in the form of light. Light, therefore, is an essential of life, and all that pertains to it, to the making of a human body, and to human functional activities, as well as to the growth of vegetable organisms.

If a child or an adult is deprived of light, he suffers as do plants when thus deprived. If one will recall the growth of a potato in a cellar, its blanched and feeble appearance, its struggles towards the few rays of light that gain entrance, he can see how a human being will suffer when deprived of this life-giving agent.

The human body, like plant life, requires the full ray—white light—for health, though some diseases of the one as of the other are relieved by parts of this ray, as the blue, yellow, red, etc. The human body also requires the direct ray, and can not maintain a good condition of health if obliged to depend upon reflected light.

It has already been remarked that air deprived of sunlight becomes dead, the plant deprived of sunlight becomes sickly and eventually dies, and the child deprived of sunlight sickens, and may die from the deprivation. Light is a stimulant, and we employ it in diseases of debility. If a child is suffering from asthenic disease, and is confined to a room on the north side of the house, or the light is cut off by houses or trees, we at once suggest the propriety of such change as will bring sunshine into the room a considerable part of the day. Bitter tonics and restoratives have failed, sunshine is the restorative indicated.

We have cases in which the cure comes from the free light out of doors, and we insist that the little sufferer shall have it every day. In some cases we use the sun-bath with good effect; the entire body, except the head, may be exposed to the direct rays of the sun before a window, or if a part is feeble this only may be exposed to the sun.

In some cases the white ray is broken up, and we use but a part of it—this part now has color. If we want sedatives, the color is yellow or green; if stimulation, red or blue. An alternate blue and white light exerts a peculiar influence upon the body, that can hardly be classed as stimulant or sedative. It is stimulant to a plant and improves its nutrition, but upon the human body it seems rather to increase waste.

If a patient is suffering from nervous irritation, or acute local disease, we find that yellow, green. or leaden colored window shades, and wall papers exert a markedly soothing influence. If on the contrary there is a feeble circulation and innervation, poor nutrition, and general asthenia, we find benefit from pink or red shades and papers, and sometimes from blue. It is the same with clothing. If an anemic patient is clothed in yellow, I suggest the change to pink or red.

EXERCISE. The necessity of right exercise to health of body has already been referred to, as well as the necessity of rest, in conditions of irritation. Every part is made for use, and maintains its structure because it is rightly used. If it is disused, it is eventually lost.

Right exercise is as essential to the life of the child as the adult. Nutrition of the body depends upon it, and a good body cannot be grown without it. It is the same of a part, if not used it is atrophied; and in such cases right exercise becomes a prominent restorative. If the exercise can not be taken in the natural way, we have the parts rubbed and kneaded, and use passive movement. Many deformities in childhood grow from disease of parts; they are corrected by exercise, either passive movement or movement under the influence of the will.

8

FOOD. One of the most important objects in the treatment of disease is to place and keep the stomach in such condition that a portion of food may be taken and digested every day. Foods are of two kinds—*calorifacient and histogenetic*—heat-producing and tissue-making. The first of these is of the most importance is some forms of disease, the last in others. We can not attempt to define the indications for one or the other here, but will give some simple rules which may guide us in selecting and giving food:

" 1. Solid food should rarely be given during the progress of an acute disease, as the stomach and digestive organs are not in a condition to furnish the fluids necessary for its proper comminution, and hence it does not digest, but decomposes, giving rise to irritation and other annoying results.

" 2. As a general rule, the severer the disease, and the further the system is from a condition of health, the lighter and more diluted should be the food. Thus, in a high grade of fever or inflammation, we would give whey, toast-water, thin farina, or tapioca, weak chicken or mutton broth, etc.

" 3. In states of great exhaustion the food should be concentrated, very nutritious, and yet deprived, as far as possible, of all material that can not be appropriated by the stomach. Thus we would give beef essence, concentrated chicken, or mutton tea, farina with milk, etc.

" 4. In all febrile and inflammatory diseases the food should be given at that period of the day in which there is least vascular and nervous excitement, and it should never be forced on the patient when suffering from high fever.

" 5. Never give food when the patient is suffering from severe pain, as at such times it is impossible for the digestive organs to appropriate it.

" 6. If the tongue is heavily coated with a yellowish coat, a bad taste in the mouth, and a feeling of weight and oppression at the stomach, it is better not to give food, or at least give it in a fluid form and in small quantity.

" 7. Never force food on a patient when his stomach revolts at it, or if it produces nausea, oppression, or pain. It is much better to wait until medicine or time has placed the stomach in a condition to digest it.

" 8. When the digestive powers are much impaired, and it is important to give food to sustain the strength, it should be given in small quantities, and at regular intervals, like medicines.

" 9. If there is an absolute demand for nourishment to sustain the strength of the patient, and it can not be given by mouth, it is sometimes an excellent plan to administer it as an injection.

" 10. Much care is necessary during convalescence from disease that the patient does not eat too much, or that which is indigestible. The digestive organs are now enfeebled, and, if overworked, there is not only an excess of imperfectly elaborated material taken into the system, but the exhaustion is extended to the entire body, and impairs the functions of other organs and parts."

TONICS. Tonics may well be classed with restoratives, as the end of both is to improve nutrition. The difficulty in the way of the administration of tonics with children, is their bitter, unpleasant taste. Many remedies, therefore, that we should like to use have to be excluded on this account.

QUINIA. I employ quinia internally in the diseases of children, only for its specific influence in antagonizing the malarial poison. Though it is one of our most powerful nerve stimulants, and in some cases, aids digestion and nutrition in a marked manner, its taste is so unpleasant as to prevent its ordinary use.

In some cases where this stimulant influence upon the nervous system is required, we may use the Triple Phosphate of Quinia, Strychnia and Iron, (Aitkin's Tonic Mixture). It is a powerful remedy, and in many cases it will restore the appetite, improve digestion and blood-making, and aid nutrition in a marvelous manner. As the dose for a child need not be more than five or ten drops three times a day, the child will sometimes take it without objection. It may be prescribed as follows : ℞ Comp. Tonic Mixt. ℥ss., Simple Syrup ℥iss. ; half a teaspoonful three times a day.   •

COLLINSONIA. The Collinsonia Canadensis is a very valuable remedy, possessing as it does an influence upon several parts of the body. The proportion for a child two years old will be : ℞ Tincture Collinsonia ℥ss., Simple Syrup ℥iijss., Mix. ; dose from one-half to one teaspoonful.

To improve the appetite and aid digestion it may be administered in connection with iron, cod-liver oil, etc. It is not bitter, but quite pleasant to the taste, and sits very kindly upon the stomach.

In chronic diseases of the respiratory apparatus it is a favorite remedy, having a specific influence in removing irritation and improving functional activity of all parts supplied by the sympathetic system of nerves.

Iron. Iron is a most efficient tonic and restorative in childhood. It can be so prepared as to be readily taken and well appropriated by the stomach. As a general rule, it should not be administered while there are evidences of fever, or during the progress of inflammation. And in the treatment of chronic disease it will rarely be of benefit so long as the secretions are locked up.

I prefer the following preparations: ℞ Tincture of Muriate of Iron ℥ij., Glycerine or Syrup ℥iv.; dose a teaspoonful three or four times a day. ℞ Citrate of Iron ℨss.; Water ℥j.; dissolve and add Glycerine or Syrup ℥iij., mix; dose, a teaspoonful three or four times a day.

Phosphorus. Phosphorus may sometimes be employed with advantage as a restorative. Different preparations are in use, but I think the best are the Hypophosphite of Lime, Hypophosphite of Soda, and Compound Syrup of the Hypophosphites. Of the first, the dose will be one grain; of the second one or two grains; and of the syrup, one-third of a teaspoonful three or four times a day.

The indications for the use of phosphorus are, a pale, doughy condition of the skin, cold extremities, abundant pale urine of low specific gravity, and deficient innervation.

Sulphur. Though not regarded as a tonic, or even as a restorative, I regard sulphur as a very important restorative in some exceptional cases. It is not very easy to indicate the cases in which it is most useful, but they are those in which there is a deficiency of heat and impairment of muscular power. I use it in doses of three to five grains, associated with cod-liver oil, or other fatty material.

Cod-Liver Oil. This is a very important restorative in many cases where the nutritive powers are impaired. It furnishes material for combustion, and thus saves the nitrogenized food and tissues from being burned. It also furnishes food for the nerve centers in a form readily appropriated. Cod-liver oil, when the stomach will receive it kindly, is easily digested, and increasing the temperature and the innervation of the body, all the functions are better performed.

The dose of cod-liver oil for a child two years old, will be a teaspoonful three or four times a day.

CUPRUM. *Specific Indications.*—The skin has a tawny or dirty hue, sometimes greenish, the tongue is pale, and the gums blanched, the pulse is small and soft, and more frequent than usual. Discharges from the bowels lack color, and if there is diarrhœa, they are pale and like rice water.

Dose.—℞ Tinct. Acetate of copper (Rademacher's) gtt. x., water ʒiv.; a teaspoonful four times a day. Sometimes we get the restorative action of copper by having the patient take it in the shape of a *greened* pickle—a small portion three times a day.

Homœopathic Indications, (3d to 6th decimal).—Spasmodic affections, whooping cough with spasms, long continued paroxysms of suffocation, cough, much rattling of mucus; epilepsy; violent diarrhœa with cramp in the stomach and chest, much flatus with the stool, nausea and vomiting of frothy mucus, sometimes green, metallic taste; pulse small, soft, almost imperceptible; scarlet fever when the rash suddenly disappears, followed by stupor and delirium or convulsions.—*Ehrmann.*

SILICA. *Specific Indications.*—The skin is rough, the epidermis desquamates, and the hair is harsh; teething is slow, and the tissues feeble.

Dose.—I employ the third decimal trituration as a remedy; usually I order an oatmeal or cracked wheat diet, or the use of bread from unbolted flour.

Homœopathic Indications (3d to 6th decimal).—Headache with great sensitiveness to a draft of air. Must have the head covered. Great inclination to perspire. Hardness of hearing. Otalgia with stitching pains. Nasal catarrh with offensive odor and loss of smell. Inflamed glandular swellings. Fistulous ulcers. Ulcers burning, putrid, indolent. Carbuncles. The skin heals badly; a slight injury inclines to suppurate. Bones swollen, inflamed. Caries. Weakness of the joints, especially the ankle. Feet perspire very much, and smell very offensively. Phthisis pulmonalis, with great sensitiveness to a draft of air, and profuse night sweats.

SODA. This remedy has been studied in two or three places, and we only need say here that when the tongue is continuously pallid and broad, it will be found a good remedy. It improves the appetite, digestion and blood-making, and the nutrition of tissues.

LIME. *Specific Indications.*—Acid fermentation in the stomach, food does not digest well, colic, green acrid discharges from the bowels, inflammation of cellular tissues, boils.   ·

Dose.—When we wish to influence the digestive tract we use lime-water with the milk. If it is for inflammation of the cellular tissues, or boils, we may use lime-water, or minute doses of sulphide of lime.

PART II.

CARE AND MANAGEMENT OF INFANTS.

CHAPTER III.

THE young of man is the most helpless of all created beings, requiring constant care from the hour of birth until its second or third year. But it is especially for a short time after birth that it is thought to need extra attention, and when it very frequently gets an officious interference that renders it uncomfortable, and frequently leads to disease.

Many of the prejudices and whims of the olden time still remain, and some of them, indeed, receive the tacit support of the physician. It is well to educate the public mind to the truth in these things, and though the physician may find it a thankless task at first, it will bring its return in an increased confidence and respect.

Works on obstetrics point out the things to be done, and how to do them. It will be our business here, at least to some extent, to point out some things that should not be done, and other things that are not done properly.

WASHING THE CHILD.—A physician should not have the least hesitation in telling a nurse how to wash the child, if she shows ignorance in regard to it. A very common plan is, to rub the child with lard freely, and then with an ordinary cloth, and poor soap and lukewarm water, to wash off the grease and secretions upon the skin. It is a difficult job, and the child is rubbed until its skin is irritated, and still not cleansed, and it is rendered fretful and uneasy for days.

(119)

The true way is, let the person washing the child rub it thoroughly with the lard, and then with a piece of soft flannel, or even cotton, wipe the child clean. Then with soap and water, and a *soft* cloth or sponge, it may be gently yet thoroughly washed. The drying had better be done by gently pressing the towel or cloth upon the skin, rather than by rubbing. This makes all the difference possible in the feelings of the child and its future comfort. And as the skin has a very close sympathy with other parts, in addition to being an extremely sensitive part itself, we can readily see that, in some cases at least, it will be a means of avoiding serious disease.

CLOTHING OF THE CHILD.—I object most decidedly to many parts of the infant's dress, as usually made; and I have no hesitation in giving my objections words, whenever and wherever I am brought into contact with these errors.

The *bandage* of the child is *always* pinned too tightly. The child no more needs this swathing than the adult; its abdominal muscles are just as strong, and just as capable of performing their functions. But there is something worse than the mere suffering which follows such confinement. The bandage extends upward so as at least to embrace the floating ribs, hence the insertion of the diaphragm, and often much higher. Here there is a more or less serious impairment of the respiratory function, as the band is more or less tightly applied. We not only have suffering and restlessness upon the part of the child from this, but occasionally such impairment of the lungs as will lead to a fatal result.

There is another lesion that may be directly attributed to this. I allude to infantile *constipation*. It comes on gradually; it is attended with colic, and sometimes with infantile dyspepsia. True, in many cases it does not produce any present marked inconveniences, but it lays the foundation for future constipation, hemorrhoids, prolapsus ani, and, as I believe, for hernia.

The child needs such band only for the purpose of retaining the dressings of the cord, and to give warmth. One thickness of soft and elastic flannel, not so wide as to extend upward on the thorax, or down to be wet with the excretions, will answer this purpose. Being elastic, it maintains its place without being applied so tightly.

Its skirts are usually as badly made. The waist is made of cotton stuff, usually stiff and harsh, and so wide as to prevent any warmth from the flannel attachment. As frequently put on it chafes the child under the arms, and is unpleasantly bulky and cumbersome about the body.

The diaper is as bad or worse than the remainder. If the parents are in good circumstances, and it is a first child, it will be of linen diaper; at any rate it will be harsher than there is any need for. Let this be wetted and dried several times without washing, and we have a cause of irritation that would make the adult wince if obliged to wear it. I have seen scores of cases in which irritation and chafing about the genitals, the perineum, nates, and thighs, was the result of such treatment. Yet the mother could not imagine *why* her child should be so affected.

While I do not insist that they should be washed after each discharge of urine, though it would be better and pleasanter for the child if they were, I object to their being used until they smell and are as foul as it is possible for so much linen or cotton to get. If children are delicate, the absorption of these excretions must prove very detrimental.

When a child shows such irritation as I have named, recommend that the diapers be made of canton flannel, which is soft, unirritating and pleasant, and that they be rinsed in clean water every time they are soiled, before drying and using.

How often should the Child be Washed ?—The question might be put in a different way—not so much how often, but how much should the child be washed? Some are washed a very great deal; indeed, some poor innocents have their lives soaked out of them.

It would be a mercy to the children to put a few monthly nurses, and not a few mothers, through a similar course of soap and water, as afterward they would hardly be so free with it.

I have seen children washed out of the world in a few months, and sometimes in a few weeks, and have had to put a stop to it a number of times. The child is very sensitive to cold, and being entirely divested of clothing, and having the evaporation of water from the surface for some minutes, is too much for many.

To wash a child properly, the nurse should have a blanket or other woolen material of sufficient size to wrap the child entirely up. Then a portion of the body should be washed at a time, the child being kept well covered in the meanwhile. Even with this care it is not necessary that the child be washed every day.

Washing is done for cleanliness, and need not go further than that; and when children are feeble it had better fall short of it than exceed it.

ATTENTION TO THE CORD.—The question is usually asked, how long will it be before the cord falls off? and what shall we do with it in the meantime? It usually requires from five to nine days for the separation of the cord, and cicatrization of the umbilicus, and occasionally it is prolonged beyond this.

It is important that the cord be tied sufficiently tight to prevent any hemorrhage. True, it is but rare that there is any severe hemorrhage, yet I have had such in my practice. But a small amount of blood discharged causes the cord to adhere to the cloth, and the mass becomes hard and unpleasant, if not irritant to the child. I always recommend the young physician to examine the cord before it is dressed, and if not certain in regard to this point, to apply a second ligature.

The cloth with which the cord is dressed should have its upper surface oiled (in obstetrics we say oil when we mean lard), so that it may not adhere if it should become necessary to remove it. Every day when the child is dressed, it should be placed in such a position that there will be no strain upon it. And in washing the child, the cord and cloth should be removed, so as to wash beneath it.

In the majority of cases, irritation and inflammation about the umbilicus is the result of tension upon the cord, upon the dressing becoming hard and irritant from bleeding, or other cause, from a too tight band, or from the excretions passing up so as to soil the dressings of the cord, and subsequent want of cleanliness.

No traction or other means should be used to remove the cord, as it will be separated in due time by the natural process of absorption.

When separated it is well to apply a soft cotton cloth, folded, that the recent cicatrix may not be disturbed by the rubbing of the child's clothes. If the part seems somewhat tender, a cloth spread with mutton suet or simple cerate may be applied.

ULCERATION OF THE UMBILICUS.—Occasionally we are called to prescribe for ulceration, or rather a want of cicatrization of the umbilicus. Sometimes there is considerable redness and evidences of inflammatory action, and the child is feverish and fretful.

I generally prescribe one of three things. In the majority of cases, that the part be powdered with subnitrate of bismuth twice daily. If there is much inflammation, that it be dressed with stramonium ointment. Or, if there is an erysipelatous redness, order a solution of permanganate of potash, one grain to the ounce of water.

If the child is feverish and restless, we would give Aconite, one drop to the ounce of water, one-half of a teaspoonful every hour.

EXCORIATION AND CHAFING.—Various parts of the infant's body may be chafed, and become excoriated, and painful. I have already mentioned the chafing that comes from want of attention to the diapers of the child, and want of cleanliness. It is quite easy, by inspection, to determine the nature of the difficulty, but it is not always so easy to give immediate relief.

The use of Colgate's Glycerine Soap to wash the child, and strict attention to cleanliness, will sometimes be all that is necessary. Occasionally dusting the inflamed parts with *scorched flour* will answer the purpose. The common idea is that something " greasy " is necessary, and frequently ointments will have been used before the physician is spoken to. Mutton suet, if well prepared, answers as well as any ointment, if we except the glycerole of starch, which is an excellent application in such cases. The most certain, as well as the speediest cure, has followed the use of subnitrate of bismuth in fine powder, dusted over the parts, two or three times a day.

DOES THE CHILD NEED MEDICINES IMMEDIATELY AFTER BIRTH?—Many persons believe that the young child needs something to act upon its bowels. And for this purpose they propose to give it urine and molasses, lard and molasses, or sometimes castor oil. Others think it should have a whisky or rum mixture, to work the mucus out of its throat.

The truth is, the less the child is interfered with the better it will get along. Nature makes provision for all of its wants, and an action of the bowels is provided for in the first milk of the mother. The child will have a sufficient opportunity to take medicine, without commencing thus early, and it will be good policy to impress this upon the mother and nurse.

There is no need of saffron tea to keep off the jaundice, or make the child a better color. No need of German chamomile to strengthen its stomach, or of pap or panada, or of any of the hundred and one things that people like to force upon the child. What it wants is warmth and quiet, and not food or medicine. Nature furnishes food at the proper time, and furnishes it exactly adapted to its wants.

DIFFICULTIES IN NURSING THE CHILD.—Occasionally we find a child that will not nurse, at least it persistently refuses, and accompanies its refusal by energetic remonstrances. This seems, in some cases, an excellent illustration of *original sin*, though, doubtless, there is some good cause, other than this, if we but knew it.

If the child is fed the difficulty increases as time progresses; for, becoming accustomed to the spoon or bottle, he is less inclined to take the breast. The proper way is, not to feed the child, but, by kindness and perseverance, get it to nurse. The child should be applied to the breast when in a good humor, and not when crying; sometimes it will be an advantage to wet the nipple with the milk, as an extra inducement.

In some cases the difficulty is with the mother, not with the child. The nipples are retracted, or not properly formed, and the child is unable to obtain a sufficient hold to draw the milk. In this case a breast-pump, applied for a moment, will draw the nipple out so that the child can nurse. If a breast-pump is not at hand, it may be accomplished with an ordinary pint bottle; fill it with hot water, and pouring it out quickly, apply to the breast.

The *sore* nipple that is so frequently met with, and that proves such an annoyance and source of difficulty in nursing, is best cured by wearing a leaden nipple shield, and small doses of Phytolacca with or without Aconite. If sheet-lead can be readily obtained, this may be easily fashioned into a cup that will accurately fit the nipple. If not, a piece may be hammered from ordinary bar or other lead. We want it to fit accurately and pleasantly, so as to bring it in contact with every part of the nipple. It should not be too heavy, as it is to be constantly worn, except when the child is nursing. It has often seemed strange to me that with so simple and certain a means of cure, such suffering should be permitted.

FOOD OF THE CHILD.—The natural food of the child for the first eight months, is the mother's milk, and it is rarely the case, when the mother is healthy, that any other food will be required.

If the mother is healthy, and still fails to supply the amount necessary, the probabilities are that there is a deficiency of some kinds of food. Frequently it will be of farinaceous foods, which, if prepared in palatable form, give an increase of good milk. Occasionally the fault will be found in a lesion of digestion, and the employment of the bitter tonics and restoratives will answer the purpose.

When the child can not have its natural food we have to advise a substitute. That most readily obtained and easily prepared is the milk of the cow. In its natural condition it is too rich in some of the elements of food, too poor in others. The constituents of human, cow's and goat's milk are as follows:

	HUMAN.	COW.	GOAT.
Water	883.6	861.0	868.0
Butter	25.3	38.0	33.2
Casein	34.3	68.0	40.2
Sugar of Milk and Extractive Matters	48.2	29.0	52.8
Fixed Salts	2.3	6.1	5.8

To prepare the milk of the cow for food for the child during the first weeks of life, one-third the quantity of water should be added, and one drachm of sugar to the pint of milk, and the whole boiled. After the first month, the proportion of water may be lessened to one-fourth, and after the child has passed its sixth month it may be given pure.

Those who have had occasion to use artificial food in raising children, prefer the milk of the goat to the cow, and there is no doubt but that it is more easily appropriated by the child. In France, goats are frequently trained so that they will lie down and permit the child to nurse directly from the udder.

Liebig's food is a most excellent substitute for the mother's milk, being much better than the milk of the cow. It contains the elements of human milk in a form easily digested by the child. It is composed of Wheaten Flour, two ounces; Barley-malt Flour, two ounces; and Bicarbonate of Soda, thirty-one grains. In preparing it for use, the following directions are observed:

Take a heaping tablespoonful of the powder and mix it with two tablespoonfuls of water; then add to it ten tablespoonfuls of milk, and heat the whole over a gentle fire (do not boil). When the mass begins to thicken, remove it from the fire, and stir for five minutes. Then place it over the fire, and heat again, with stirring, until it becomes quite fluid, and then suffer it to boil.

In the preparing of the food, do not use too much heat, or haste, as it will prevent the necessary chemical change taking place. Use a gentle heat, and only allow it to come to a boiling temperature at the end of the process.

In employing artificial food, whether milk or that just named, it it better that it be given with a nursing bottle, rather than with the spoon. There is more than one reason for this direction, but one good reason is sufficient. The act of sucking calls into use the muscles connected with the salivary glands, and excites those glands, hence there is good insalivation of the food. *Maw's* nursing bottles are decidedly the best, though they require more care to keep them clean and in order. But the ordinary nursing bottle will answer if it is well looked after.

Milk should never be allowed to stand in the bottle, which should be emptied and rinsed every time it is used, and kept in a vessel of water, or at least partly filled with water. In this way it is kept clean and sweet.

Everything connected with the food of the child should receive the greatest care, and should be frequently scalded.

A very slight decomposition in the milk, from sourness of a vessel, will give the child an indigestion, and may seriously derange the functions of the stomach and bowels. These facts can not be too forcibly impressed upon those who have the nursing of children in this way.

The food of the child should always be given at the temperature of the body, and this will require the employment of heat at all periods of the day. A nursery lamp answers this purpose very well, but there is a much simpler means available wherever coal-oil or gas is employed for light. It is a nursery attachment, which may be attached to any lamp or gas-burner, and which will hold a cup steadily over the blaze, where it will quickly heat. These are made by Messrs. Tifft & Howard, 406 Pearl street, New York, of whom they may be ordered by mail, if they can not be obtained nearer.

It seems very important, when trying to raise a child by the milk of the cow, to be able to determine whether the milk obtained is good. There is a very simple method, which will give this information to any one. Allow the milk to stand for six hours, skim it, and if, upon holding it to the light, the skimmed milk has the natural color, it is good. But if it is blue, or green, or varies in appearance in any way, the milk should be rejected as not fit for the purpose, and another cow's milk should be tried. I need not say that it is important that the child should have *one* cow's milk, and I am satisfied that it is much better if the cow is fresh.

As the child attains to the end of its first year, it will require additional food to the mother's milk. If it is strong and healthy it may be gradually accustomed to the ordinary diet of the family; farinaceous food being preferred to animal. If, however, the digestive organs are not strong, it will be better to confine it to a milk diet, or to food prepared with flour and milk, or with sago, tapioca, etc.

A WET NURSE.—Next to the mother's milk as food for the child, the milk of another woman is the best, and when a choice can be had, preference should always be given to a wet-nurse, Mothers have a prejudice to overcome in agreeing to this; yet, as it is manifestly for the child's benefit, and as its chances of living are one hundred per cent. greater, it is a duty to make the sacrifice for the child's benefit. When we have really tried it, having a wet-nurse in the house, and having the care of baby, will not be found nearly so objectionable as anticipated.

The nurse should be selected with reference to her health, the age of her milk, and with reference to her character and associations. We do not expect perfection in a woman who is obliged to take this position, and must make many allowances.

As a general rule, we will find the health of persons in the lower walks of life much better than those who enjoy more of this world's goods and comforts. The general appearance of the patient will usually tell the story at the first glance. Such persons are rarely scrofulous, at least to that extent that it would be objectionable, and if they are, they carry the marks in old cicatrices, and enlarged glands.

What we are most fearful of is a syphilitic taint, secondary syphilis. We may not be able always to determine this, for it may exist in the system, without having as yet shown itself in the usual symptoms. A person suffering from secondary syphilis in the early stage will show maculæ, and if the tongue is examined we will find the evidence in recent or old indurations.

The danger of getting a person afflicted with syphilis is not great, except in large cities, and even here the pleasures and profits of a life of shame are so much greater than attends the position of a wet-nurse, that they are not frequent applicants.

When there is doubt of the capacity of the applicant to furnish a sufficient supply of milk for the child, the breasts may be examined. If round and firm, of usual size, and a well developed nipple, there will be but little doubt in this respect. It is important that the nipple be free from sores or abrasions, as this not only prevents free nursing, but it is the source from which vices of constitution are most frequently transmitted.

It is best that the milk of the wet-nurse be as young as the mother's. When other things are right, three or four weeks, or even as many months, may be permitted. I do not see any objection to the milk being too young, if it has passed the first week.

As regards the character of the wet-nurse, we would like to have it as good as possible, as well as her associations. Still, beggars can not always be choosers; we want to save the life of the child, and wet-nurses are not too plenty, and we may, by kindness and good offices, improve both character and associations.

WEANING THE CHILD.—Almost all the works that speak of weaning the child, name twelve months as the period the child should nurse. This, however, is English and continental authority, and is not always adapted to this country and our habits. If a child is well developed, and has cut its teeth thus far without trouble, there is no reason, so far as it is concerned, why it should not be weaned. Of course it is essential that for a couple of months it has been accustomed to eating, and has digested its food well. I do not think weaning is prudent until we have the conditions named.

If, as in some cases, the milk of the mother is found to disagree with the child, it may be weaned at a much earlier period, and will thrive better on the milk of the cow, or the other foods named, than if permitted to nurse. Not unfrequently the mother's milk undergoes a change from the eighth to the fourteenth month, that renders it poorly adapted to the wants of the child, and gives rise to various symptoms of infantile dyspepsia, and to irritability of the bowels.

The re-appearance of the menstrual discharge sometimes produces a very marked effect upon the milk, and in nearly all cases impairs its value. But as this, in the usual course of events, does not appear until the child has attained the age of one year to eighteen months, it is not often that our attention will be called to it. When, from such reason, there is evident change in the milk, and it gives rise to gastro-intestinal or other disease, we will advise weaning the child.

When the mother becomes pregnant while nursing, the milk very generally suffers after a time. It is necessary in this case that the child be weaned, not only for its own sake, but for the mother's; for nursing a child would interfere with the natural progress of gestation.

The objections to weaning the child are of two kinds—with reference to the mother, and with reference to the child. The first objection would seem trivial, if we did not so often witness the health of the mother completely broken down by frequent child-bearing. It is the popular impression, and to some extent it is true, that so long as a woman nurses her child, she will not have her menstrual periods, and so long she will not be in danger of becoming pregnant. Frequent pregnancies are a very great burden to a woman, especially if she has not the proper assistance in the care of her family, and we need not wonder that she prefers to nurse one child for two or three years, rather than have another added to her cares.

The objections upon the part of the child are with reference to its health, and its powers of digestion and assimilation in maintaining a separate existence. In some families every child has to pass through a second summer of cholera infantum, which requires great care, especially with regard to its food. In such cases it is best for the child to nurse eighteen months, or even two years.

In other cases the infant is far from robust, and its development is slow; it cuts its teeth tardily, does not digest food well, and is liable to attacks of gastro-intestinal irritation. Here, also, nursing the child should be prolonged.

When it is determined to wean the child, there is much discussion of *how* it is to be done, as it is no easy matter in many cases, and, as usually managed, very troublesome in most.

Trying to " taper off," as a drunkard tries to quit drinking, giving the child a little now and then, or depriving him of it in the day-time, giving it to him at night, is just about as efficient in the one case as in the other. At the end of a week the child

9

is no nearer weaned than when it commenced; but in most cases is farther from it, for it has been worried and tantalized until it has a stronger desire for the breast than before the weaning commenced.

It is also a poor plan to put bitter or nauseous things on the nipple, thinking to disgust the child, and make it reject the breast. While it answers occasionally, it fails in the majority of cases. It is objectionable, also, because it makes a bad impression upon the child's mind, and it feels that it has been deceived, and wronged.

The only true way is, to stop giving the breast at once, and convince the child that this is right. In the day there will be but little trouble, and if the father or some friend will take care of the child for two or three nights, the trouble will be all over. The moral effect of such a course upon the child will be a decided benefit, and will give the mother a power of control that she could not otherwise possess.

SLEEPING.—The young child should be accustomed to regular periods of sleep during the day, indeed, all its habits should have great regularity. As it grows older it does not need so much sleep, and after it has passed its first year, it will not usually require more than one sleep in the day-time.

At night the child should have its separate bed. I think it very objectionable for the child to sleep with its parents. Experience has proven that the young yield vitality to the older, and the sound to the unsound. This would be a very cogent reason if either of the parents suffered from ill health. But beyond this it is not well for the child to breathe the emanations from the bodies of other persons, as it must, if thus confined, and frequently covered so as to almost exclude the air.

Another, and a very important reason is, that if allowed to lie with the mother all night, it will obtain a habit of frequent nursing that will prove detrimental to its own health and to the health of the mother. Many times an infantile dyspepsia may be traced to this source.

Let the child lie in its crib or cradle at the bedside, and when it requires nursing the mother can take it in the bed with her and give it the breast, placing it in the cradle again afterward. A child cared for in this way will be much less trouble to the mother, in addition to the advantage to its own health.

REGULAR HABITS.—There are few persons that can be convinced how readily a child may be trained to regular habits, in all things, unless they have seen it. Very frequently they will not believe the evidence of their own eyes, but account for it by saying, that it is owing to a difference in children, some being good, others being bad. The truth is, that a child is just as sensible to impressions as the adult, and is just as ready to yield obedience if it is asked in a proper manner. Beyond this is the fact, that habits in childhood are very readily formed, and very persistently maintained. This is as true of bad habits as of good.

During the first three months the child should be accustomed to taking the breast about every three hours, and at very regular intervals. As it grows older it will not require it more than four times during the day, and once at night. One-half the slight illnesses, and the irritability and fretfulness of childhood might be avoided in this way.

Nursing the child whenever it cries, simply to keep it quiet, is a most pernicious habit, and will very certainly lead to ill health, usually in the form of infantile dyspepsia and colic. And yet this is a very common habit, and will require much effort upon the part of the mother to break it up.

The time of getting up in the morning, of being washed and dressed, of taking the morning sleep, of the afternoon sleep, should all be as regular as the clock. I know that the physician will be met by many objections when he urges this ; the most common being, that the mother has other duties besides taking care of her child, and the attention it requires can not always be given at the minute. The fact is, however, that duties regularly performed become light by the side of those irregularly done. No one ever accomplishes much, unless the work is pursued by system. So in this case, it is a saving of time to do things promptly to the minute, and have the child in good humor and take its long refreshing sleeps, rather than from neglect, to have it irritable and fretful, difficult to get to sleep and hard to keep asleep.

There is no necessity of rocking a child when sleeping, indeed but little necessity of their being rocked at any time. If they are accustomed to go to sleep at a certain time, all that is required is to give them the breast, and put them in the crib or the bed.

Americans might learn a valuable lesson from our German citizens, in regard to the management of young children. They put them in the crib and cradle and cover them warmly, and keep them there, except when nursed. There seems to be no trouble about it, the child is comfortable, is accustomed to that method of living, and takes it as quietly as if it were its natural mode of existence. When it becomes old enough to play, a tassel, with some simple toys, is suspended within reach of its hands, and with them it will amuse itself for hours at a time. When old enough to sit alone, it is accustomed to amuse itself in a similar way.

MORAL GOVERNMENT.

As the health of the child will depend to a considerable extent upon the condition of the nervous system, especially of the brain, the physician will frequently have to give the mother instructions in regard to its moral government; especially will this be the case when the child is suffering from chronic disease, which is aggravated by irritability, displays of temper, and the various cryings and frettings so common in badly-regulated households. True, in these cases he is not able to effect as much by good advice as he would like; yet, there are other cases in which he may, by a few judicious words of counsel, direct the mother, who is willing, how to train her child, so as to conduce to her happiness as well as its own.

In this respect, the experience of a mother has higher value than any other person, and I will give the short chapter from Chambers' Infant Treatment, by the lady editor of that work:

"During the first few weeks of life, happiness is solely derived from the healthy operation of the bodily functions. Until the senses begin to act so as to convey impressions to the brain, there can be no pleasure drawn from external circumstances. The activity of the senses, and the enjoyment produced, will be in proportion to the state of the health. An infant who is continually in pain, either crying, moaning, or in a state of repletion, or of exhaustion from the consequences of suffering, will be but little attracted by the light, sound, or motion which first engages the senses of infancy. In no other instance, perhaps, are the influences of the physical condition so immediate and so evident. An infant, even of three weeks old, will exhibit a haggard, grief-worn countenance, sunken eyes and shrunken face, painful to those whose experience tells them what these signs indicate. But the fair, plump, contented look of the healthy babe, speaks a language of comfort, prophetic of the approaching dawn of intellect.

"The general irritability caused by disordered functions, renders the impressions upon the senses even more painful than pleasurable; the disposition for enjoyment bestowed by the feeling of health is denied ; the mother's voice, her smile, are associated with pain as much as with pleasure, and the affections are imperfectly and tardily aroused. As weeks pass on, habits form, and instead of a habit of contentment, there is one of fretfulness. An infant so constituted is either reared with an indifference to its continual crying and fretfulness, or with the apprehension which causes its nurse to be continually seeking how she may quiet or prevent its cries. At the age when food alone appeases it, the babe is always eating or sucking ; as it grows older, sugar, cake, etc., are superadded, with the addition of noises or rough exercise, and but too frequently some sedative or composing draught, which the mother believes herself obliged to adopt in order to procure the child needful repose, or the servant surreptitiously administers to relieve herself from incessant fatigue. When the time arrives that restraints and guidance should be adopted, the fear of farther irritation by contraction leads to a system of bribes, deceit, and coaxing; all the lowest sentiments of human nature are appealed to ; and at two years old we have a selfish, willful, ill-tempered child, with violence apportioned to its strength, and intelligence prompted by ill feelings. It is not to be supposed that these moral disorders belong exclusively to bad health. A healthy child may be selfish, willful and ill-tempered at two years of age, if injudicious treatment have cultivated the lower sentiments ; but the healthy infant is predisposed to receive happy impressions, and enjoys the condition called good temper—a term which in infancy is synonymous with good health. The nurse has fewer temptations to mismanagement, and the affections and intelligence being more healthful and active, moral mismanagement actually produces less permanent injury.

"There can not, then, be too much value attached to the physical condition of an infant; to the condition of the parent while pregnant and while nursing, and to the regulation of every particular connected with the health of her offspring. This being the first object, both in point of time and importance, the next consideration is the means of developing the moral and intellectual faculties."

PART III.

DISEASES OF CHILDHOOD.

CHAPTER IV.

THE diseases of childhood do not differ in their pathology from the diseases of the adult. In each disease, there is a depression of the vitality, either of the whole or of a part, and though there may be increased functional activity in some particular functions or part, we must never take it as an indication of an increased vitality or of a real functional increase.

The animal body in early life is immature in all its parts, and what change there may be in the character of diseased action, is in part dependent upon this. The tenacity to life is just as strong in the child, in many cases even stronger than in the adult. Possibly, this increased tenacity of life, will counterbalance the immaturity of organs, in all children who possess vitality sufficient to reach adult years.

There is this fact to be taken into consideration in all our estimates of infantile disease: That there is a certain proportion of children, who are not born with sufficient vitality to live longer than the second year, and some, indeed, who have not the capacity to live this long.

It is not worth our while here to discuss to any considerable length why this is so, but it concerns us to know that it is a fact, and to know how we may determine it.

It is a well-known fact in vegetable physiology that imperfect or unsound seed will not produce perfect plants; and that when the seed-bearing plant requires to be fertilized by the pollen of another, an unsoundness of this will lead to deterioration; that in breeding animals, sound and robust young are not expected

(134)

from immature or unhealthy parents; and that peculiarities and imperfections will be increased from generation to generation, by propagating from animals thus impaired.

In our civilization there is a continual breeding in, with reference to physical and vital imperfections, and hence to save the integrity of the species, it is necessary that the imperfect young should die before they have arrived at an age to transmit their physical imperfections.

PHYSIOLOGICAL MARRIAGE.—Prof. William Byrd Powell advanced the theory, that persons of similar temperaments could not have healthy offspring, or in some cases, would not be capable of begetting children; even two very healthy persons, physically and mentally well-developed, and with each the capacity to live the allotted time of man, might have children that would not live to the age of two years, or in some cases would not have the capacity for independent existence at all.

The principles of this discovery, if discovery it is, are thus briefly stated by Prof. Powell:

"LAW I. When the constitutional similitude of the respective sexes is such that a qualified observer can not detect an appreciable difference, sterility will be the result of their marriage. *Illustration:* Washington and his wife were, respectively, sanguine, and it is known that sterility was the result. Between General Jackson and his wife there was a nominal difference of constitution; he was bilious sanguine, and she was bilious; nevertheless they were physically the same, both being exclusively vital, and it is known that sterility was the result. The first Napoleon and Josephine were, in person, greatly different, and in constitution they were nominally as different, and yet there was no physiological difference. He was sanguine encephalo-bilious lymphatic, and she was bilious encephalic, consequently they were, respectively, compounded of equal varieties of vital and non-vital conditions, and it is known that sterility was the result of their alliance.

"LAW II. When the constitutional similitude of the respective sexes is less than complete, or is appreciably different, progeny will result, but it will be dead-born, imbecile, scrofulous, deaf, blind, or otherwise imperfect. *Illustration:* I can furnish three hundred examples of this law, but as they are not histori-

cally known, they would be of no value in this relation. I can
cite one, however, which is historically known, viz.: the first
Napoleon and his second wife. Her temperament was bilious·
encephalo-sanguine, and his temperament I have indicated.
There was between them an appreciable difference of constitution, and the result of this difference was one son; but the difference was too small to secure to him a normal viability, for he
died of a scrofulous affection of the lungs, at the age of eighteen
years. · It is most indisputably the fact, that a considerable difference of constitution must obtain between the respective parties
to a marriage, to secure to offspring a soundly viable constitution. To discover the least difference consistent with a physiological marriage was indispensible, but before discovering this the
conviction became forced upon me that my discovery could not
become of general utility without the discovery of a law of
universal application. By a great amount of observation and
study, I succeeded in discovering the desired law, and it is of
easy application, and will universally secure a physiologically
legitimate offspring, and the greatest possible happiness to the
parents. Those, therefore, who make domestic happiness, and a
really useful progeny, conditions of marriage, must observe the
following law:

"LAW III. One of the parties must be exclusively vital—that
is, must be either sanguine, bilious, or sanguine-bilious (the last
being a compound of the two former, is also vital), and the
other party must as certainly be more or less non-vital, that is,
more or less lymphatic or encephalic. All marriages, in contra-
vention of this law, are physiologically incestuous, and the consequences will be vicious in proportion to the delinquency.

"LAW IV. The greatest dissimilitude of constitution that can
obtain between the sexes, when they are respectively of the same
species, is that which obtains between a vital and a non-vital
temperament—and this is the most favorable to progeny. But
marriages of this character are greatly impracticable in any
country. It is a very remarkable fact in the physiology of human procreation, that a high degree of constitutional dissimil-
tude is about equally unfavorable to progeny. It has been seen
that a high degree of similitude entails a scrofulous diathesis, and
with a high degree of dissimilitude, as when one party is white and
the other negro, the progeny is invariably scrofulous, I believe."

LIFE LINE.—To determine whether a child has the power to live to adult years, is of much moment if it is practicable. Most persons judge by the general appearance, and by the functional activity of the important organs. A physician in active practice for a number of years, will be able to form a very good opinion in most cases, but will occasionally be seriously in error.

I have been accustomed to place much reliance upon a measurement of the base of the brain, which determines its depth. Let the meatus auditorius externus represent the floor of the cranium, which it does only approximately, and a line drawn from the occipital protuberance to the external angular process of the os-frontis where it articulates with the malar bone, represent the superior part of the basilar brain, as marked by the *tentorium cerebelli*. The measurement between the two is approximately the depth of the base of the brain.

If the measurement is one-half inch or more, the child has the capacity to live to adult years, and will pass safely through any ordinary sickness. If the measurement is from three-fourths of an inch to an inch, it will live when it would seem hardly possible. But if the life line fall below the measurement of one-half inch, the power of living is feeble, and the child will succumb to the slightest ailments, and with one-fourth or even one-third of an inch, we may safely say that life is not possible.

As a physician's success will, to a certain extent, depend upon his powers of prognosis, such knowledge as this will prove of much importance. It tells us when we may rely on treatment to effect recovery, and when the best treatment will fail of success.

For the general pathology of disease, and the principles upon which we base our practice, the reader is referred to my " Principles of Medicine."

TEMPERATURE. PULSE. RESPIRATION.

In an abnormal condition of the system, the first things to be considered are, the temperature, pulse and respiration. Pulse and respiration are more readily affected, and to a greater extent, than the temperature, but the latter affords a more correct diagnosis of the patient's condition than the former. The pulse may deviate from a healthy standard, from some trivial cause as that of anger, slight indigestion, and sometimes sleep will produce a change, but under such circumstances the trouble will be of no moment and soon subside. However, should a change of temperature occur and remain for a time, a general abnormal condition will prevail. Too much importance, then, can not be attached to the study of the temperature, and a thorough knowledge in disease of the body thermometer.

TABLE
Showing the Temperature, Pulse and Respiration at various Ages.

				PULSE.	RESPIRATION.
Temperature above normal age, at birth,			1.5	120 to 150	40 to 60
"	"	"	1st month, .5	120	40
"	"	"	1st year, .25	105 to 130	30 to 40
"	"	"	15th year, .21	75 to 85	20
"	"	"	21st year, .0	70 to 75	16 to 18

Deviations from this table have been found without any constitutional disturbance, but they were few, and so far as tests have extended it is accurate.

TEMPERATURE.—Exercise and diet, as well as climate, may affect and cause the temperature to vary, but it will be but a temporary change when thus disturbed. Should it, however, deviate, and remain so for some hours, we may expect an abnormal condition of the system. If, on the day following we note an increase in the temperature, running as high or higher than that of the day preceeding, the question of disease is beyond all doubt. In measles there is usually a high temperature three or four days preceding the eruption ; in scarlet fever it is less a day, perhaps; in variola two days; in whooping-cough, the first indication of the disease is, a change in temperature usually running quite high ; in diphtheria it precedes the soreness of throat. *Increase* of temperature calls for sedatives and cooling remedies. *Decrease* of temperature demands stimulants and sustaining treatment. The following things are to be taken into special consideration : The hour at which the temperature rises or falls. *An increase* of temperature, beginning a little earlier each day, is cause for alarm, beginning later is encouraging. *A decrease,* beginning earlier each day is a good indication, beginning later, it is to be closely watched; also the duration of temperature. Should a high temperature prevail and remain without a decline for some time, it is to be regarded as very unfavorable. Likewise a low temperature, remaining so for a time or continuing to decrease, is to be considered a most dangerous symptom, being of greater danger than a corresponding rise would be—the danger lying

in the great depression, and probable want of vital force necessary to accomplish a reaction. The daily fluctuations of temperature should be carefully noted, going frequently from one
point to another, and not remaining stationary any given time.
The food of the child has an influence on the temperature; after a
good nourishing meal, there will be a rise in the index of the
thermometer; if there is no rise, the food does not contain
sufficient nourishing properties. This rise will not exceed .5,
unless in a case nearing starvation, when it may be increased.
A too simulating food will produce a rise of one degree in the
temperature; after digestion there will be a decrease of temperature until the normal standard is reached.

PULSE.—As previously said the pulse is more easily disturbed
than the temperature, and an increase may be occasioned by the
most trivial cause. The position of the body influences the
pulse, it being augmented when the body is in a standing position, and decreased when lying down. Joy, grief and anger will
vary the pulsations of the body; food likewise affects it. Hence,
an accelerated pulse must not always be regarded with alarm;
but when such a condition supervenes with general constitutional
disturbance, it demands investigation. In studying the pulse we
want to note the frequency, regularity and fullness. Referring to
the table we find that in a healthy adult there are from 70 to 75
pulsations per minute; we note the fact on the living body, and
finding it to correspond, we next observe the regularity, whether
the pulsations are uniform in volume and time, or whether there
is an irregular action of the heart, causing the pulse to intermit.
Fullness of pulse is ascertained by the sensation imparted to the
finger from the current of blood as it passes through the artery.
If a large portion of the pulsation is felt, we get fullness, but if
the finger discovers but a small portion of it we have the reverse;
likewise we detect a *hard* pulse, *soft* pulse, and a *wiry* pulse, and
measure them according to the standard of increase, decrease and
perversion. With children it is more difficult to obtain accurately the number of pulsations to the minute, but the general
character of the pulse is to be taken into consideration in order
to determine the departure from a normal condition. If possible
observe the pulse while the child is sleeping. The pulsations
will be more accurate than when the little sufferer is awake and
irritable.

RESPIRATION.—In disease we not only desire to know the number of respirations per minute, whether there is an increase or decrease, but we must as fully understand the character and source of respiration; whether it occurs with regularity, or whether it is panting and short; whether it is abdominal or thoracic, and whether a full inspiration can be taken without causing pain. Note, if any, the difference in the character of respiration when the child is sleeping and when awake; also the sounds, if any, produced by respiration. Observations made during sleep are generally more accurate than made when the child is awake, as any exertion quickens the respiration.

CLASSIFICATION OF DISEASE.

It is most convenient, and at the same time the most successful plan to study disease as it influences particular parts or functions. We thus get all the diseases of a part in a group, and are able to learn better the differential diagnosis between them, and also between them and others of different parts.

There are certain diseases that are general in their character, and have their cause principally in the blood, or in functions intimately associated with it—as blood-making, nutrition, and retrograde metamorphosis. This will naturally form the first division.

We may then study diseases of the digestive organs in a group, next, diseases of the respiratory organs, then diseases of the kidneys, of the skin, of the locomotive apparatus, of the nervous system, and of the organs of special sense.

CHAPTER V.

FEBRILE DISEASES.

FEVER is a disease involving to a greater or less extent every function of the body. Continuing for a time, it impairs the tissues, in some cases to such an extent that the further continuance of life is impossible.

The functional derangement is studied by the rule of *excess*, *defect* and *perversion*, though in all cases the change is in reality an impairment, and is associated with depression of vitality.

In analyzing a simple fever, we find it composed of: 1st. An *excess* in the frequency of the pulse, an *excess* in the temperature of the body, and an *excess* of innervation. 2d. A *defect* of excretion by the skin, the kidneys, and the bowels, and a *defect* of digestion, assimilation and nutrition. 3d. A *perversion* of the blood and of some functions.

PATHOLOGY OF FEVER.—The doctrine of the *humoralist* that fever is a disease of the blood, offers the best explanation of the phenomena of fever, and leads to the most certain methods of treatment. We must not forget, however, that with such derangement of function, especially of waste and excretion, and its complement nutrition, the solids of the body are being diseased from day to day.

In some fevers, the cause of the disease has been influencing the functions of nutrition and waste for a considerable time before the outbreak of the fever, and during this time all the tissues which have been formed are more or less diseased. In such a fever, complete recovery does not take place until all such tissue is removed. Thus, in *typhoid* fever, we find the disease continuing until the soft tissues have been removed to a very considerable extent; and in those exceptional cases, where the patient does not lose flesh during the progress of the disease, a perfect recovery does not take place; the reason is obvious—the imperfect tissues remain in their place, and to that extent the nutritive process fails. The old tissue is incapable of functional activity, and hence the imperfection with which all the processes of life are performed.

If we regard the cause of fever, as acting in and from the blood, as consisting of some *morbid material*, either generated within the body, or introduced from without, but always acting to impair the quality of this fluid, and consequently depressing every functional activity, we have a plain and very simple pathology. We can readily see how such condition of the blood should occasion the disturbance of the circulation that we witness, and how this abnormal circulation should be the cause of the diminished functional activity of every part of the body.

Based upon this view of the pathology of fever, the treatment becomes also simple and straight-forward. The rapid but imperfect circulation of the blood is corrected by certain special medicines, called sedatives. The excretory organs can then be stimulated by remedies that act directly upon them to increased activity, and any morbid material thus removed. We are then in a position to call the digestive organs into action, and obtain normal nutrition of tissue. Lastly, we have a class of remedies which directly influence zymotic processes and products, and in some cases at least, are antidotal to the cause of disease. In other words, they are agents which, when introduced into the blood, combine with and modify or destroy all those activities which are unnatural to and not under the control of the vital forces.

As before remarked, we must not lose sight of the structural changes which go on during the incubation and progress of the fever, and which impair the integrity of the tissues. But the treatment for this is as distinct as for the other. This imperfect tissue must be removed, and the nutritive process must be stimulated to its repair. The remedies which increase excretion favor retrograde metamorphosis, and the two go on together. A tonic and restorative treatment, with a good diet, completes the treatment, and the two go on harmoniously together. As the old tissue is removed by the first, new material is prepared, and formed into tissue in its place by the second means. We have thus, as Dr. Chambers tersely expresses it, a real " renewal of life."

CLASSIFICATION OF FEVERS.—We may divide the fevers of childhood into four varieties. The evanescent, or *febricula*; the *periodic* or malarial fevers, divided into two varieties, intermittent and remittent; *continued* fever, and the *eruptive* fevers.

FEBRICULA.

Occupying a prominent place in the diseases of childhood, if we regard the percentage of cases, is this *little* fever. I call it little on account of its brief duration, and the little danger that attends it. In the apparent symptoms it may be, and frequently is, a very high or active fever, and gives rise to much uneasiness on the part of parents.

CAUSES.—The causes of febricula are numerous, but why they influence the system in this way is not well known. Probably, the most frequent cause is the action of cold arresting secretion from the skin. The similar arrest of secretion from the kidneys or bowels may produce the same effect. In other cases, an indiscretion of diet, followed by imperfect digestion, introduces into the blood an imperfectly-formed material, which must be removed.

PATHOLOGY.—Taking the humoral view of this disease, we regard it as being produced by a morbid material in the circulation. In the first case, from retained excretions; in the second, from mal-digestion. This view is borne out by the history of the disease—always terminating by the re-establishment of secretion, and by the treatment, which is always successful when it attains this end. We do not wish to ignore the existence of the nervous system in this view, for it is influenced by the cause of fever, and in some cases its morbid action is capable of generating such a blood-poison, but usually this is but a minor element.

SYMPTOMS.—The symptoms of *febricula* are those common to all fevers. The patient has a chill of longer or shorter duration, but not usually of much severity. Preceding this chill, it will be observed, sometimes for a few hours, that the child is dull and languid, cross, or that it wants to sleep more than usual. Following the chill, the pulse increases in frequency, and the temperature becomes higher, the secretions are checked, and there is more or less excitement of the nervous system. The febrile reaction comes up rapidly, and usually attains its greatest intensity in from two to four hours. It may continue to the termination of the disease as a continued fever, or it may be broken up into exacerbations and remissions.

The natural duration of the disease is from one to three days, the fever gradually declining for the last third of the time, and finally terminating by the establishment of free secretion.

Occasionally, during its progress, the excitation of the nervous system becomes very great, and the child suffers severely from this, and in some cases, may have convulsions. In other cases there is determination of blood to the brain.

DIAGNOSIS.—There is no difficulty in determining that the child has fever, for all the symptoms are very clearly defined. But it is not so easy, always, to determine the character of the disease. The points that I rely upon to effect this diagnosis are these : though the skin is hot, it is never dry or harsh ; the tongue is uniformly moist, and the mucous membranes of natural color ; though the pulse is frequent, it is not otherwise much changed, but gives a sensation to the finger of normal fullness, and freedom of circulation. As the reverse of this is shown in the more persistent fevers, the diagnosis is quite plain.

PROGNOSIS.—The prognosis is favorable whether the patient has treatment or not, and we can safely say that the disease will terminate favorably by the close of the third day.

TREATMENT.—If secretion is arrested, we want to look to that, and usually but little additional treatment will be necessary. For the excited circulation and to favor diaphoresis, add Aconite gtt. v. Asclepias, gtt. x. to xv., to half a glass water, and give in teaspoonful doses every hour until a good action has been obtained. If there is irritation of nervous system, alternate Gelseminum with the sedative, gtt. x. to xv. to water 3iv., a teaspoonful every two hours.

If the child is vigorous, a warm bath may be given once a day, but if delicate the dry rubbing or the employment of some oily preparation is far better.

INTERMITTENT FEVERS.

It is only in those sections of county where the malarial poison is intense, that young children have ague; and during the first six months, they rarely have it, even in these localities. After the child has passed the age of two years, the system seems more easily impressed, though not so easily as after the age of puberty.

CAUSE.—The cause of intermittent fever is generally conceded to be a poison generated by the decomposition of vegetable matter, and receives the name of " vegetable malaria."

PATHOLOGY.—There is no doubt but that the cause of intermittent fever acts upon the blood, and from it upon other parts and functions. There is always a period of incubation in which the morbid material, whatever it may be, is increasing. Finally it produces depression of the vital powers, and a chill. Then the vitality of the system seems concentrated for its removal, and we have the febrile reaction, succeeded by the sweating stage, in which secretion is re-established.

The system thus freed from the poison enjoys a period of comparative health (the intermission), but the germs of the poison still remaining in the body, it is gradually reproduced, and there is another period of depression, reaction, and secretion. Thus it may continue for an indefinite time, the system never freeing itself entirely from the cause of the disease.

SYMPTOMS.—An intermittent fever in the child, presents the same symptoms, and in the same order as in the adult. There is a period of depression for a short time, in which the child is dull, its face pallid, the extremities cold, and the lips and finger nails bluish. The chill is thus introduced, and increases in intensity for a longer or shorter time. The child is seen to be cold, has more or less rigors, its pulse is small and increased in frequency, and the secretions locked up.

Continuing thus for a short time (from fifteen minutes to two or three hours), the symptoms of the chill gradually pass away, the surface becomes warm, loses its bluish appearance, the pulse increases in frequency, and the symptoms of nervous depression pass away. But going beyond the point of healthy reaction, the surface becomes hot and more or less dry, the pulse increased in frequency from twenty to forty beats per minute beyond the healthy standard, and is also changed as to freedom, and is more or less hard, and the nervous system is excited, with a gradual increase in the febrile symptoms. The fever attains its maximum intensity in from two to six hours, and there is then a like gradual decline to the *sweating stage.*

This stage of an intermittent is announced by the diminished frequency of the pulse, and disappearance of nervous excitement. The temperature of the body is reduced, and the child breaks out into a free perspiration, and there is also increased secretion from the kidneys.

There is then a complete intermission, the child presenting no evidences of disease for twenty-four, forty-eight, or seventy-two hours, as the fever is *quotidian, tertian,* or *quartan* in type, except in some rare cases in which it assumes the form of a double quotidian, there being two revolutions of the fever each day.

This period being passed, there is again a recurrence of the phenomena of chill, febrile reaction, sweating stage, and intermission, until finally the fever wears itself or the patient out, or is arrested by medicine.

WITH NERVOUS COMPLICATION.—In some cases the nervous system suffers severely. There is very great depression during

10

the chill, with tendency to congestion. And during the febrile reaction there is dullness, with tendency to coma. In other cases there is very great excitation of the nervous system, with determination to the nerve-centers. And this will sometimes continue during the intermission. In some cases the ague is attended with convulsions.

WITH INFLAMMATORY COMPLICATION.— In other cases we notice a very high febrile reaction, the fever being prolonged beyond the usual time. The sweating stage is imperfect, and during the intermission the child still suffers, the skin is dry, the urine scanty, and the pulse somewhat hard.

WITH GASTRIC COMPLICATION.— In still other cases, the stomach and associate organs seem to suffer severely. In some the stomach is very irritable, and there is nausea and vomiting during the chill and early part of the febrile reaction. And even in the remission, the irritability of the stomach remains, so that digestion is imperfectly performed, and medicine is not received kindly, and is absorbed slowly and with difficulty.

In another class of cases the tongue is pale, broad, and coated with a white, or yellowish-white, tenacious fur. The appetite is capricious, digestion imperfect, and nutrition of tissue is not well performed. Medicine is received with difficulty, and if absorbed at all, slowly, and does not act kindly, or produce its usual influence.

In still another class of cases the mucous membranes are darker, even dusky, and the coatings of the tongue are also dark-colored. Digestion is imperfect, and of course nutrition is to a considerable extent suspended. In the severer cases there is a tendency to septic decomposition. And in all, the usual antiperiodic treatment fails to arrest the disease.

TREATMENT.—During the chill the child must be kept warm; this may be aided by giving warm drink occasionally, and there is nothing better, especially if the child craves drink, than hot lemonade, sweetened just enough to make a pleasant drink, and let it be taken at pleasure. It will prove a grateful and efficient remedy. As soon as reaction takes place, give the required sedative with any other remedy that may be indicated. Aconite, if the pulse is small and rapid and there are no complications. Veratrum, if the pulse is full and bounding and with gastric complication. Asclepias is the child's diaphoretic and an essential part of the treatment, when the skin is harsh and dry; Gelseminum is always to be thought of when the pupils are contracted and face flushed, with irritation of nervous system. If there is dullness with a tendency to sleep, indicating congestion, Belladonna in small doses, gtt. iv. to viii. to water ℥iv., a teaspoonful every half hour or hour, according to the severity of the case, will prove a fine remedial agent. As soon as the pulse and temperature begin to decline and secretion is established, the antiperiodic is to be given. In uncomplicated cases carbazotate of ammonia is preferable to quinia; it may be given even during the fever with perfect safety. The dose is small, from one-eighth to one-sixth of a grain, triturated with sugar of milk and given every three hours. When complications exist, and particularly if there is considerable depression, quinia is preferable to any other antiperiodic. It may be given in from one to two grain doses, repeated every two hours during the intermission. After the chill is broken, and especially in malarial localities, it is best to continue the antiperiodic for three or four days, giving it less frequently, however, in one grain doses twice daily, repeating it on the seventh day.

Quinia is *the* antiperiodic, though there are other remedies that act in the same manner, but less efficiently. I prefer to give a sufficient quantity in a single dose, rather than in broken doses. This for a child two years old, will be from two to three grains. It is given best in a cold infusion of green tea. If the chill is broken, it may be repeated for two or three days in one-grain doses.

For some days afterwards, it will be well to continue the Veratrum or Aconite, and in a malarial country repeat the antiperiodic every seventh day.

In the nervous complication, with dullness, and congestion of the nerve-centers, I use Belladonna, associated with Aconite, as in this formula: ℞ Aconite, Belladonna, aa. gtt. v. water ℥iv.; a teaspoonful every hour during fever, and every three hours during the intermission.

Rhus tox. will be found indispensable in some nervous complications; give it in alternation or with the Aconite.

Stimulant frictions to the skin, as with a combination of some of the essential oils with lard, is an aid to the treatment. This means is continued until the symptoms are no longer marked during the intermission, and then quinine is given as before.

In the second form—with irritation of the nervous system and determination of blood—I prescribe : ℞ Tincture of Gelseminum gtt. x. to xx.; Aconite or Veratrum gtt. v.; water ℥iv. A tea-spoonful is given every hour while there is fever, and every two hours during the remission. If the symptoms have been very severe, the acetate of potash may be given for a day or two before the administration of the antiperiodic.

In the case of *inflammatory* intermittent, the administration of the special sedatives, until a complete sweating stage and inter-mission is produced, will frequently be all that is required. In the more persistent cases, gently opening the bowels, and the use of a solution of acetate of potash, will be necessary. The im-portant point in all of these cases is, not to give the quinine until the system is thoroughly prepared for its reception.

The *gastric* complication gives the greatest trouble, as there is not only imperfect digestion and nutrition, but medicines are not absorbed from the stomach. In the first case named, I should give the patient Nux in small doses every hour, and the usual doses of Aconite with Gelseminum. The child may have a hot foot-bath once or twice a day, or sometimes a general hot bath. Cold applications over the stomach will usually answer a better purpose than counter-irritants.

In severe persistent cases the following local application will relieve when all other means fail : Take of cinnamon, cloves and alspice aa., sufficient quantity, cover with good whisky and water, equal parts, and make an infusion, strain and keep it hot, wring flannel cloths out of this and apply to the epigastric region; it will quickly relieve and refresh the little sufferers.

When the tongue is broad, and coated with the pasty coat, I advise the administration of sulphite of soda in doses of two to four grains, every two or three hours. When this condition is very marked, and the disease has continued for some time, a thorough emetic is the shortest road to a cure. In the meanwhile, the child should have Veratrum in small doses regularly, and if there is nervous irritability, Gelseminum.

When there is the dusky discoloration of the mucous mem-branes, the patient wants Baptisia and Aconite every hour, and the quinine should be given in the form of an elixir, prepared with muriatic or nitric acid and simple syrup. This will supply the required acid ; and render the antiperiodic ready for immediate absorption.

Where quinia can not be given by mouth, I frequently arrest the disease by its endermic use. Especially in the nervous and gastric forms of the disease, this method will be found beneficial. I order : ℞ Quinia Sul. ʒj., Adeps, or warm olive oil, ʒij. Let the child be thoroughly rubbed with this twice daily; brisk friction while it is being employed is of great importance. I have not failed of success in any case, with this method, where the internal treatment had been pursued as I have recommended.

MASKED INTERMITTENT.

In the diseases of children, as with the adult, we meet with many cases, in which, if the periodic influence is not the cause of the disease, it continues it, and prevents success from the ordinary treatment. Many of these cases are very obscure, and it will require that the practitioner be impressed with the importance of closely looking for the malarial complication, that it be detected.

Almost all of the inflammatory diseases may be thus complicated, and we frequently find it at the bottom of persistent gastric and intestinal disturbance, and of lesions of innervation, and especially of nutrition. In the adult we find the majority of masked intermittents in the form of neuralgias.

The treatment of these cases will be the same as for an ordinary intermittent, and quinia, one to two grains in the form of an elixir, may be given with an acid ; or carbazotate of ammonia or Alstonia may be given, in the following doses : grains two of the first every two hours ; from one-sixth to one-fourth grain of the second every three hours, and one to three grains of the last four times a day.

REMITTENT FEVER.

The fevers of childhood are almost always remittent in form. A continued fever, being of very rare occurrence. Thus we will have remittent fever from the ordinary malarial cause of this disease in the adult, and also from the various causes which may give rise to fever.

The disease is divided into two varieties, *regular infantile remittent*, and *slow infantile* remittent fever. The first possesses all the elements of a fever, and runs a regular and uniform course; while the second is defective in many of its symptoms, and is irregular in its course and duration. We will consider the regular form of the disease first.

CAUSES.—As just remarked, the causes of infantile remittent fever are of two kinds. In a certain proportion of cases, it is as distinctly malarial, as in the same form of the disease in the adult. I know of no means of determining this cause, other than that it prevails in localities and at times when the adult remittent fever is noticed. The ordinary cause of continued and typhoid fever in the adult—animal malaria—will produce a species of remittent fever in the child. The ordinary causes of simple fever, cold, arrest of secretion, imperfect digestion, deficient waste of tissue, whatever may be their origin, will give rise to this disease.

PATHOLOGY.—The febrile poison, whatever may have been its source, seems to act primarily upon the blood. During the incubation of the disease, which occupies a longer or a shorter period of time, the solids of the body become to some extent involved through impaired nutrition. Thus, when the febrile symptoms are fully announced, there is the double lesion, a morbid material in the circulation and impaired vitality of the blood, and imperfect material through all the tissues of the body, from the depraved nutrition.

For the first days of the fever, the lesions seem principally of function, but as the disease is protracted, the solids of the body become more and more affected. This is in three ways—by the impairment of the vitality of tissue, by the febrile-reaction—by imperfect waste, leaving worn-out and dead tissue in its place— and by imperfect and faulty nutrition, whereby feeble and imperfect tissues are formed.

Post-mortem examination shows a deterioration of the blood, and in some of the worst cases, the red-corpuscles of the blood are much broken down. There is usually some evidences of congestion and infiltration, especially of organs which have suffered from irritation during the progress of the disease. The tissues are but little softened in the majority of cases. The local lesions are principally inflammatory in their character, and owing to complications.

SYMPTOMS.—In the majority of cases there is a forming stage of from one to six days. During this, the child seems dull and listless at times, at other times is cross and fretful. Sleeps at unusual times during the day, but ,not soundly, and is restless and uneasy at night. The appetite fails, the bowels are irregular, and occasionally slight febrile exacerbations occur.

The chill is not usually so marked as to attract notice. The child seems very quiet and dull, draws up to the fire, and wants to drink more than usual. Lasting but a short time, febrile reaction comes up quickly, and in a couple of hours presents its most marked symptoms. The skin is hot and somewhat dry, the pulse frequent and hard, the mouth dry, and the tongue coated white; the urine is scanty, the bowels constipated, and there is considerable nervous irritation.

The fever varies considerably, as regards the remissions. In some cases there is but one remission, and that usually in the morning; in others there are two, and in others three, and in still others the fever is broken up into short febrile exacerbations of irregular duration, so that there may be a dozen or more in the course of twenty-four hours.

As the disease progresses, the symptoms increase in severity, the febrile reaction is higher and more prolonged, and the remissions less marked. The important functions are also involved to a greater extent, and the patient more and more debilitated.

Remittent fever is frequently complicated, and, as a general rule, the greatest danger to life is from this. The more common complications are, of the brain, the digestive organs, and of the respiratory apparatus.

WITH DISEASE OF THE BRAIN.—There are two principal lesions of the brain, noticed early in the disease. These are—irritation with determination of blood, and congestion.

In the first case, the child is noticed to be more than usually restless and fretful, its eyes are bright, and it is continually wanting drink, and various things that it sees. In a short time it is noticed that the face and scalp are flushed, the head is hotter than usual, the eyes are bright, with contracted pupils, and the restlessness has increased. The disturbance thus commenced may go on to the development of inflammation, or after lasting for some time—three or four days—may terminate in congestion and coma; or in other cases, the excitement may eventuate in the production of convulsions.

In the second case, the child is dull, and has a tendency to sleep much, but sleeps with its eyes partly opened. The eyes are dull, pupils frequently dilated, and the face has a heavy, expressionless appearance. As the case progresses, coma comes on, and gradually deepens until it becomes impossible to arouse the child from it.

WITH DERANGEMENT OF THE DIGESTIVE ORGANS.—The common lesions of this apparatus are—irritation of the stomach, and atony of the stomach with morbid accumulations. The irritable stomach is readily recognized in most cases. The child can not take food, drink, or medicine, without nausea and retching. The tongue is elongated and pointed, tip and edges more or less reddened, and the coating—usually white—confined to its center.

In the second case there is also nausea and efforts at vomiting, food is not digested, and medicine produces sickness and is not absorbed. The tongue is unusually pale, broad, and covered with a pasty white coat.

In both of these cases, the disease runs its course more rapidly. Not only on account of the sympathetic derangements that flow from such gastric' disturbance, but also because digestion is arrested, and all our means of cure are inefficient because not absorbed.

WITH DISEASE OF THE RESPIRATORY APPARATUS.—In the winter and spring, infantile remittent fever is frequently complicated with disease of the respiratory organs. In the majority of cases, it is nothing more than an irritation, with a more or less troublesome cough. In others, a well developed bronchitis, and in others, a lobular pneumonia is set up.

The symptoms are usually very plain. The child has cough, with increased frequency of respiration, and sometimes slight difficulty in breathing. At first there is dryness of the bronchial mucous membrane, afterward increased secretion. Where these symptoms are marked, a physical examination of the chest should be made, to determine the character of the trouble.

DIAGNOSIS.—The diagnosis of an infantile remitttent fever is easily made. From *febricula*, we determine it by the *dry* skin, *hard* pulse, and *dry* or broad and pale tongue. From continued fever, by the regular appearance of remissions. From inflammatory diseases, by the absence of local symptoms; for even when complicated, the local symptoms are distinctly secondary.

PROGNOSIS.—The prognosis is favorable. Unless the complications are severe, there is little danger of a fatal termination in viable children. As heretofore noticed, a certain proportion of children are *non*-viable, and will die of any disease, or even without any disease, before they have passed the age of childhood. The mortality to be expected in the course of ordinary practice will be from two to five per cent., depending upon the fact just stated.

TREATMENT.—The treatment of a remittent fever is usually very simple, and also very successful. The first object is to reduce the pulse to the normal standard, and get a free and equal circulation of blood. The second, to remove any irritation or other derangement of the nervous system. The third, to estabish secretion, and get a good condition of the digestive apparatus. The fourth, the employment of agents to antagonize the malarial or other cause of disease.

The first indication is very surely accomplished by the use of the special sedatives, and the accessory means—a general bath once or twice daily, and the hot foot-bath. I usually prescribe : ℞ Tincture of Aconite or Tincture of Veratrum Viride gtt. v. to gtt. x., water ℥iv. ; a teaspoonful every hour. The action of these remedies should not be looked for at once ; they require time, but it is noticed that under their use there is a continuous amendment in all the symptoms. In the course of forty-eight hours, the febrile reaction has to a considerable extent subsided, the pulse is but little above the normal standard, and the circulation is free and uniform; and we are ready to look after the second indication.

As a general thing, we will not need to use any special remedies to relieve the nervous system, for any irritation or other disturbance will pass away with the febrile reaction. The special cases that demand treatment will be named hereafter.

The third indication is also frequently fulfilled by the action of the special sedatives, without other means. But when secretion is not established as we should like, we put the little patient on tincture of Asclepias ℨi. to water ℥iv., a teaspoonful every hour, which, with a hot mustard foot-bath, will soon establish secretion from the skin. More frequently we will want to stimulate the action of the kidneys, as being the most important excretory organ. For this purpose I prefer a solution of acetate of potash.

Secretion from the bowels should be obtained by minute doses of Podophyllin, as named in the first part of this work, or other laxative. Or, if there is simply retention of feces, it may be overcome by the use of mild laxatives or enemata. Purgatives should not be used unless there is a special indication for their employment.

The fourth indication of cure is accomplished in the strictly malarial cases, by the administration of quinine in doses of one to two grains every two hours, given with an acid. I never give quinine to child or adult until I have so prepared the system for its reception that it will act kindly, and as a curative agent. With the action of the sedatives, and secretion beginning to be established, it may be given with a considerable degree of certainty.

With the subsidence of the disease, and the establishment of convalescence, but little medicine will be required. I think it advisable to give the sedatives in quite small doses, for two or three days after the fever has entirely passed away, as the child convalesces more rapidly under their influence. In some cases, a restorative aids recovery. Elixir of Guarana, gtt. x. four times daily, or Comp. Spirits of Lavender, answers a very good purpose. Or in some cases, dilute muriatic acid ʒij. to simple syrup ʒij., half a teaspoonful in water as a drink, every three or four hours, will answer a good purpose.

Determination of blood to the brain is arrested by the administration of Gelseminum, which, in this respect, is specific in its action. I usually give it in combination with the sedatives, preferring, when the fever runs high, to dispense with the Aconite, and increase the quantity of the Veratrum. In the majority of cases, ten drops to the four-ounce mixture, is sufficient. But when the determination is active, and likely to progress to inflammation, I prescribe twenty drops to four ounces of water. When there is danger of convulsions from the same cause, Gelseminum may be relied on with great certainty to prevent it.

The dull, congested condition, with tendency to coma, is antagonized by the administration of Belladonna, which is also specific in its action. I generally prescribe it with Aconite, in the following proportion: ℞ Tincture of Aconite, Tincture of Belladonna, aa. gtt. v., water ʒiv.; a teaspoonful every hour. The hot stimulant foot-bath, and occasionally counter-irritation to the spine, will be beneficial. But the dry rubbing, in both mild and severe cases, will frequently be all that is desired. It arouses the inactive condition of the system, restores the capillary circulation, and with the internal use of Belladonna, the cure is complete. The rubbing or dry bathing must be done gently but thoroughly, five minutes being sufficient time to operate, repeating it every hour and a half or two hours, the hand movements being from the base of the brain down to the sacrum, the arms, thorax and abdomen operated on in the same manner. I have removed the congestion in very severe cases with no other remedies than these. The Belladonna wants to be given in small doses. Children being very susceptible to its influence, from gtt. iv. to viii. added to half a glass of water, and given in teaspoonful doses every half hour will give excellent results.

The *irritation of the stomach* is met by the use of cold or warm applications to the epigastrium, and heat and stimulants to the extremities as the external means. Internally, Nux or Ipecac may be given, and they will do the work best if given with as little fluid as possible. To accomplish this the preparation of Nux should be made stronger than usual, about gtt. v. to the ounce of water, and give gtt. xv. every half hour. Ipecac will operate charmingly in severe cases, if given in doses varying from one-tenth to one-sixth of a grain according to age of child. If there is nausea with flatulence, Colocynth alternated with the Ipecac will be the remedy; or subnitrate of bismuth, in mint water. These are continued until the irritation is removed. At the same time, small doses of Aconite may be given for its sedative effect, and also for its influence in controlling the irritation of the solar plexus of nerves, which attends the gastric irritation.

Much care will be required in these cases, to prevent a renewal of the gastric irritation, as the disease progresses, especially to give our remedies in such way that they will not be likely to produce it.

In the second case, with atony and *morbid accumulations,* if the symptoms are marked, the shortest method will be to give a prompt emetic. This is not a pleasant means with children; but when the symptoms are grave, it is the most successful method.

In malarial localities the emetic is usually a first consideration, as it is worse than useless to give the child other remedies until the stomach has been placed in a condition to receive them. The acetous emetic tincture or an infusion of gentian-root is preferable to use. In the same cases it will be absolutely essen-

tial to use :　℞ Podophyllin gr. 1-20, Leptandrin gr. 1-15, sugar
of milk sufficient to make one grain of the powder.　Dose, one
every three hours during the first day, and three, one at morning,
noon and night, during the second day.

In other cases, sulphite of soda in five grain doses, may be
administered every two or three hours, or if the tongue is red
and dirty, sulphurous acid.

As a general rule, disease of the *respiratory* organs will yield
readily to the treatment for the fever.　If an inflammation is
developed, I advise the mush-jacket to the chest.　And if there
is much cough, some simple remedy to relieve this.　℞ Cam-
phorated Tincture of Opium, Syrup of Lobelia *aa:* ℥ss., Simple
Syrup ℥iii. M.　Dose from x. to xx. gtt. three or four times a
day.　Or, ℞ Chlorate of Potassa ℈ss., Tinct. Aconite gtt. x.,
Tinct. Opium ℈ss., Simple Syrup ℥iv. M., and give a teaspoonful
four times a day.

SLOW INFANTILE REMITTENT FEVER.

This disease is not only slow in its progress, but also obscure
in its symptoms, so that many times it is very difficult to make a
correct diagnosis.　As before remarked, more or less of the
symptoms of fever are wanting, and others are quite irregular.

CAUSES.—In a part of the cases that come under our notice,
the cause is undoubtedly the malarial poison that produces other
forms of periodic disease.　This may be taken for granted, when
it occurs in a malarial region, or if the child has been exposed to
this cause, however temporarily.

In other cases, the disease is produced by derangement of the
digestive organs, by imperfect nutrition and waste, by deficient
excretion, and indeed by any of the many causes of diseased
action.

PATHOLOGY.—As with the other fevers, this must also be regarded as primarily a disease of the blood, but as it advances, all
the functions are impaired, and there is also a continued deterioration of tissue.　An arrest of nutrition is a prominent feature,
so that waste continuing, the child becomes very much reduced
in flesh.

There are no special evidences of diseased action, found upon
post-mortem examination, unless there has been some local com-

plication, which generally manifests itself during life, by well pronounced symptoms.

SYMPTOMS.—For a week or two the mother notices that the child is not so well as usual. Its appetite is variable, its temper capricious, is fretful and irritable, is sometimes dull, wants to sleep at unusual times, but does not sleep easily or well at regular times, is restless and uneasy at night, tossing about in the bed, and wants to drink frequently. These symptoms slowly increase, and the parents finding it is not the temporary ailment from cold or teething that they supposed, call a physician.

He hears the history of the case as above, and an examination confirms the symptoms. He also finds that the pulse is not natural, but has a sharper beat, is a little more frequent, and has an unusual hardness in some cases, is softer and more easily compressed in others.

The tongue shows evidence of disease in its change of form and its coating. In a few cases it is contracted, and presents a dusky redness, and the coating, if any, has a shade of brown. In the larger number of cases it is more or less pallid, frequently broader and seemingly larger than common; in both cases the coating is a dirty or yellowish white.

The bowels are irregular, and the feces unnatural. In some, the feces are dark-colored, brownish, or greenish, in state of effervescence, and have a peculiar fetor. This is more frequent with the dusky-red tongue, and is generally said to indicate acidity of the *primæ viæ*; whether this is so or not, it is best remedied by the administration of an acid. In other cases the feces are light-colored, papescent, and have not the natural feculant odor.

The skin is dry, and in bad cases gives a parchment-like sensation to the hand. Frequently, the extremities are cold, and in some cases the temperature of the child is lower for a considerable portion of the day.

The urine, if it be observed, will usually be found free, pale, and of low specific gravity, although the waste of tissue is going on rapidly. In some cases it is highly colored, stains the clothes, and possesses in an eminent degree, the urinous odor.

At some period of the day, slight febrile symptoms are developed, and continue for one or two hours. But they are not very well marked, and frequently escape notice. As soon, however, as the attention is drawn to it, the periodicity of the disease will be noticed.

Continuing on thus for weeks, the child becomes much reduced in flesh, and it no longer has the strength to sit up, or make an effort at play. Finally, if not arrested, or it is not naturally removed, some local complication is set up, which runs an acute course, and causes a fatal termination.

DIAGNOSIS.—The diagnosis of slow infantile remittent fever is made by *exclusion*. The evidences of disease are very marked, but *what* the disease is, is obscure. The patient is examined with reference to local lesions. Is there disease of the brain or nervous system, of the digestive organs, of the respiratory system, or of the excretory organs? We determine there is not. It is then a disease of the blood, and a lesion of nutrition and waste, and the symptoms observed will readily determine its periodicity.

PROGNOSIS.—The prognosis is favorable if the disease is diagnosed, and a proper treatment adopted. But in its advanced stages it will not bear a harsh or very active treatment, but, on the contrary, requires time and patience. The vitality of the patient is so exhausted, and the functions are so impaired, that they require gentle and judicious stimulation to obtain the desired action.

TREATMENT.—I place the patient upon the use of Aconite alone, if the pulse is soft and feeble, giving it in the usual doses. If the pulse is hard, I associate Veratrum with it in the proportions already named. I regard the action of the sedatives as very important in these cases, improving the circulation, both as to frequency, freedom, and an equal distribution of blood. No disease will better illustrate the fact that these remedies called sedatives, are really stimulant and tonic to the heart and blood-vessels.

The sedatives are continued for one, two, or three days alone, or with but little other medicine, and aided by the general bath and hot foot-bath. A salt-water bath answers a very good purpose to stimulate the surface. We find that the circulation becomes better, the skin less dry, bowels more regular, innervation better, indeed, that every function has improved slightly under their use.

If there has been marked dullness and hebetude of the nervous system, the combination of Belladonna and Aconite should be employed. If there is irritation of the nervous system, Gelseminum, or Rhus, as indicated, should be substituted. These means of meeting special indications, should not be neglected.

With the influence upon the system named above, we may give remedies to call the excretory organs into action, if it seems necessary. The administration of small doses of Podophyllin, triturated as heretofore named, with one-fourth or one-half a grain of Sulphate of Hydrastia, three or four times a day, increases secretion from the intestinal canal, and improves the appetite. Acetate of potash, to the extent of ten grains daily, will give free excretion by way of the kidneys.

Whether we use these means to increase secretion or not, we find a very important aid to the cure in quinine. I consider the patient prepared for its administration whenever the pulse becomes soft, the skin soft and moist, and the tongue moist and cleaning. Usually, I give from two to three grains at one dose, and afterward repeat it, one grain each day, for three or four days, or may be a week.

When quinine by mouth is objected to, or there is any other reason for not giving it in this way, I use it by inunction. I am satisfied that in some cases this use is decidedly preferable to the other, being more certain, as well as pleasant. I have the child thoroughly rubbed once or twice a day with an ointment of one drachm of quinine to two ounces of lard.

In some cases we find the tongue broad, pale, and covered with a pasty coat. If these symptoms are very marked, the mouth being nasty, and the breath fetid, I think the speediest method is the administration of a thorough emetic, and its repetition if necessary. In place of this I give the sulphite of soda in the usual doses, repeated every two hours. Rhus, Belladonna, Gelseminum, Apis, Arsenicum, and other remedies, will find a place in this disease according to indications.

When the mucous membranes present the dark-red or dusky appearance, a very marked effect follows the use of muriatic acid. Some very persistent cases, that have been intractable to the ordinary method, will yield readily to the same treatment with the acid addition.

As a general rule, when the disease is soon arrested in its course, convalescence is steady, and tolerably rapid. Where it is not, we may put the patient upon the restoratives named in the treatment of the preceding form of fever.

CONGESTIVE FEVER.

There is a form of fever in childhood that may be properly called congestive, resembling in many of its symptoms the congestive remittent fever of the adult. It might have been passed by with the brief notice given in remittent fever, complicated with congestion of the brain; but it would leave the student of medicine at a loss, when he was called to cases which, though distinctly febrile, presented the predominant symptoms of congestion.

PATHOLOGY.—The causes of this may be the same as produce any form of fever; the congestive character of the disease depending upon the condition of the patient, rather than the causes of the disease. In the majority of cases, however, there is something in the cause which seems to paralyze the vegetative or sympathetic nerves, and which occasions the congestion. In some seasons, and in some localities, we find a tendency to this form of disease.

Post-mortem examination evidences the local engorgements of blood, in discoloration, and transudation. In the severer cases, the blood is more broken down, and the tissues more softened, than in ordinary fevers.

SYMPTOMS.—As a general rule the chill is protracted and better marked than in the fevers described. Indeed, in some cases it will last the greater part of the twenty-four hours, not presenting so much a coldness as a dullness and hebetude.

As the febrile reaction comes up, the child is not excited, but sleeps or dozes with its eyes partly open, and a full and expressionless appearance of the face. This continues until the fever commences to decline, when the child arouses up and seems more lively and better. This condition of the nervous system continues throughout the disease, gradually increasing as the fever becomes more intense, becoming a profound coma in the severer, and especially in the fatal cases.

The pulse is full and somewhat labored, or in some cases the casual observer would notice but little change from its normal state, beyond its frequency. There is distinct evidence of an impaired capillary circulation, in the fullness of loosely connected tissues and the dusky redness or lividity. The circulation becomes more free during the remission, and the symptoms of congestion pass away.

I have noticed the lesion of the nervous system, which is so constant as to be a part of almost every fever. But there are also local congestions, which become very marked.

The *respiratory apparatus* suffers in this way in many cases. It is noticed that the child has a slight cough, attended with some difficulty in breathing, and occasionally removing a frothy mucus. In a few hours the breathing becomes labored, there is a marked rattling or blowing sound in the chest. The cough fails to raise anything, though it is evident there is increased secretion, and the evidences of imperfect aeration of the blood become well marked. Continuing on in this way, we find all the symptoms increasing rapidly, and sometimes death will ensue in twenty-four to forty-eight hours.

In other cases there is evident congestion of the abdominal viscera, and arrested function of stomach, liver, intestinal canal, etc. The symptoms are not so distinct as in the preceding case; yet the prostration of the little patient, the fullness and evident uneasiness in this region, are pretty good evidence. Add to this that the respiratory function is oppressed, and the breathing principally thoracic, with an absence of the physical signs of disease of the lungs, and the diagnosis becomes quite plain.

Congestion of the kidneys is announced by a rapid increase of coma, attended by a scanty or arrested secretion of urine.

DIAGNOSIS.—There is usually but little difficulty in making the diagnosis of this disease. The evident impairment of the circulation is manifested by the pulse, and by the appearance of the patient. The dullness and hebetude, and the development of coma, are very characteristic. The local lesions are very readily determined, as named above.

PROGNOSIS.—If not properly treated, the prognosis is not so favorable as in other forms of fever. It frequently runs its course to a fatal termination by the fourth or fifth day. But if attention is given to this feature of the disease at the commencement, with our special remedies, the mortality will be but little greater than in other forms of fever.

TREATMENT.—There are two methods of overcoming this congestive condition, both of them good, though differing entirely in their character and mode of action. The first of these methods is by active emesis, the second by the specific action of Belladonna.

11

There are some cases in which I should prefer the employment of an emetic. In a case where the symptoms of congestion had developed rapidly, and especially where there was marked coma, the action of the emetic would be the most speedy, and probably the most certain. The acetous emetic tincture is the best emetic in such cases, being very thorough in its action, and also giving the stimulant influence upon the circulation.' I administer it in doses of gtt. x. to xx. every ten minutes, with some stimulant or aromatic infusion, and continue to repeat it until very thorough emesis is induced, and a free circulation of blood follows. The nervous prostration and coma are usually overcome by this action, and in many cases, after such thorough emetic, the disease will run the usual course of a simple remittent fever.

Should the symptoms of congestion return, the emetic may be repeated every day, or even twice daily. In these severe cases I like the action of the hot blanket pack, as an adjuvant to the other treatment. To a kettle of hot water add an ounce of mustard, wring a blanket out of it, and wrap the child up in it, placing it in bed, and covering warmly. In half an hour the wet blanket may be replaced by a hot dry one, in which the child can remain until the means employed are successful.

Another very efficient means of revulsion is—heat three, four, or five bricks on the fire until they are hot enough to vaporize water, yet not to burn the clothing; wrap them in cloths wrung out of equal parts of vinegar and water, and place them at the feet, legs and body of the child, covering the whole loosely with a blanket, which should be well tucked in at the feet and around the child's shoulders, to prevent the escape of the vapor. It is very efficient as described, but may be rendered more powerful by the addition of tincture of Capsicum to the vinegar and water in which the cloths are wet.

The above may be called the *indirect* method of treatment, and, as will be seen, requires considerable skill, and much attention from the nurse. For, when improperly used, it may do as much harm as it would do good if used rightly.

The *direct* method is wholly different, but to persons in the habit of active medication will hardly seem sufficient for the purpose. The remedy with which I propose to overcome the congestion, is Belladonna, in small doses. I generally administer it with Aconite. ℞ Tinct. Aconite, Tinct. Belladonna, aa. gtt. v., water ℥iv.; a teaspoonful every hour. Had I not seen the *specific* action of Belladonna in such cases so frequently, I would not be willing to give it this recommendation. But if my experience is to be depended upon, the remedy is most reliable, and will give entire satisfaction.

In this connection a very important part of the treatment will be stimulants and restoratives, and I shall refer to Elixir of Guarana, Comp. Spirits of Lavender and beef essence. They have filled the requirements so often that I consider them indispensable in these prostrating diseases.

Usually, in these cases, I continue the same remedies throughout the treatment, as there is still some tendency to congestion or feebleness of the capillary circulation, especially of the brain.

In congestion of the *respiratory apparatus,* in addition to the means named, or when it is a special lesion, other parts not suffering so much, I depend principally upon small doses of Lobelia internally. For a child three to six months of age we may prescribe: ℞ Tinct. of Lobelia (seed) ʒj., Compound Tinct. of Lavender ʒij., Simple Syrup ʒij. Dose, half teaspoonful every fifteen minutes or half hour.

For the local application I direct a cloth, sufficiently large to cover the anterior surface of the thorax, spread with lard and dusted with Comp. Powder of Lobelia and Capsicum. This is changed once or twice daily until the complication is removed.

In congestion of the abdominal viscera I order hot stimulant fomentations to the bowels, occasionally preceded by dry cupping if the child is of some age, and an enema of an infusion of Bayberry, to which is added one drachm of tincture of Lobelia. The enema as named, is of especial importance in those severe cases in which the danger is imminent and speedy relief essential.

In congestion of the *kidneys* I apply hot fomentations across the loins, sometimes preceded by dry cups, and move the bowels freely by the use of an enema of Comp. Powder of Jalap and Senna. In those cases in which there is a constant tendency to congestion of these organs, but in which the urine is not suppressed to any considerable extent, Santonine, gr. one-sixth to one-fourth every three hours, will prove effectual.

In all other respects the fever should be treated in the same manner as named for infantile remittent. In those cases in which the abdominal organs are principally affected, the sulphate of quinia combined with warm olive oil, and the abdominal organs thoroughly bathed with it and rubbed gently three or four times daily, will give prompt relief. Too much importance can not be attached to this kneading and friction process, it supplies a long needed want in medical treatment, particularly in congestion and some other diseases, as marked beneficial results, after repeated trials in practice, have proven. In those cases where an acid is required, I would always combine the quinia with it, adding aromatic elixir to form a pleasant preparation. The sulphite of soda will also prove a very important remedy in some of these cases—the tongue presenting that peculiar broad, pale, pasty condition, heretofore named as indicating it.

CONTINUED FEVER.

As heretore noticed, continued fever is not of frequent occurrence in early life; and, up to the fifth year, the common fevers of children are those already described. Indeed, with but exceptional cases, we will not meet with continued fever, except in those years in which it is the prevailing disease of the adult, being endemic or epidemic. At such times the cause of continued fever is very intense, and influences the child as well as the adult. A continued fever is never *sthenic* in childhood. In the majority of cases it will present the symptoms of *synochoid*, or *common continued fever*, and only when the cause of *typhoid* fever is intense, do we have this form of the disease with the intestinal complication. Typhoid symptoms, *i. e.*, the evidences of depression and sepsis of the blood, is, however, of tolerably frequent occurrence.

CAUSES.—We accept the usual theory, that this class of fevers is caused by an animal poison, or malaria. What it is exactly we do not know, but that animal matter in certain states of decomposition will produce fever of this kind, is very clearly proven. The fever poison may be generated by decaying animal matter, or by a person suffering from such disease, and being introduced into the circulation through the lungs, or the ordinary methods of absorption, sets up a process of change which gives us the phenomena of fever.

PATHOLOGY.—As just stated, the cause of continued fever is an organized body undergoing decomposition. It may be perceptible to the senses, or it may be in gaseous form, but it possesses the property of setting up a like change in any material that possesses its elements with which it may come in contact.

The theory of Liebig is, that the fever poison resembles yeast in this respect, and that it not only sets up a process of decomposition in the blood, but develops a material similar to itself out of its elements. The evidence of this action is so marked and so well known that it is not now a matter of controversy.

The fever poison possesses varying degrees of intensity, just as yeast is weak and active. Like this, also, it possesses the power of setting up these changes in minute as well as in large quantity. When the fever poison is active we find it producing fever of a grave character, which either runs its course rapidly to a fatal termination, or is prolonged and attended with great prostration. On the contrary, when the fever poison is mild, the fever, though it may be protracted, is not of a grave character.

These fevers are characterized by the length of their forming stage, or period of incubation. During this time the fever poison is insidiously undermining the powers of life. Digestion is influenced by it, as is assimilation and the formation of blood. Every molecule that is formed into tissue during this period, receives the impress of the poison in an imperfect development; and it is not until the patient is divested of all these imperfect tissues that real convalescence commences.

Post-mortem examination reveals the same lesions that are found in the adult. There is marked change in the blood, both of the albumen, fibrin, and red globules. It does not coagulate firmly, does not change readily when brought in contact with oxygen, and the microscope shows the red globules more or less broken down. There is no special lesion of structure, except in those cases in which the glands of Peyer are involved, and these have presented the characteristic symptoms of typhoid fever during life.

SYMPTOMS.—One of the most marked features of a continued fever, is the long duration of the forming stage, which frequently embraces from one to three weeks. During this time the child will be dull, uneasy and fretful at times, will sleep more during the day, and will be restless and wakeful at night. Its appetite will be impaired, the bowels irregular, and the skin dryer than usual. The tissues become soft and flaccid, and it loses strength in a marked degree.

The chill is usually protracted. In many cases an entire day, sometimes, indeed, two days will be occupied with alternate slight chills and febrile exacerbations. Slowly the fever becomes established, and, to the ordinary observer, it does not seem so severe as in the simple *febricula*, or infantile remittent, but the experienced practitioner sees in the prostration of the nervous system, the feeble circulation of the blood, the parchment-like skin, and the marked debility, evidences of a grave lesion.

Day by day, as the fever advances, the strength becomes more impaired; the pulse is more frequent, is smaller and harder, or more frequently soft and easily compressed; the skin is dry and harsh, and manifestly inactive; the urine is tolerably free, but is pale, frothy, and has an unpleasant odor; the stools are also frequently papescent, and have an unpleasant odor; the nervous system is markedly implicated. In some cases the patient is very restless and irritable, the eyes bright and the pupils dilated; but in the majority the patient is dull, the face is expressionless, the eyes dull, and the child dozes with its eyes half open.

Passing on toward the eighth or the tenth day, we notice that the child's mind wanders, though there is never the manifest delirium that we witness in the adult. In other cases the dullness has passed into coma, which gradually deepens until it terminates the life of the patient.

The tongue shows distinctly the character of the disease. In a majority of cases it is contracted, its movement is impaired, and the coating upon it has a tinge of brown. In this case the tongue and mucous membrane of the mouth are of a dull-red or dusky hue, which once observed will always convey to the mind the evidence of depravation of the blood. In a much smaller number of cases, the tongue is broad and pale, and covered with a pasty white coat.

In the first case, as the disease advances, the tongue becomes dryer, less mobile, fissured, and the coating of a deeper brown, until finally, in the severer cases, it is dry, black, and fissured, can not be protruded or hardly moved in the mouth, and bleeds when it comes in contact with anything, or even when movement is attempted. I am of the opinion that this condition is due rather to the harsh medication so frequently adopted, than to the disease.

When the symptoms above are most marked, the passages of the nose and the throat are also involved, and in consequence there is a dry whistling respiration, sometimes with much difficulty. The lips also are dry, fissured, bleed, and form dark unpleasant-looking crusts.

The symptoms of *typhoid* fever have the addition of the intestinal lesion. Early in the disease the bowels become lax, which by the fourth or fifth day is a marked dirrrhœa; there is also the evidence of pain in the bowels, and tenderness on pressure. A peculiar pinched appearance of the face is frequently seen, and may almost be regarded as pathognomonic of this condition.

As the disease advances the intestinal complication is the most marked feature. The diarrhœa is very intractable, with evident pain and uneasiness before the evacuations. The discharges present an unpleasant appearance and an unusual fetor. Occasionally there is tympanitis.

The natural duration of the disease is from three to four weeks, though in some cases it may terminate fatally by the end of the first week, and by appropriate treatment it may be aborted in the early part of its course. During this period it may be complicated with local diseases, as named under the head of infantile remittent, but as they present very nearly the same symptoms, and possess the same pathological character, they need not be further described.

DIAGNOSIS.—We determine the character of this fever, first, by the long duration of its forming stage, and the nervous prostration at the period of chill and commencement of febrile reaction. As it progresses, the continued febrile reaction, without remission, gives it a distinctive character.

The *typhoid* form of continued fever is diagnosed by the appearance of diarrhœa early in the disease, with tenderness on pressure about the umbilicus, and evident intestinal uneasiness, if not pain. In these cases, also, the fever is of an asthenic type, and there is especial prostration of the nervous system.

PROGNOSIS.—The prognosis is not unfavorable if the disease is properly treated; but under the old antiphlogistic treatment the mortality was sometimes large. Indeed, if patients are carefully nursed, and have proper food, the disease being allowed to run its regular course, the mortality will not be more than ten per cent.

TREATMENT.—The object of treatment in this disease is twofold—to shorten its duration, and relieve unpleasant symptoms.

In some few cases we may *save* life, but they must be few, as the mortality is but small, and some of the deaths are unavoidable, as the disease involves all the tissues of the body, which must be removed before convalescence is completely established. With simple functional lesion we might expect to arrest the disease at once; but we are doing well to accomplish it by the seventh to the ninth, or sometimes the twenty-first day.

As heretofore named, the first indication of cure is to correct the functional disturbance, for this must precede the removal of structural lesions. The derangement of the circulation of blood is manifestly first in order, and if we can correct it, other indications of treatment will be readily accomplished.

I prescribe in this case, as in other fevers, the special sedatives, as: ℞ Tinct. Aconite, or Tinct. Veratrum, gtt. v., water ℥iv., a teaspoonful every hour. This is continued to the complete establishment of convalescence, whatever additional means may be employed.

If there is irritation of the nervous system, with flushed face, restlessness, sleeplessness, and other evidences of determination of blood, we give Gelseminum in combination with the sedatives; but if on the contrary there is dullness and hebetude, the patient being drowsy, sleeping with the eyes part open, and as the disease progresses tending to coma, we use Belladonna instead, as: ℞ Tinct. of Belladonna, Tinct. of Aconite, *aa*. gtt. v., water ℥iv., a teaspoonful every hour.

The patient has a sponge bath once or twice daily, always used carefully so that the surface will not be chilled. Sometimes we find it better not to employ the bath after the first two or three days, on account of the prostration that follows. The hot foot-bath is both an aid to the action of the sedatives, and valuable to relieve irritation of the nervous system and produce sleep. When there is great wakefulness we render it more stimulant by the addition of mustard or capsicum.

There are some cases, though in most seasons but few, in which decided benefit will follow the use of an emetic in the early stage of the disease; indeed, occasionally it offers the only safe course. The cases are those in which the tongue is broad, covered with a heavy yellowish white fur, especially at its base. The breath is frequently fetid; there is nausea and retching, and medicines are rejected by vomiting, or, if retained, are not absorbed. In such cases deterioration of the blood goes on rapidly, and typhoid symptoms are developed early. A thorough emetic of acetous emetic tincture, or Ipecacuanha, removes accumulations from the stomach, stimulates it to better action, and prepares the way for the successful administration of other remedies, and the digestion of food.

There is yet another condition in which an emetic can be employed with marked advantage. I allude to those cases in which there is marked nervous prostration, with dull eyes, expressionless countenance, tendency to sleep, which will soon pass into coma. In these cases the action of the emetic arouses the nervous system, and at the same time gives a better and more equal circulation of blood.

When the tongue is pallid, usually with a whitish pasty coat, I like the action of the alkaline sulphites. The sulphite, or hyposulphite of soda is in most common use, and may be given in doses of five grains every three hours to a child two years old. It may be commenced at the beginning of the treatment, and used with the special sedatives. It is discontinued when the condition for which it has been given is removed. Frequently a day will make such a difference in the patient, especially after the remedy has been given for three or four days, that we will find it necessary to put the patient upon the use of an acid.

When the mucous membranes are dark-red, or dusky, the tongue being in a majority of cases dry and contracted (occasionallly it is moist), I give the muriatic acid from the commencement of the treatment, using the formula heretofore given: ℞ Dilute Muriatic Acid, ʒij., Simple Syrup, ʒij. Add it to water so as to make it pleasantly acid, and give as a drink. The child can hardly take too much; usually about one teaspoonful of the mixture every two hours will be the proper quantity.

There are some cases in which *irritation of the stomach*, will be an unpleasant feature in the early part of the disease. Not only so, but it will aggravate all the symptoms, prevent the action of remedies, and cause the disease to run a more rapid course, and increase its fatality.

This irritation should be arrested early. So important is this, that it will take precedence of all other treatment. I generally prescribe a cold application—cloths wrung out of cold water, or its opposite—hot fomentations applied over the whole abdomen. An infusion of the bark of the young limbs of the peach-tree, in small doses, is very efficient in quieting this irritation. In other cases, the subnitrate of bismuth, in mint water, may be employed. An infusion of the compound powder of rhubarb, given in small doses, has answered a good purpose, sometimes continuing the remedy until it produces a slight laxative effect.

In this case, Aconite alone of the sedatives may be used, with the addition of Gelseminum, if there is irritation of the nerve-centers, or Ipecac if there is intestinal irritation. The dose is five drops of the Aconite to four ounces of water.

Cathartics are not employed in this fever, unless there are special indications for their use. This would be, evident accumulations in the intestinal canal, and irritation from this cause. The usual evidence from the tongue, is its uniform yellowish coat, extending from base to tip.

With the means already named, judiciously selected to meet the indications of the case, we find the patient progressing favorably. The disease yields, day by day—the pulse becomes less frequent, but is greater in volume and in freedom of circulation—the irritation or depression of the nervous system passes away, and the secretions are gradually established. In a majority of cases, no other medicines will be required, as with proper nursing and food all symptoms of febrile disturbance will have passed away by the seventh to the fourteenth day.

In a few cases we will find it necessary to stimulate secretion from the skin and kidneys, after the sedative has produced its effect. The Comp. Powder of Ipecac and Nitrate of Potash, will be found a good remedy for this purpose, and will also act very kindly on the nervous system, and promote rest and natural sleep.

Occasionally we are called to take charge of cases of fever which have run the first part of their course, and have sometimes been aggravated by the medicines employed. All forms of fever will, if severe, or badly managed, present much the same lesions and symptoms; these being of an asthenic or typhoid type.

In diagnosis, as well as in our therapeutics, we group them in four classes—with reference to the circulation—with reference to digestion and the formation of blood—with reference to the nervous system—and with reference to waste and excretion.

The action of the heart is rapid, but imperfect, and though the blood apparently moves faster, there is deficient circulation in the capillaries. Here small doses of the special sedatives exert a most marked influence. As the pulse becomes less frequent under their influence, the action of the heart is stronger and more regular, and the blood circulates more freely, and is more equally distributed. In the majority of these cases we prefer the use of Aconite, but are governed even in this advanced stage by the rules already laid down.

It is of very great importance to the success of treatment at this advanced stage, that the stomach be placed in condition to receive and digest small portions of food, for the real danger is, many times, from starvation—a failure of vital power, from a want of nutrient material to sustain it.

One of two conditions will usually present—the tongue will be contracted, dry, fissured, and coated brown or black; the lips, also, will be dry, and occasionally covered with an unpleasant crust. The mucous membrane of the mouth, tongue, nose, etc., is dark-red or dusky. In this case, we administer dilute muriatic acid, as heretofore named, giving it as a drink, and as freely as the child chooses to take it. In the other case, the tongue is broad, pallid, and moist, and covered with a pasty fur. Here we give sulphite of soda, in the usual doses. For the use of the anti-zymotics the reader is referred to page 95.

To improve innervation, and get its general influence upon the system, I like the action of quinine used by inunction. Internally, it is not kindly borne at this stage of the disease. We find also, that this fatty inunction improves the condition of the skin, and also increases the general vital power of the child.

The food of the child will be boiled milk, with sufficient salt to make it pleasant. It is more kindly received and better digested, if taken quite warm, the heat acting as a stimulant to the stomach. Children take food, because they have an appetite or desire for it, and, not as the adult sick may, because their reason shows the necessity of it. In disease, when there is no appetite, but on the contrary, disgust for food, they can not sometimes be persuaded to take it in the usual manner of taking food. In such cases, instead of letting them suffer from the want of it, I have the milk given as a drink, even sometimes mixed with one or two parts of water. The necessity of this attention to giving food should be forcibly impressed upon the mother.

If the child will not take milk, we may have preparations of arrow-root, sago, tapioca, maizena, etc., or occasionally a beef essence will answer a good purpose. A good way to make a beef tea for a child, or even an adult, is as follows: Cut a lean and tender piece of beef in small pieces, rejecting all the fat; put in a tin or porcelain-lined vessel and cover with cold water; set on the stove and bring it slowly to the boiling point, letting it boil for half an hour. We thus get the elements of the beef in a fluid form, and with the necessary salt it will be quite palatable.

The attention to the skin, by the quinine inunction, or an occasional stimulant bath, with brisk friction, will be all that we can do to get secretion from this organ. As a general rule, it will not be good policy to stimulate the kidneys by special means, at this advanced stage of the disease. Occasionally, when convalescence is being established, we may administer a saline diuretic in small quantity, to facilitate the removal of waste. The tincture of muriate of iron, recommended to aid convalescence, strengthens and increases the action of the kidneys. As a general rule, the bowels had better not be interfered with. But where there is irritation from accumulations in the intestinal canal, a gentle purgative of compound powder of rhubarb will answer a good purpose. In those cases in which the tongue is broad and pale, minute doses of Podophyllin, triturated as heretofore described, may be used with advantage. Occasionally we associate one-fourth to one-half grain of hydrastis with it.

In some few cases, there is tardy passage of urine, or complete retention, and if this is not seen to, it will lead to a fatal termination. The use of diuretics, as generally advised, is bad practice. Instead, I prescribe santonine, triturated with sugar, in doses of one-sixth to one-fourth gr. every hour, until the patient is relieved. Its action in this respect is specific.

In some other cases, the urine is *suppressed*, producing at first irritation of the nervous system, then coma, which gradually deepens until the child dies. In other cases, convulsions ensue, and the patient dies of them. In this case, also, I do not depend upon diuretics. Instead, I have hot fomentations applied assiduously across the lions. Usually the fomentation is of vinegar and water; but if there is evident congestion, it may be rendered stimulant.

SPOTTED FEVER.

Though not a disease of childhood exclusively, yet it has prevailed among children more than adults, and deserves consideration here: Indeed, if we were to reject all the diseases which attack the adult as well, we would have but a small list remaining.

Spotted fever has prevailed in this country to a very considerable extent since 1862. In some localities it seemed endemic, while in others it was decidedly epidemic. Whether or not it is contagious has been in dispute; some contending that it is, others that it is not. My own opinion is that spotted fever is, in ordinary cases, contagious, like typhoid fever; that when very malignant, with marked symptoms of putrescency, a fever poison is evolved which will affect persons who come in contact with it. There is, again, an epidemic form in which the contagion is as marked as in *typhus* fever.

Many have supposed that spotted fever was a new disease, bearing no relation to diseases known and described by authorities, and have been at a loss how to treat it. The fact is the disease has appeared several times previously, and bears a very close relationship to typhus fever, with the addition of a subacute cerebro-spinal meningitis.

This fever first made its appearance in this country in the town of Medway, Massachusetts, in the year 1806, and prevailed to a considerable extent in New England from that time up to 1815. At this time it presented the same symptoms, was as malignant, and attended by the same mortality.

The same disease is noticed by historians as having prevailed over the greater part of Europe in 1505, 1528, 1556, and at various times up to 1805; and we can trace it under the names of hospital, jail, putrid, or spotted fever, in almost all parts of the world during the last century, following in the train of the great European armies, among which it made the most destructive ravages, and by whom it was spread over the greater part of Europe.

CAUSE.—The cause of spotted fever is undoubtedly an animal poison of very great activity, resembling in many respects the fever poison or malaria of typhus. We do not know definitely how it was produced in this country; yet in all previous epidemics it has been traced to crowding, bad ventilation, and especially to the decomposition of human excreta.

PATHOLOGY.—In its pathology this fever does not differ very materially, save in its malignancy and rapidity, from that last considered. The fever poison, whatever it may be, when once introduced into the blood, reproduces itself more or less rapidly, and finally causes the death of this fluid. In some cases the virulence is such, that within forty-eight or even twenty-four hours, the blood is completely broken down, dies even before the vital functions have ceased.

Post-mortem examination reveals a breaking down of the blood in greater degree than in typhoid fever. There is also extravasation of blood, and especially of its coloring material, so that parts which were congested during life are much discolored, as are the most dependent parts of the body after death. Parts that have suffered from local congestion are also softened, sometimes so much so that they may be readily separated and broken with the handle of the scalpel.

The spleen is engorged with dark grumous and broken down blood, and is frequently enlarged. The liver is also dark-colored, swollen, and friable. The lungs seem to have suffered in like manner, are filled with blood, and the bronchial tubes contain a dark-colored offensive mucus.

The surface of the body presents in some cases a remarkably spotted or ecchymosed condition. The discolorations are purplish or almost black, and most numerous on the most dependent parts of the body. A close examination reveals that they are true ecchymoses or vibices.

SYMPTOMS.—The symptoms vary in different cases and in different localities, but may be divided into two prominent classes, as follows:

First. For two or three days the patient is listless, dull, and stupid, the face is flushed and dusky, eyes tumid, some pain in the head and back, loss of appetite, tongue dusky-red and coated with a dirty-white mucus, skin dry. This is the forming stage of the disease, and instead of lasting as long as named, will, in the severer cases, not be longer than twenty-four hours.

Following this there is a tolerably well-marked chill, lasting for two or three hours, and attended with great prostration. Febrile reaction follows, sometimes high, at others not very well marked. In the one case the surface becomes intensely hot and flushed, the pulse 120 to 140 in the adult, and so that it can scarcely be counted in the child, sharp and hard, with great irritability and restlessness, though there is marked dullness of the intellectual functions. The thermometer marks a. temperature in these cases of from 104 to 109 degrees. The urine is scanty and the bowels constipated. Frequently there is difficult respiration, some cough, and sibilant rales.

In from two to six days an eruption appears upon the surface, very closely resembling measles, but more clearly defined. If the patient recovers they commence fading out by the end of the first twenty-four hours, but do not disappear entirely for some days. If the disease progresses unfavorably, they become dusky, and at last livid and associated with vibices. As their color becomes darker the nervous system of the patient becomes more oppressed, his mind wanders; and, becoming livid, he sinks into a stupor from which he can not be aroused, and which in a short time terminates in death.

In the second case there is but little reaction, the pulse running up to 90 or 100 in the adult, 110 to 120 in the child, and oppressed. There is tendency to coldness of the extremities, the skin being harsh and dry, or sometimes moist and atonic. The eruption appears the first, second, or third day, and is a dusky-red, not readily effaced by pressure.

There is marked dullness and hebetude from the commencement, and frequently the patient is almost entirely unconscious a few hours after the appearance of the eruption. It runs a very rapid course in most instances, terminating fatally by the third to the sixth day. The eruption becomes dusky and livid, petechiæ appear, the tongue is dry and brown, sordes on the teeth, urine and fæces very offensive, coma or low muttering delirium, and gradually increasing difficulty of respiration.

DIAGNOSIS.—The diagnosis is not always easily made at the commencement of the disease; yet this is not so important, because the symptoms show it to be a grave form of disease, of a congestive and malignant type. The extreme febrile reaction in the one case, associated as it is with dullness and hebetude of the intellectual functions, the dusky discoloration of the tongue and mucous membrane of the mouth, are characteristic symptoms.

In the other case, the great prostration, feeble reaction, dullness and tendency to coma, the appearance of the dusky eruption, etc., show the nature of the disease.

PROGNOSIS.—The disease varies in malignancy and mortality in different sections of country and at different times. So that while under one class of circumstances we should regard the prognosis as favorable in a large majority of cases, in another it would be unfavorable.

Much will depend upon the time when the patient is seen. If quite early in the progress of the disease, very severe cases may be conducted to a favorable termination. While, if the disease is allowed to progress unchecked for a day or two, or is aggravated by injudicious medication, the more mild cases will be rendered unmanageable.

TREATMENT.—There are three plans of treatment that I think may be relied upon, and I will state them plainly, endeavoring to point out the special cases where a preference should be given to one over another. Taking the majority of cases I think I should value them in about the order in which they are stated, relying upon the first plan especially in very bad cases.

Make an infusion of Capsicum one part, Bayberry six parts, having it as strong as the patient can take it with comfort. Then give the acetous emetic tincture in doses of from one-fourth to one-half teaspoonful every ten or fifteen minutes, with as much of the infusion as the child will drink. We do not wish to produce immediate vomiting, but desire to get the general influence of Lobelia upon the system, so that if it is not well tolerated by the stomach, we lessen the dose, and apply a stimulant fomentation over the epigastrium, to aid its retention. Continuing it in this way for one hour or more, we notice that the depression of the fever is being replaced by the influence of the Lobelia, and when this becomes marked, we carry it to free and thorough emesis.

In the meanwhile, the child being placed in bed between blankets, hot bricks wrapped in flannel wrung out of an infusion of Capsicum with vinegar and water, are placed at the feet, by each thigh, and by each side of the trunk, at such distance as to be in no danger of burning the child. Have the blanket loose over the body, but well-tucked down at the feet and around the neck, to prevent the escape of the vapor. After a free and vigorous circulation is established, the body may be rubbed dry with a flannel, and wrapped in a dry blanket.

In the second method of treatment we desire to obtain the stimulant influence of Lobelia or Ipecacuanha upon the circulation. I would order them in this form: ℞ Lobelia Seed (powdered), gr. xx.; Capsicum ℨj. Mix and divide in eighty powders. These may be administered as often as every fifteen minutes, every half hour, or hour. Occasionally they will produce slight nausea, which is not objectionable, providing it does not go so far as retching or the rejection of the remedy. Ipecacuanha is given in doses of one-fourth to one-half grain every hour, always less than will produce vomiting, and is thus continued until reaction is well established. I like the action of the Lobelia the best, though the Ipecac has been used with considerable success.

To aid these, an enema of an infusion of Bayberry, Lobelia and Capsicum, may be employed with good advantage. I am satisfied that I have seen the patient aroused by this means, so as to obtain the influence of other remedies by the stomach, when without this, it would have been impossible.

Dry cupping to the spine, is sometimes of advantage, but in the majority of cases, I would prefer the stimulant vapor-bath, as first named, and friction with Comp. Tinct. of Capsicum.

The third method of treatment is based upon the specific action of Belladonna to overcome congestion and stimulate the circulation. This has not been as thoroughly tested as we would wish, yet some very favorable reports have been made of it. In California, the Atropia has been employed; we have always used the the tincture of Belladonna. For a child two years old, I should order: ℞ Tinct. of Belladonna gtt. v. to x.; Tinct. of Aconite gtt. v.; water ʒiv. Of this, give a teaspoonful every half hour, until its influence in arousing the nervous system and overcoming the congestion is noticed, afterward every hour, until the circulation becomes free, and the dangerous symptoms have passed away.

In either case I should employ the sulphite of soda, or sulphurous acid, in the usual doses, as soon as this first influence was established, and continue it until the dirty coat was entirely gone. But if at any time, the tongue becomes dry and dark, I would substitute the dilute muriatic acid.

The Comp. Powder of Leptandrin may be given in from one to two grain doses four times a day with remarkable efficacy. It acts mildly, but efficiently, upon the glandular system, and removes morbid and worn-out material from the intestinal tract, leaving it free from irritation.

Quinia may be used by inunction, early in the disease, and its use in this way continued to the complete establishment of convalescence. I do not think that its internal administration in the early stages of this disease, has been attended with any good effect, but, on the contrary, has frequently increased the depression of the nervous system.

Tincture of muriate of iron may be occasionally given alternately with the sulphite of soda, so as to obtain the good influence of both. In a majority of cases, it will be well to put the patient upon its use for some days after other remedies are suspended.

EPIDEMIC CEREBRO-SPINAL MENINGITIS.

Closely associated with spotted fever, and also with diphtheria, is the disease known by the name of epidemic cerebro-spinal meningitis. These three bear a very close relationship, in that each presents very similar lesions of the nervous system, both during the progress of the disease, and in the sequelæ. In all three there is the evidence of the action of a blood poison, and the breaking down of the blood, and in each, death may be the result of the lesion of the nervous system, or of the lesion of the blood.

In this country, within the last ten years, the three diseases have prevailed in an epidemic form, one succeeding another, and in some cases seeming to merge into one another. We had first the epidemic diphtheria, next the epidemic spinal meningitis, and last the spotted fever.

PATHOLOGY.—The profession are not agreed as to the pathology of epidemic spinal meningitis, though it is now generally regarded as bearing a very close relation to typhus and typhoid fevers, epidemic dysentery, etc. I think there is no doubt but that it is produced by an animal poison, which, gaining entrance to the blood, gives rise to the phenomena of fever, and acting from the blood, specially affects the cerebro-spinal centers, producing the peculiar lesions that characterize this disease.

What this poison is, we are unable to say, neither are we able to account for its origin or propagation, in many cases. That the disease is contagious, in its severer forms, I am well satisfied, but this is also true of typhoid, especially of typhus, and, at times, of nearly all diseases which present that grouping of symptoms called *typhoid*. In some instances, it has seemed as if there were an endemic influence causing the disease. In others, it has been distinctly epidemic. There is yet much mystery in regard to epidemic influences, and until the subject has been more thoroughly studied, it will be useless to theorize upon it.

SYMPTOMS.—The cases of cerebro-spinal meningitis may be divided into two classes, the distinction being very marked.

Occasionally, we will find it prevailing in both forms, at the same time, in a locality; but more frequently it will maintain the one form in all the cases, at one place, or during one season. We may call these two classes the *rapid* and the *slow* cerebrospinal meningitis, as this expresses the greatest difference in the symptoms.

In the first, or *rapid* form of the disease, there is but a short period of incubation, rarely exceeding a day. The patient feels dull and prostrated, and if old enough to complain, it is of pains in the back, head and limbs. The chill is usually well marked, the extremities being cold, the surface shrunken and pallid, and occasionally, severe rigors. It will be noticed that during the chill there is greater dullness of the mind and prostration than should attend an ordinary chill, and the patient seems to suffer severely.

In the course of one or two hours, the chill passes away and febrile reaction succeeds. The surface becomes flushed, and the temperature is increased. The pulse increases in frequency to 120 or 140 beats per minute. The face is flushed, the eyes injected and suffused, and the head is warmer than other portions of the body. The tongue and mucous membrane of the mouth are usually dusky, and the tongue coated with a yellowish-white pasty fur.

During the first few hours, sometimes for a day, the patient complains of pain in the back, and muscular pains in various parts of the body, and though there is great dullness of intellect, yet the patient is restless and uneasy. It will also be noticed that moving the child increases the suffering, sometimes so much that we are obliged to let it remain in the one position.

By the second day, the patient has sunk into a stupor, from which it is difficult to arouse it. The surface is markedly flushed and dusky; the pulse very frequent and wiry, in the majority of cases, but in some it is open but oppressed. Respiration is difficult, and the patient shows marked evidences of imperfect aeration of the blood. The characteristic symptom is a fixure of the spinal column, the head being drawn backward, or to one side; or the disease may show itself in the same way in the dorsal or lumbar regions. Occasionally, we notice evidence of partial paralysis this early in the disease. At other times, convulsions come on early, or the disease is announced by them.

Thus it progresses rapidly to a fatal termination, the patient rarely lasting longer than four or five days, if not relieved, and sometimes it terminates fatally within forty-eight hours.

In the other form of the disease, the forming stage may last from one day to a week, presenting the usual symptoms of dullness and hebetude, and arrested function. The chill is not very marked, though it may last for the greater part of a day, or be made up of slight chills and febrile-reactions, of short duration.

Reaction comes up slowly, and is not fully established before the end of the first twenty-four hours. The temperature is increased, the pulse increased in frequency and hardness, the patient restless and fretful, and if old enough, complaining of pains in the back, head, limbs, or not unfrequently, simply of a *hurting*, without being able to locate it at any one point. The secretions are arrested; the skin becomes dry and harsh, the urine scanty and high colored, and the bowels constipated. The tongue in some cases is contracted and reddened, with a coat having a shade of brown ; in others it is broad, pallid, and covered with a pasty white coat.

Usually at first, the face is slightly flushed, the eyes bright, the pupils contracted, and the mind active. The patient is uneasy and restless, and does not sleep well. The child maintains one position, and the fixure of the back is marked, and if moved, so as to bend the spinal column, it cries out with pain.

Thus, day after day, the fever continues, sometimes presenting the symptoms of a remittent, at others of a continued fever. There is a gradual increase in its severity, and necessarily an increasing debility of the child. Occasionally the fever will run very high about the sixth to the tenth day, and there will be evident delirium. Passing into the third week, the symptoms assume a typhoid condition, which gradually increases as time passes.

I do not think that at this time there are any distinctive symptoms, but the tenderness on pressure over the spine, and the pain when the child is moved, except, possibly, the greater excitement of the nervous system, which in this case replaces the dullness of typhoid. But sooner or later in the disease, this excitement is replaced by coma, which sometimes becomes a marked feature in fatal cases.

A peculiarity of the disease is, that having run the course I have described, for two, three or four weeks, the symptoms gradually give way to treatment; the fever is arrested, secretion established, the child sleeps well at night, takes food and seemingly digests it, but further than this there is no advance to recovery. There is no increase of the strength, indeed, no increase of flesh, and thus week after week will pass by without an appreciable change. After a time, however, it will be noticed that the child is failing, and in two or three weeks it dies—but of what it is impossible to say. I have known of many cases that had a duration of three or four months, and an exceptional case that terminated fatally at the commencement of the eighth month from the date of attack, there being no time during which the child was able to sit up.

DIAGNOSIS.—In the first form of the disease, the symptoms are of a very grave character from the commencement, and we are able to trace their relationship to spotted fever and the more malignant cases of diphtheria. The pains in the back and head, the severe muscular pains, are characteristic, and even where the child is too young to describe its sufferings, its appearance will evidence it. The pain, or expression of suffering, upon moving the body, is the evidence of spinal disease.

In the second class of cases, the disease comes on insidiously, and there may be but little, if anything, to arouse the suspicion of the practitioner that he has more than an ordinary fever to treat. But after awhile his attention is attracted to the pain when the child is moved, and the greater irritability of the nervous system and restlessness than is common in ordinary fevers.

PROGNOSIS.—I do not regard the prognosis as unfavorable if the disease is seen in time, and a proper treatment is adopted. Taking the disease as it ordinarily prevails, it is probable that the mortality will vary from ten to twenty per cent. We must not forget the fact that in some situations the cause of the disease is very intense, and it exhibits very great malignancy. Indeed, in some localities, many times, death would have commenced before the physician was called to the patient.

TREATMENT.—In the first form of this disease, I would strongly advise that the treatment be commenced with a thorough emetic, of some preparation of Lobelia. In this case as in spotted fever, we desire the general influence of Lobelia, as well as the act of emesis, or in other words, we want the act of emesis as the result of its general action. The acetous emetic of our dispensatory, or the Comp. Powder of Lobelia and Capsicum, are good preparations for this purpose. Give in five to ten drops doses with warm water, repeated every five or ten minutes, and when the system is brought fully under the influence of the remedy, which will be in one or two hours, then more freely until emesis results.

In many cases it will be necessary to follow emesis with a cathartic. This is particularly requisite in malarial localities. The intestinal tract is in the same morbid condition as the stomach, and must be aroused ere we can hope for a satisfactory response from our sedatives or other indicated remedies, when quick action is desired. I like the action of Rhamnus P. best, prepared as follows : ℞ Tinct. Rhamnus P. ℥j., Tinct. Taraxicum ℥j., Simple Elixir ℥j. M., and give from gtt. xv. to xxx. every two hours, until there has been good action. If slow action is desired we can use the Leptandria compound gr. ij. every two or three hours. Then comes our sedatives, Aconite or Veratrum, whichever is indicated ; Veratrum, if the patient has a full bounding pulse, and Aconite if it is small in volume and easily compressed. Of the former, gtt. x. to half a glass of water ; of the latter, gtt. v. to water ℥iv. Marked cerebral irritation or coma will generally be present in a major portion of cases, and in the irritative ones, Rhus tox. is a prominent remedy ; given, a case with excited circulation, frontal pain, prominence and redness of papillæ, and tip of tongue, and we have the true indications for the remedy. In these cases, then, we would think of Rhus and give it curatively. We have another class of symptoms that present great vascular excitement, contracted tissues and pupils, face flushed, and extreme restlessness ; here we would substitute Gelseminum for Rhus and get its specific action ; gtt. v. of the Rhus to water ℥iv., and xv. of the Gelseminum to the same quantity of water, will be the right proportion ; give in teaspoonful doses every hour. In the opposite cases we have dullness and hebetude, patient sleeps with eyes partially unclosed, tissues are sodden and the whole system is in an atonic condition. From five to ten drops of Belladonna added to half a glass of water and given every half hour or hour, according to the severity to the case, will remove the congestion and leave the system in better condition for the action of other remedies.

If there was large pulse, tongue broad with dirty coat, sweetish taste in the mouth with pallor of mucous membranes, I should prescribe sulphite of soda, grains two to four every three hours; but with deep redness of mucous membranes, slick tongue, and typhoid symptoms, the patient wants muriatic acid. In such cases I should associate Baptisia with the acid. In septic conditions they are definite agents and as such produce definite results. With the system prepared for the antiperiodic, we may give it either by mouth or by inunction. Where the soda is indicated I prefer giving the quinine by inunction. In the other cases I should give an acidulated elixir of quinine.

Where there is a torpid circulation of the skin, I should associate it with a stimulant, as follows: ℞ Quinia Sulphas ʒj., Capsicum gr. x. to ʒss., Adeps ʒij.; mix thoroughly. This may be used with brisk friction two or three times daily, and if the temperature of the extremities is lowered, apply dry heat or friction with the hand.

In the second case, the treatment need not be so active. I am not certain, however, but that in many of these cases we might obtain much advantage from the action of the emetic in the first two or three days. Indeed, I am satisfied that in two cases I arrested the disease by this means.

But we would think more particularly of Rhus and Macrotys, plus the required sedative, in these irritative cases, though even the sedative is not always necessary, Rhus doing the work nicely. Indications for Gelseminum will be met with that remedy substituted for Rhus. If we get an indication for a certain remedy in a large number of cases, we then have an epidemic remedy, and with the sedative is all sufficient. Two or three remedies rightly chosen, have, in some of the severest types of disease of this character, always given me good results.

The general sponge-bath, sometimes with hot water, two or three times a day, will be found an important aid. If the bowels are torpid they may be stimulated to action by a gentle laxative, or by an enema. If the secretion of urine is scanty and high colored, some diuretic infusion, as of mentha viridis, with small portions of sweet spirits of nitre, may be given.

DIPHTHERIA.

Diphtheria was the first of these epidemics, making its appearance in some sections as early as 1855. From then, up to 1864, it prevailed in most parts of our country. It was thought by many to be a new disease, though a reference to authorities will show that several epidemics of the same have occurred before, and that it was well described, aud received its name from French observers in the last century.

CAUSE.—The cause of diphtheria is undoubtedly a specific animal poison, though how generated or propagated we are unable to tell. It prevails as an endemic or epidemic disease, and is rarely if ever found in isolated cases. I have no doubt that it becomes contagious, like other similar diseases, when it occurs in its most malignant form. There seems, sometimes, to be a very close relationship between diphtheria and scarlatina, and cases have been recorded where an eruption attended it. In this, as in some other diseases, the anomaly may have been a mistake in diagnosis, rather than a difference in the disease.

PATHOLOGY.—I have not changed my opinion of the pathology of diphtheria, first published in 1861, and upon which the treatment of the majority of our physicians has been based. I quote it as then written, desiring to keep it on record as being the first announcement of a doctrine that is now generally admitted as correct.

" I hold diphtheria to be a general as well as a local disease, as is proven by the languor, listlessness, torpor of the nervous system, and derangement of the excretory organs, which, as a general rule, precede all local disease; all being symptoms of perversion of the blood, and almost invariably indicating the establishment of febrile reaction. We also find the evidence of the perversion of the blood in the heavily coated tongue, which is always more or less discolored at the commencement of the disease, and always in severe cases, exhibiting the brownish tinge, with more or less sordes upon the teeth as it progresses ; in the diphtheritic deposit which is markedly different from the exudations from highly-vitalized blood; in the secretion of urine in severe cases being abundant, in all cases discolored, frothy, more or less clouded with a peculiar, somewhat cadaverous odor—what the ancients would have termed *illy-concocted*; in the evacuations from the bowels, obtained by cathartics, which are frequently large, dark, and almost invariably fetid; and especially in the condition of the blood itself, when the disease has attained its maximum, which is dark, is not changed by exposure to air, forms a loose and easily broken-down coagulum, or does not coagulate at all.

" Post-mortem examination in those cases that have run a regular course, *i. e.* that have not been terminated by an extension of the disease to the larynx, shows us the blood broken down to a considerable extent, more or less discoloration of tissues from extravasation of the coloring matter, and softening of the tissues. These facts, it appears to me, prove conclusively the opinion given above."

There are some cases in which the disease seems almost wholly local, yet these are mild. Other cases will present the evidences of local disease first, and it will only be after some days that the serious character of the general lesion will be manifest.

SYMPTOMS.—As above named, the symptoms of the forming stage are similar to those of fevers and inflammations generally. For a day or two, sometimes for a week, the patient is listless and languid, does not play with the usual zest, is fretful at times, does not sleep well, especially at night, drinks frequently, and has a variable appetite.

Following this is a slight chill, lasting one or two hours; not unfrequently it is so light that it is not noticed by the parents. Following this, febrile reaction comes up slowly, and varies greatly in different cases. In some the fever is acute, and is a marked feature of the disease. In others the symptoms of fever are but slight—an accelerated and soft pulse, arrested secretion from the skin, kidneys, and bowels, and an increased temperature of the body, as marked by the thermometer, though it is not so perceptible to the hand.

As the disease progresses the fever assumes an asthenic or typhoid character, and there is evidently a serious lesion of the blood. In a few cases the fever is high from the commencement, and continues to present sthenic symptoms during its entire progress.

The patient complains or shows signs of sore throat at the commencement of the disease. There is difficulty and pain in deglutition, the child swallows frequently to moisten the throat, and there may be slight difficulty in breathing. On examination we find the mucous membrane of the fauces, tonsils and pharynx somewhat swollen, sometimes of a vivid red color, at others dusky or livid, and occasionally presenting a blanched appear-

ance. On some of these parts we will notice the characteristic
exudation—spots of an ashen-gray or white lymph upon the surface of the mucous membrane. They are usually small at first,
not larger than a grain of wheat, or at farthest a three-cent piece.
They are usually grouped together, two or three or more at a
point, which is more swollen or discolored than adjacent parts.
There may be but one of these points of exudation, or several.
As the disease progresses the swelling becomes more marked, and
the points of exudation more numerous. The patches likewise
increase in size, sometimes coalescing, so as to uniformly cover
quite a large surface.

For two or three days, in the majority of cases, the throat is dry,
sometimes, indeed, during the entire progress of the disease.
Then secretion is established from the mucous follicles, and some
patches of exudation being removed, there is a free secretion
from the denuded surface. The salivary glands also become more
active, and the saliva is tenacious, thick, and ropy; and altogether the secretion is large, and requires frequent efforts at
removal. Occasionally cases present themselves in which this
seems to be the most unpleasant symptom.

In the later stages of the disease, we may distinguish two
classes of cases. In the first the dryness continues, and the parts
become stiff and immobile, so that after a time deglutition becomes almost impossible, and respiration is rendered very difficult
and labored. Extending upward to the posterior nares and nasal
cavities, these are closed by the swelling; and descending to the
inferior portion of the pharynx and epiglottis, these and associated parts are swollen and rendered incapable of motion, and
the child dies, partly from want of food and drink, and partly
from imperfect aeration of the blood.

In the second class of cases, secretion commences about the
second or the third day. By the fifth day it is quite free, some
portions of the exudation are being detached, and the exposed
surface secretes pus. In very severe cases this ulceration progresses in every direction, but is mostly superficial. The tissues
seem to have lost their vitality, and the muscles their power of
contraction, and they hang feeble and pendulous and infiltrated
with serum, when the connective tissue is loose. Thus we have
paralysis of the throat in the second as well as in the first case.

This also extends upward to the nose, sometimes presenting
the distinctive characteristics of diphtheria throughout. In some

of these the discharge will be profuse, in others it is retarded, becomes dried, and thoroughly closes up the passages.

In other cases the disease extends downward and involves the pharynx. Here the child presents all the symptoms of croup— the whistling respiration, croupal cough, loss of voice, and gradually increasing difficulty of breathing. The occurrence of the laryngeal complication is sudden, and it runs a rapid course. Thus, if not relieved by remedies, it will usually terminate fatally within forty-eight hours.

DIAGNOSIS.—Diphtheria is readily diagnosed by the specific character of the sore throat. Where there is the peculiar ashen exudation upon the free surface of the mucous membrane, there is a case of diphtheria, no matter what the other symptoms may be. When there is no such exudation the disease is not diphtheria. I admit that in some exceptional cases the patches of exudation may be thrown off very early, and when the patient is first seen there will be simply an ulcerated sore throat, but in all there is the exudation at some period.

PROGNOSIS.—As is the case with all endemic and epidemic diseases, it prevails with different degrees of severity in different places and at different times. Thus one physician may meet with it in a form so malignant and running its course so rapidly, that a majority of the cases will prove fatal. While another will see it in a mild form, and with but simple treatment a large majority recover. Thus with some there has been a mortality of twenty to fifty per cent., with others of not more than two or three per cent., both pursuing the same treatment. Of course, in laryngeal complication the prognosis will be more doubtful.

TREATMENT.—Diphtheria being a disease of the blood, the treatment will in all cases be general, with the addition of local treatment in severe throat complications. Should nausea prevail, with heavily coated tongue, I would preface the treatment with the administration of a thorough emetic ; this will arouse the system and place the stomach in condition to receive other remedies.

In my experience with this disease, in the greater number of cases taken from the commencement, I find Phytolacca to be the most frequently indicated. It has a specific influence upon the blood, the glandular system, and the throat. We may use it with the indicated sedative, or, as I prefer, in alternation with

Tinct. Aconite or Veratrum gtt. v. to x., water ℥iv.—Tinct. Phytolacca gtt. xx., water ℥iv.; dose, a teaspoonful alternately every hour.

If there is an excess of saliva, make a solution of potassa chlorate ℨss. to half a glass of water, and give the child a teaspoonful four times daily. If there is fetor with ulceration, Thymol may be given both internally and used with the spray. ℞ Thymol gr. viii., Glycerine, water, aa. ℥ii. M.; a teaspoonful once in three hours, and spray the throat three or four times a day.

With enlargement of the glands a local application of Phytolacca and glycerine will be excellent treatment. Two drachms of the first to an ounce of the last; saturate a flannel bandage and apply to the throat, renewing two or three times a day.

Dusky color of face and mucous membranes, with typhoid symptoms, indicates Baptisia, which may be given in the usual doses with Aconite. Children of scrofulous habits will improve nicely on Lachesis, and Lycopodium may be given when there is a low grade of fever, with diminished secretion of urine.

Cases may arise where, in the latter stage of the disease, it will be necessary to use an emetic, but it will be in laryngeal complications, and must be used with care. We may use the acetous emetic tincture in small doses, just sufficient to produce nausea and gradual emesis.

Following the sedatives and antiseptics must be the sustaining treatment. The temperature must be carefully noted at different periods during the day, and if it decreases, stimulants must be administered frequently, and in small quantities. Should high temperature prevail, we want to think of cooling remedies in addition to our sedatives. In malarial localities, with the system prepared for the antiperiodics, quinine should be given early in the disease. If there is no gastric irritation, but an alkaline condition, the better way will be to give the acidulated elixir. With an atonic condition, or congestion, I prefer the olive oil preparation with thorough friction.

In the matter of local applications, there are two remedies that I prefer to all others: they are Phytolacca and compound Stillingia Liniment. As previously alluded to, the Phytolacca will be an essential part of treatment in all cases with enlarged glands, using it in combination with glycerine. The Stillingia liniment is a stimulant and an excellent counter-irritant, and may be used in both throat and lung complications. Saturate a flannel cloth, and apply three times a day. It exerts a marked control over the croupal symptoms, when applied over the larynx. In mild cases we may use the hot vinegar pack.

If there is stupor and tendency to coma, or pain in the throat with difficulty in swallowing, give Belladonna. Sepsis with deep redness of mucous membranes and slick tongue, wants muriatic acid. If the tongue has a dirty coating, with pungent breath, and the excretions look as if fermented, the patient wants sulphurous acid. We may use the first as an acidulated drink, adding two or three drops to half a glass of water, and let the child drink at pleasure. We get a much better action from the acid in this way than we should if syrup were added. Of the sulphurous acid, 3ss. to water 3iv., and given in teaspoonful doses every two hours, will give marked benefit. Occasionally we have a case calling for sulphite of soda. We recognize it by the pasty-white coat on the tongue, pallidity of mucous membranes, and a general atonic condition of the whole system. The dose will be three grains every two hours.

Now come our cases of unusual severity: the patient is debilitated, and the throat ulcerating with a sickening fetor. The disease has extended up into the nares, and the lips are parched, and may have sordes upon them. Treatment must be prompt and decided, and we turn with confidence to bichromate and permanganate of potash. The former we would use more particularly in those cases of a croupy nature, accompanied by difficult, wheezy breathing, and great depression. From ten to fifteen grains may be dissolved in half a glass of water, and given in teaspoonful doses every two hours, alternating with Veratrum. Of the latter salt, we will add from five to ten grains to four ounces of water, and give a teaspoonful every two hours.

As adjuvants, I would recommend gargling, if the child is old enough, and spraying with the atomizer, or inhalations. Water medicated with chlorate of potassium, salicylic or carbolic acid, thymol, chloride of calcium, with muriated tincture of iron, and the vapor of hot vinegar, may be employed with success. Cloths must be kept at hand to absorb all discharges from the mouth, and burned immediately; and the lips and mouth sponged with a solution of thymol or chlorate of potash. The little patient must be kept as quiet as possible, and the food will necessarily be of the most sustaining character, as beef essence, milk, and custards of sea-moss farina.

With this, as in other similar diseases, the general bath and the hot foot-bath are employed. In the majority of cases a stimulant bath will be preferable, as of salt water with brisk friction, or the addition of capsicum or mustard to the water. The hot mustard foot-bath is a very valuable means, and may be used two or three times daily, and for half an hour at a time. It is especially important in severe cases, indeed in any case, that the extremities be kept warm. I have often thought that in some it was just the difference between death and recovery. Let the feet and legs be cold for six hours, in a severe case, and the throat disease will have advanced beyond the reach of medicine. In the application of stimulants I prefer Capsicum, and the use of dry heat will be the best means. It is a very good plan to say to the nurse, "If complaint is made of any part, sponge it with hot water, and cover at once with hot dry flannel."

Where the larynx becomes involved, and the symptoms of croup developed, the treatment must be prompt and thorough, if we expect to save life. To give temporary relief, I direct the inhalation of vinegar and water sufficiently often to give ease, using at the same time hot fomentations assiduously applied to the throat. When the case does not seem to be progressing rapidly, I place the patient upon the use of Aconite and Phytolacca alone, using the Stillingia liniment as a local application over the larynx. If, however, it is progressing rapidly we may give the patient small doses of acetous tincture of Lobelia and Sanguinaria, so as to keep up continuous slight nausea; and when the patient is brought fully under the influence of the remedies, it is carried to free emesis. So far as my experience goes, however, we have much better results if we follow the direct indications, and give the small doses.

ERUPTIVE FEVERS.

The eruptive fevers are caused by a specific contagion generated during the progress of the disease, and propagated by contact, by inhaling the gaseous exhalations, or by the absorption of the poison from the clothing or the excretions. The poison is capable of being disseminated by the atmosphere to a limited extent in ordinary cases; but when severe, and in numerous cases, it sometimes affects the atmosphere for considerable distances, especially in the direction of the prevailing winds; in these cases the disease assumes an epidemic character.

In the most of these diseases the virus may be transmitted by inoculation. Indeed, in any way by which it is brought in contact with the body, so that it may be absorbed into the blood, the disease may be disseminated.

The causes of these diseases vary in intensity in different seasons, and in different localities, so that at one time they will be remarkably mild, and at another time very severe, even malignant. It is pretty evident that the condition of the atmosphere, as regards heat and cold, moisture or dryness, and especially its electrical condition, has much to do with the intensity of these, as well as other diseases.

It has been noticed that the eruptive fevers, as well as some other diseases, prevail in cycles, the periods being usually about seven years. This has been accounted for in the case of the class under consideration, by the fact that all the material for such disease is used up in its progress of two or three years, and its reappearance depends upon the presence of a new population not protected from it.

A peculiar feature of these diseases is, the immunity from them produced by one attack. It seems that in each of these cases, the cause of the disease destroys all the material capable of being influenced, or all susceptibility to the influence in the first attack. As a general rule, the immunity from the influence of the cause a second time, is perfect after the first attack, and the person may be exposed without the least danger. There are some rare cases, however, in which the person is not thus protected, and in which it may recur one or more times.

These fevers differ from all others, therefore, that an attack protects the individual from ever having the disease again, even though being exposed to the same cause. They differ also in the specific character of the eruption, which is a part of the fever.

Liebig thus accounts for the disease and its protective influence : " When a quantity, however small, of contagious matter, that is, of the exciting body, is introduced into the blood of a healthy individual, it will be again generated in the blood, just as yeast is generated from wort. The condition of transformation will be communicated to a constituent of the blood ; and in consequence of the transformation suffered by this substance, a body identical with, or similar to the exciting or contagious matter, will.be produced by another constituent substance of the blood. The quantity of the exciting body newly-produced must constantly augment, if its further transformation or decomposition proceeds more slowly than that of the compound in the blood, the decomposition of which it effects. * * * In the abstract chemical sense, reproduction of a contagion depends upon the presence of two substances, one of which becomes completely decomposed, but communicates its own state of transformation to the second. The second substance thus thrown into a state of decomposition is the newly formed contagion. * * When the constituent removed from the blood is a product of an unnatural manner of living, or when its formation takes place only at a certain age, the susceptibility of contagion ceases on its disappearance. The effects of *vaccine* matter indicate that an accidental constitution o f the blood is destroyed by a peculiar process of decomposition, which does not affect the other constituents of the circulating fluid."

As in the case of sthenic fevers, the cause, in these diseases, seems to influence the blood almost in direct ratio to the functional disturbance. Thus, with an active fever, and arrested secretion, the poison seems to be generated in increased quantity.

In some cases the influence of the contagion does not cease with the formation of the specific virus, but originates a septic decomposition of the blood, giving rise to putro-adynamic symptoms, which frequently result in death, sometimes even before the appearance of the characteristic eruption.

VARIOLA—SMALL-POX.

Small-pox, like other of the eruptive fevers, is caused by a specific contagion, generated during the progress of the disease in the human body. This contagious matter may be propagated by contact with the person or with articles of clothing, or indeed anything that has been in contact with him, and by inhaling an atmosphere impregnated with the poison. This atmospheric cause varies in different cases; in some extending but a short distance from the person, frequently confined to the room; in other seasons extending considerable distances in the direction of the prevailing winds.

The disease is, so far as we know, always produced by the one cause, the specific virus of small-pox, and is never generated

anew or without this.   While it is thus contagious, it also becomes endemic and epidemic by peculiar states of the atmosphere.
It also prevails in cycles, as heretofore stated, some of these being
mild, others severe.   More commonly, perhaps, we have one or
two years in each cycle, in which the disease is severe, then
others of a milder type.

The virus of small-pox acts upon the blood, giving rise to such
changes in this fluid as produce all the phenomena noticed,
and generating within it a virus similar to itself.   Finally, this is
thrown upon the surface in the form of pustules, and the blood
being freed from it convalescence ensues.   We may, in this disease, study the pathology of all fevers, as this presents all the
phenomena of a fever, with a cause tangible to our senses.   We
may trace the incubation of the virus in the body, as it gradually
increases and influences all the functions of life.   The gradually
increasing prostration terminating in a chill; the febrile reaction
following this, the eruption of the small-pox pustules, and finally
the subsidence of the disease when the virus is thus wholly
removed.

SYMPTOMS.—Small-pox is divided into two classes, the *discrete*
and *confluent,* though these are differences in its severity, not in
its character.   In discrete small-pox the pustules are not so numerous but that they have room for full development upon the
skin.   In confluent small-pox the points of eruption are so
numerous, that as they are developed into pustules they crowd
upon each other and run together, so that on large portions of
the surface nothing is seen but the eruption, and, as it matures,
the formation of a single crust.

In the *discrete* form the premonitory symptoms are not severe.
The child will be fretful, restless in its sleep, and, if old enough,
will complain of feeling tired.   The chill is tolerably well
marked, though not severe, and lasts from one to two hours.
The febrile reaction comes up pretty actively, though it bears a
closer resemblance to febricula than to the severer fevers.   The
eruption generally makes its appearance during the latter part of
the second and during the third day, and at this time the fever
has attained its highest point.   By the fourth day the fever will
have subsided, to a considerable extent, and from this to the
tenth or twelfth day, is so slight as to give but little annoyance.
At this time, when maturation is complete, a secondary fever is
developed, which, for a few hours, is pretty active.   This passing

13

off, secretion is established, the appetite and digestion are restored, and the child convalesces rapidly.

In the *confluent* form the premonitory symptoms are noticed earlier. The child appears.depressed, its appetite is poor, it is restless and fretful during the day, and does not sleep soundly at night. As we approach the period of chill these symptoms are more marked, and the child presents evident symptoms of suffering. The chill, in this case, is usually well marked and protracted, the temperature is really lowered, the pulse increased in frequency, the skin contracted, presenting the cutis anserina, and in many cases there are marked rigors.

The febrile reaction comes up rapidly in most cases, and runs high. The surface is hot and dry, the pulse frequent, full and hard, the urine scanty, the bowels constipated, the mouth dry, the tongue coated with a white fur, the face flushed, the eyes bright, with great restlessness and irritability. The child expresses evidences of suffering that can not be mistaken, and we suppose it has all the aches and pains that are pathognomonic of small-pox in the adult.

The fever continues on without abatement, rather increasing than otherwise, until the eruption makes its appearance. The symptoms of suffering pass away after the first day, and there is usually not so much restlessness, though occasionally the irritation of the nervous system continues to be a marked feature, and there may be convulsions and occasionally evident delirium.

About the time the eruption is making its appearance the throat seems to be stuffed up, and there is free secretion from it and the mouth. In some cases this becomes quite severe, the throat is swollen, and the secretion profuse and tenacious, which render deglutition and even respiration difficult.

The eruption appears on the third day, at first on the face, then the trunk, and finally the extremities, usually not being fully out before the end of the fourth or the fifth day. In most cases there is a slight abatement of the fever after the eruption has made its appearance, presenting frequently a morning remission and evening exacerbation. It varies greatly in intensity in different cases. In some it is very active in all its symptoms, and children are frequently delirious. In others it presents the usual symptoms of an infantile remittent. In the more malignant cases it is asthenic in character, and the symptoms are of a typhoid type.

About the twelfth day from the chill, or ninth from the eruption, a secondary fever makes its appearance, and lasts from one to three days. Usually it is very active, and the child suffers more from the fever than at any time during the progress of the disease. Maturation being complete, when this fever declines, the secretions are re-established, and the child slowly convalesces.

Having described the general symptoms of the disease, we must now study the eruption. When it first appears it simply presents a small red spot, resembling somewhat a flea or mosquito bite. These look as if slightly elevated, and when we place a finger upon one it feels hard, as if a small shot were imbedded in the skin. On the second day the redness has increased in size, the elevation is more perceptible to the eye, and a minute point is seen in the center from which the future pustule is to be developed. On the third day the red spot has increased in size still more, and in its center is seen a small vesicle, distinctly rounded in form, and filled with a clear, limpid serum. On the fourth day this has increased in size, is flattened, seems to be tied down in its center—umbilicated—and the lymph is becoming yellowish and opaque. On the fifth day it has attained about one-half its size, is yellowish and opaque, distinctly umbilicated, and stands on a swollen base, and presents a red areola.

Continuing to increase in size it attains its maturity by the ninth day, when it is three-eighths to one-half inch in diameter, the red areola being about as much larger; it is yellowish and opaque, and distinctly umbilicated. The tumefaction of the base, when the points of eruption are numerous, is such as to give a uniform swelling of the surface. Thus, in the severer cases, the eyes are closed, and the features so effaced that the person could not be recognized.

When maturation is thus completed, some of the pustules burst and discharge a portion of their contents. In all it desiccates, and forms a crust or scab, which is retained in contact with the skin by the epithelial investment of the pustule. In from three to nine days this gives way and the scabs are thrown off. The parts, where they are developed, present a bluish or livid appearance, which slowly passes away in some five or six weeks, though it can be noticed for as many months when the person is exposed to the cold.

In some cases a process of ulceration is set up at the base of the pustule, and the true skin is destroyed, to a greater or less extent. This causes those depressions that we call *pitting*, and which are never effaced.

Small-pox may be complicated with inflammation of any organ or part. These are not very common, and are usually recognized with ease. There are some irregularities in its progress, however, that should be studied with care. In one, the disease having progressed to the fourth day, we find the skin swollen, and assuming a dusky appearance; yet the eruption does not come out, or appears sparsely. At the same time we find the child becoming comatose, and also that this is influencing every function of life, until finally death results. In a second, the eruption having made its appearance, recedes; the skin is swollen and dusky, and coma comes on, as in the preceding case.

In malignant small-pox—that in olden times received the name of *black* small-pox—the skin seems swollen and dusky from the commencement. When the eruption makes its appearance it also is discolored, and the pustules, instead of being yellowish, have a shade of brown or black, or in some cases are black. This is associated with the brown tongue, sordes on the teeth, frequent small pulse, and other typhoid symptoms.

Diagnosis.—The diagnosis of small-pox is not difficult. The symptoms preceding the chill for some two or three days,—this being marked, and followed by high febrile reaction, in which the child evidently suffers much pain,—are sufficient to arouse our suspicions. The eruption of red points, *hard under the finger,* is the diagnostic evidence at this time. Afterward the *umbilicated* pustule and its regular and slow development, distinguish this from all other eruptions.

Prognosis.—I regard the prognosis of small-pox as favorable, except in those exceptional cases last described. Of course, some children will succumb to even its milder form, because they have not vitality to live through any disease. The mortality attending small-pox seems to have lessened in the ratio that the people are protected by vaccination, thus proving that the virus gains intensity as developed from a large number of persons at the same time.

Treatment.—It is contended by the majority of writers that small-pox is a self-determined disease, and has a regular course to run, and that this can not be modified or shortened. In this respect it is said to resemble typhoid and other continued fevers. I admit the resemblance, and contend that its course can be both modified and shortened, as we have proven to be the case with continued fever.

We have the evidence of this in the history of inoculation. Here the small-pox virus was introduced into a system not protected, but by a system of diet and means to promote the secretions, it was caused to run a very mild and also a short career.

If the small-pox virus is generated within the blood, and this is the cause of the disturbance we witness, that disturbance will be less, just in proportion as this is developed slowly and removed by the excretory organs. There is no doubt but that the virus of small-pox may be removed by the excretory organs in very considerable quantity. We have already seen that the development or activity of such blood poison was in the ratio of the frequency of the circulation and other functional derangement. If, therefore, we control the circulation, and establish function, we will lessen the rapidity with which the virus is generated, and arrest the septic decomposition that accompanies it; and if, at the same time, we stimulate excretion by the skin, kidneys, and bowels, we will remove considerable portions of it, and thus lessen the eruption upon the skin.

It will be observed that the theory is good, and is based upon facts constantly witnessed in other diseases. Experience, also, which is the test for every new doctrine, confirms these views. The treatment, then, will be the same as for any other form of fever.

We put the patient upon the use of Aconite or Veratrum, in the usual doses, aiding their action by the general sponge-bath and the use of the hot foot-bath. If there is much irritation of the nervous centers, with determination of blood, we add Gelseminum; but if the patient is dull and torpid, and inclined to coma, Belladonna is used instead.

The most pronounced symptom in the first stage of small-pox is the *severe pain*, and this suggests remedies that relieve pain. Macrotys has been employed for a half century in these cases, and so marked is its action that many have called it a specific in small-pox. Bryonia will be found an excellent remedy when the patient complains of pains in the chest, with severe cough, or uneasiness in the bowels with tenderness, or pain in the articulations. Rhus is an admirable remedy in some cases, and will be indicated by burning pain, by pain in the forehead, and also by pain in the neck and occiput, the tissues being very sore and painful when pressed. Among the remedies to bring the eruption to the surface, when it is tardy, there are none better than Belladonna when the patient is dull and inclined to sleep, and Rhus where it is restless, uneasy and wakeful.

By the second or third day of the treatment we may add a portion of the tincture of Asclepias to the sedative mixture, and commence the administration of a solution of Acetate of Potash ℥ss., to water ℥iv.; a teaspoonful every three hours. This is used when the fever is active and the disease sthenic in character.

The bowels are kept in a soluble condition by the use of laxatives, but cathartic medicines of an irritant character are very objectionable. Citrate or carbonate of magnesia, manna, phosphate of soda, or castor oil, may be used, or in place of them we may depend upon enemata.

In some cases we find the tongue broad and much furred, the breath bad, and the stomach in such condition that it can neither digest food nor receive remedies kindly. Generally this may be regarded as an unfavorable commencement of small-pox, and demands prompt treatment. When marked, I would strongly advise the use of an emetic, so given as to obtain a thorough action, then continuing the treatment as above named.

Where this condition is not so marked, the sulphite of soda may be employed, in doses of five grains every two hours until removed. Even when an emetic has been used, it will be well, in some cases, to follow it with the sulphites.

In the case named, where the eruption is tardy in its appearance, we may use the hot-pack or the emetic, or both. In giving a hot-pack, in this case, I have an ounce of mustard added to a couple of gallons of water, and when hot wring a blanket out of it quickly and envelop the child, covering warmly with dry blankets or comforts. In the majority of cases it will be more convenient to use the stimulant vapor by means of hot bricks, as heretofore named. The emetic of Lobelia is the most powerful means we have, if properly used. It is given so as to produce its general influence, and then carried to thorough emesis, as was named in congestive and spotted fever.

So also in those cases in which there is a retrocession of the eruption. The emetic arouses the nervous system, throws off the cause, and re-establishes a free and equal circulation of blood. It may be used alone, or associated with the hot stimulant bath.

If the disease shows evidence of a septic condition of the blood, no class of remedies will give better results than the sulphites. The sulphite of soda is probably the best, and may be used freely, the indication being a broad pallid tongue, coated with a dirty white fur. If the tongue is red, moist, and covered with a glu-. tinous nastiness like fecal matter, sulphurous acid will be the remedy, giving it in teaspoonful doses every three hours. Sulphur may also be burned in the room to purify the air, and the bed utensils may be disinfected in the same way, or by the use of chlorinated soda.

In some cases, and these very severe ones, Baptisia will be indicated·from the commencement. The full leaden or purplish face, like one who has been exposed to severe cold, is the best indication for the remedy. Sometimes the patient shows this fullness and unpleasant color over the entire surface, and even the mucous membranes will bear the tint. Baptisia is added to the sedative mixture in the proportion of gtt. x. to gtt. xv. to water ℥iv. It is an admirable remedy, and I should hardly know how to treat these unpleasant cases without it.

Occasionally muriatic acid, or a good sharp cider, or whey, will serve a good purpose; the last making a very good diet drink in these cases.

Chlorate of potash will sometimes be indicated in small-pox, but in this, as in other cases, we follow our nose—the bad odor being the indication.

To relieve the unpleasant condition of the throat during the first days of the eruption, we order a gargle of sulphite of soda, chlorate of potash, or Baptisia, when the child is old enough to use remedies in this way. With younger children these remedies may be used with a spray apparatus, or by inhaling the vapor from them when heated. Usually, when no apparatus is at hand for using inhalations, we will find the vapor of vinegar and water to answer a good purpose.

To prevent pitting I depend wholly upon cleanliness, keeping the room darkened to obviate the irritant action of light. Using glycerine soap, I have the exposed parts washed with a soft sponge several times a day, so as to keep them clean and soft. Occasionally a soft and thin cloth, wet with equal parts of glycerine and water, may be used when parts show considerable irritation. With quite a large experience in the treatment of the disease, the results of this method have been very satisfactory.

The sedatives are continued during the whole progress of the disease, so as to control febrile reaction ; and generally where the treatment as given is pursued, there will be but slight secondary fever. When this comes up, however, it is controlled by increasing the frequency of the dose.

Throughout the progress of the disease, which is exhausting in its character, especial pains should be taken to prevent irritation of the digestive organs, and to support the strength by appropriate food. When the child is nursing the mother's milk is sufficient, and is the best food. At a more advanced age, boiled milk, slightly salted, taken hot, will be more kindly received and better digested than most other food.

In some cases it will be necessary to support the strength of the patient by the use of the bitter tonics and iron. Quinine with Hydrastine, in doses of one-half grain, is generally our best prescription. This may be aided by the muriated tincture of iron with glycerine, as heretofore named. When the mucous membranes present a dark-red or dusky appearance, the dilute muriatic acid will answer the best purpose.

VARIOLOID.

Varioloid is small-pox modified by the vaccine disease. In some persons the cow-pox is but partially protective, or it may be wholly protective at first, but as time passes, the susceptibility to the virus of small-pox is gradually reproduced. Upon Liebig's theory, the *material* in the blood, upon which the small-pox poison acts, is reproduced in smaller quantity.

Varioloid is small-pox in every sense. It is produced by the specific contagion of small-pox, and in turn it generates a virus which will give rise to the fully developed disease in a person not protected by vaccination. It differs only in that the symptoms are milder, and it runs a shorter course. The symptoms are usually those of the discrete form of small-pox. The febrile action subsides with the appearance of the eruption, or in the severer cases is remittent in character, and not severe. The period of maturation is generally but seven days, and the secondary fever is short and mild. The desiccation and removal of the crusts or scabs is also rapid, so that by the twelfth to the fifteenth day the surface is pretty well freed from them.

It is not necessary to give the treatment of varioloid, as it would be but a repetition of that just given under the head of small-pox.

VARIOLA VACCINA—COW-POX.·

Prior to the year 1721, no prophylaxis against small-pox was known in Europe, if we except the practice of *buying* the small-pox that prevailed at a very early period both in Wales and Scotland, and which, in fact, was inoculation. The practice of inoculation, according to the statements of the Jesuits, was employed immemorially in China, and by the simplest method; that of passing a needle charged with the virus through the skin when it was pinched up by the finger.'

The practice of inoculation was introduced into England in 1721, by Lady Mary Montague, who had witnessed its success in Turkey, and had a son successfully inoculated. The practice met with much opposition, as all innovations upon established customs and prejudices do, and it was not until about the year 1750, that it received the cordial support of the medical profession. At first a stimulant and heating treatment was pursued, as was the case in small-pox, and the mortality was considerable. Afterward a " refrigerant and reducing" plan was adopted, which was attended with much success, and from this time the practice came to be regarded with favor.

The vaccine disease or cow-pox, as a prophylactic against small-pox, was discovered about the year 1775, by Dr. Jenner, though he did not publish it until 1798. He noticed, while first studying medicine, that in the dairy districts of Gloucestershire, there was a current opinion that persons who had been affected with a peculiar eruption known as cow-pox, were protected against the contagion of small-pox, and might go among it and nurse persons affected with it, with perfect immunity. His mind was strongly impressed with these facts, and he commenced their investigation. It was not until 1796, however, that he became sufficiently convinced to attempt the propagation of the disease by inoculation, or as we say, by vaccination. His first case was entirely successful, the disease was transmitted, the pustules formed as described, and two months afterwards, upon being inoculated with small-pox virus, it was found not to have the slightest influence.

Dr. Good, in his study of medicine, writing about the year 1822, gives the following account of the discovery, which, as a matter of historical interest, I quote : " The disease attracted attention in the county of Dorset, about forty or fifty years since, as a pustular eruption derived from infection chiefly, showing itself on the hands of milkers who had milked cows similarly disordered. It had been found to secure persons from the small-pox ; and so extensive was the general opinion upon this subject even at the time before us, that an inoculator, who attempted to convey the small-pox to one who had been previously infected with the cow-pox, was treated with ridicule. A formal trial was made, however, and it was found that no small-pox ensued. About the same time, a farmer of sagacity, of the name of Nash, duly attending to these facts, had the courage to attempt artificial inoculation on himself ; and the attempt is said to have succeeded completely. Similar facts and numerous examples of them were accordingly communicated to Sir George Baker, who, having engaged not long before in a most benevolent, though highly troublesome controversy respecting the cause of the endemical colic of Devonshire, was unwilling, notwithstanding his triumph, to tread again the thorny paths of provincial etiology. Gloucestershire, however, another dairy county, had witnessed the same disease with similar consequences ; and the same opinion generally prevailing in distant districts of both counties, afforded proof that the power thus ascribed to cow-pox was not wholly visionary.

" Dr. Jenner, then resident at Berkley, in Gloucestershire, pursued this hint with great judgment and unabated ardor. He was at first foiled by not distinguishing between the genuine cow-pox and an ineffective modification of it, or a spurious disease of nearly similar appearance, to which the same animal is subject, but which is no preservative against the small-pox ; and found another difficulty in determining the period of time within which the vaccine virus maintains its prophylactic power. Having at length made himself master of the distinctive characters of the genuine vesicle, he ventured to publish the discovery in 1798, and to recommend inoculation with the virus of vaccinia as a substitute for variola."

The discovery was not received by acclamation, and its discoverer hailed as a benefactor of his race, as many persons would imagine ; but, on the contrary, it was met by ridicule and denunciation, and Dr. Jenner lost business and property, and narrowly escaped personal injury several times at the hands of a mob. The medical profession was as slow to admit the truth of the discovery, and as fast to cast obloquy on the discoverer, as the people; and it was many years before the doctrine and the man were admitted to be orthodox.

The disease of the cow is of rather rare occurrence, even in England, and hardly ever manifests itself except when cattle are gathered together in herds. We have no authentic account of it ever having prevailed out of England, and feel sure that it has not occurred in this country. The supposition that it is small-pox modified by passing through the cow is an error reproduced every few years, generally to make a sale for vaccine virus from some particular person. Dr. Jenner thought that he had traced the cause of the disease of the cow to the *grease* of horses. In regard to this Dr. Good remarks: "It was fortunate for Dr. Jenner, and the triumph of his discovery, that a minuter attention to the subject gave sufficient proof that there was no foundation for this opinion; and that whatever be the prophylatic power of the matter of the disease called grease, this disease is by no means the origin of the natural cow-pox, and has no connection with it."

VACCINATION.—Vaccination is not so easy an operation as thought by the majority of physicians. True, it is very simple, and there should be no reason for want of success, when good virus is used; yet an extended acquaintance with physicians of all schools, proves that from its imperfect or careless performance, it fails fully one-half the time.

The causes of failure are two-fold. In one case the skin is not so abraded as to bring the virus in contact with the true skin, where it may be absorbed. In the other the punctures are so deep as to cause a free flow of blood, and the virus is washed away.

A very simple and good method is, exposing the arm at the insertion of the deltoid, make the skin tense with one hand, and holding the lancet in the other at an acute angle with the arm, make many slight punctures over a part as large as a three cent piece, elevating the epidermis without much bleeding. Now take the virus, softened to the consistence of cream, and spread on the punctures, either letting it dry by exposure to the air, or covering it with adhesive plaster, before the sleeve is drawn down.

When there is a failure with the lancet used in this way, try the Chinese method. Charge a fine needle with the virus, raise a portion of the skin between the thumb and finger and pass the needle through. An instrument has been devised for this purpose, which answers well, being speedy, certain, and not painful. It consists of a small cupped stylet, thrown by a spring; the virus being placed in the cup, and the spring drawn back, it is discharged into the skin with a single stroke.

Formation of the Vaccine Vesicle.—As a general rule, the puncture disappears the second day, but about the fourth or fifth day, a minute inflamed spot is seen. This gradually increases in size, and is swollen and hardened, and forms the base of the vesicle, which is seen first about the sixth day. At first it is spherical and filled with a transparent limpid fluid, but as it increases in size it becomes flattened, and when it attains maturity the center is lower than the circumference.

It requires twelve or fifteen days from vaccination for the full development of the vesicle, which now presents the following appearance: It is regular in its outline, being usually ovoid, though sometimes circular in form. The vesicle is uniform in its elevation, usually about one-eighth of an inch, is flattened, or even depressed at the center, and has a peculiar pearly-gray color. It stands upon an indurated and inflamed base, which forms a red areola of from half an inch to an inch outside of the vesicle.

About the twelfth day of the vesicle desiccation commences, and in four or five days is complete, though the scale or scab is not loosened for some time.

The scar left by vaccination is peculiar, yet is simulated by spurious vaccination. It remains white, the skin seeming to be deprived of its rete mucosum or colored layer, is depressed, the outlines being clean cut and well defined, and presents many little pits or depressions into the true skin.

Generally there is slight febrile reaction about the eighth day, when the vesicle has attained its maturity. Occasionally there is a marked chill, nausea and vomiting, and for a few hours the child is quite sick. Sometimes the irritation of the arm extends to the axillary glands, and these become enlarged and painful, and in some exceptional cases they have been known to suppurate.

Spurious Vaccination.—Any deviation from the course above described may be designated as *spurious*, and will not prove protective. *Regular in form, uniform and even border, flattened and of equal elevation, slight depression in the center, pearly-gray in color, uniform period of development about nine days, gradual and even desiccation, and slow detachment of the crust.* This is the vaccine eruption, and is so distinctive that I think it need not be mistaken.

Spurious vaccine matter is produced in several ways. It may be from decomposition of the vaccine lymph, by keeping, especially when moistened, and carried in the pocket, where the temperature will be high. I was acquainted with an example of this kind, in which a suppurative disease in some half dozen children was produced by such lymph. The most common cause, as I believe, is the use of second crusts from the vaccine vesicle. In many cases the vesicle having fully developed, the lymph commences to desiccate, but before completion the partially-formed scab is rubbed off or is removed by some accident. But another one shortly forms, principally from pus secreted as the result of the irritation. Of course, if this should be used for vaccination we would introduce pus instead of vaccine lymph.

If the general health was deteriorated in such person, and the blood was bad, and they were suffering from any cachectic disease, the results of such vaccination might be very unpleasant. For in some of these cases the pus possesses a peculiar septic property, that will cause extensive ulceration. I am satisfied that this was the case in the spurious vaccination (from the army) that was propagated through many sections of our country, and was regarded by some as syphilitic. In every case that I could trace, the soldiers vaccinated had been in the hospital, with health broken down by the fatigues, privations, and sometimes dissipations of the army. Even when the vaccination was good and protective, which was not often the case, a suppurative inflammation followed at the site of the vaccine vesicle, and the scabs formed contained pus from bad blood.

I had a very marked example of such spurious vaccination, in my own practice. A daughter of mine, in perfect health, was visiting her grandmother, and a case of small-pox having occurred in the village, people became excited, and almost the entire population was vaccinated. A young lady employed in the house, having a very perfect vaccine vesicle, the child was vaccinated with lymph from her arm, and it also took well, and developed regularly, and was as perfect as was possible. But, a couple of days before coming home, while playing with other children, she struck her arm, and the partly-formed scab was detached. • But a new one forming, it looked well when I examined it, and as there could be no doubt of the health of both the parties, I concluded that I would use it in my practice in place of others. The winter course of lectures had just commenced, and, as usual, I vacciuated a majority of the students, using this virus. It did not produce the characteristic vaccine disease in a single instance, but quite a number had very sore arms. Four were so severely affected as to be confined to their rooms, and in one the local ulceration and infiltration of tissue was such that I feared the gentleman would lose his arm, if not his life; but, having a good constitution, he recovered.

This unpleasant experience has caused me to be very careful in the use of vaccine, and, under no circumstances, will I employ a second scab, whether the first has been detached naturally or by some injury.

Can Syphilis be Transmitted by Vaccination?—I answer this question in the negative, to this extent—that the use of *vaccine 'lymph* will never convey the disease. To make the answer plainer, where the vaccine disease runs its regular course, and the vesicle formed presents all the characteristics of that described, the lymph from this is like pure gold, and is not influenced by any constitutional peculiarity or any disease of the patient.

But if it varies from this standard, is irregular in form and elevation, and yellowish or pus-like in color, then disease may be transmitted, for it is not vaccine, but something else. Or, if running the normal course of the vaccine disease, there is afterward irritation and ulceration, pus being formed, this may be in such quantity in the scab, especially at its base, as to be a cause of disease. Or if the first scab is detached by injury, the second one, being formed principally of pus, the use of this for vaccination may transmit disease from one to another.

For many years the medical world, following the lead of Ricord, denied that secondary lesions were transmissible. Hence, syphilis could not be transmitted as above, unless there was a primary source at the part of vaccination, which could only occur by careless handling of a person having the primary disease upon the genitals. But it is now conclusively proven, and even Ricord admits, that certain secondary lesions may be transmitted by inoculation, and among these are pustular eruptions. Thus we have an additional reason for care that the lymph we use is the normal and regular product of good vaccination. This we can always determine, while it would not be possible for us to learn or discover constitutional taints upon the part of many patients.

That our population is being slowly and insidiously infected with syphilis, through vaccination, as some would have us believe, has no foundation in fact. On the contrary, the spread of the disease depends upon the licentiousness of the people, especially the young men of the country, and the want of laws regulating houses of prostitution.

Preservation of Vaccine Lymph.—Vaccine lymph undergoes spontaneous decomposition, or at least loses its specific properties in a period varying from two to eight weeks. Hence extra precautions are required to preserve it from season to season. It is most convenient to vaccinate from the crust or scab, and this is most commonly used. For a short time it may be kept between two pieces of wax, or inclosed in this manner, additional security may be given by an envelope of tin-foil. But when we wish to keep it through the summer, we envelop the scab in tin-foil, pressing it down tightly, so as to render it water-proof, and drop it into a bottle of glycerine. I do not know how long it might be preserved in this way, but I have kept it in an active condition for a year.

RUBEOLA—MEASLES.

Measles is a disease of childhood, though it may occur at any age, and becomes more dangerous as the person is advanced in years. Like the other eruptive fevers it is caused by a specific contagion developed during the progress of the disease, and may be propagated in the same way, by contact or by breathing an atmosphere impregnated with the poison. Like the others, it also prevails in cycles of from five to seven years. In some seasons and localities it will appear in a very, mild form, while in others it will be more or less malignant.

SYMPTOMS.—From seven to fourteen days after exposure, the disease is ushered in with a chill, sometimes slight, at others amounting to a rigor. Occasionally for a day or two before this the child manifests catarrhal symptoms, has a slight cough, and may complain of pain in the head and back. Following the chill, febrile reaction comes up, but varies greatly in different cases. In some it is quite active, with a flushed hot skin, frequent, full hard pulse, and considerable irritability of the nervous system.

In all cases the catarrhal symptoms are so prominent and constant, as to be regarded as almost pathognomonic. About the time of the chill, the child seems to have taken a severe cold, sneezes frequently, stuffing up of the nose, with increased secretion and discharge, redness and watering of the eyes, increased sensibility to light, hoarseness, and a troublesome dry bronchial cough.

The febrile reaction is usually remittent, and continuing to increase gradually to the second, third, or fourth day, then declines after the eruption has fully made its appearance. The eruption comes out first on the face, neck and breast, then on the balance of the trunk, and finally upon the extremites. The single point of eruption is much the color of a musquito bite, ovoid or irregular in form, especially irregular in its border, and the color is gradually shaded off to the color of the skin. The points of eruption generally coalesce, so as to present larger patches or blotches. In very severe cases, the whole surface will be thus covered so as to present but little of sound skin. The eruption is slightly elevated, and rough when the finger is passed over it, and pressure momentarily removes the color.

It requires from twenty-four to seventy-two hours for the full appearance of the eruption. It retains about the same degree of redness for one or two days, and then slowly declines, so that about the sixth to the ninth day from the chill, it has passed away.

During the one, two or three days in which the eruption is coming upon the surface, the fever is higher than before, and sometimes the little patient is quite sick, even in the ordinary form of the disease. Then it declines, sometimes slowly, reappearing at times until the efflorescence has entirely passed away; at others, the little patient will be free from it in the course of a day.

With the appearance of the eruption the bronchial irritation and cough are markedly increased in some cases, and become very troublesome. There is also more or less difficulty of breathing, which sometimes depends upon determination to, or congestion of the bronchial tubes, and at others upon similar lesions of the parenchyma of the lungs. The part affected, and also the character of the lesion, may be determined by physical examination.

MALIGNANT RUBEOLA.—As named above, measles prevail with varying degrees of severity, from the very mild to the most malignant. We will also find that all the cases of a season resemble one another—when the disease is mild, all cases are mild—when it is severe, all cases are severe. Of the cause of this malignancy we know but little, further than that the contagious virus once attaining malignancy, propagates itself in the same way.

In one class of cases the eruption is tardy in its appearance; the fever running a pretty active course, with considerable bronchial disturbance, the third, fourth, fifth, or sixth day passes without its full appearance. The surface seems slightly swollen and flushed, and in some places the eruption is seen indistinct as if struggling to make its appearance. In the children of my own family, the eruption made its appearance, respectively, in five, eight, and thirteen days from the chill.

Necessarily, in such cases the blood must become impaired by the long retention of the rubeolous poison, and the symptoms presented will be more or less of a typhoid character.

In another class of cases, the symptoms of malignancy are manifested early in the disease. The pulse is smaller and faster, the skin flushed, but dry and husky, and the tongue covered with a dirty fur with a tinge of brown. The nervous system suffers especially in these cases. In some there is great excitement for the first day or two, even delirium, or occasionally convulsions, afterwards coma. But in the majority of cases dullness and hebetude are marked symptoms, the child dozes with its eyes partly open, and early coma comes on and gradually increases until the child can not be aroused from it.

In all of these severe cases the eruption is more or less dusky, and we may judge very closely of the severity of the disease by this. It will also be a guide in the treatment. Means that brighten the color of the eruption are beneficial, but if the duskiness increases, we may be satisfied that our treatment is productive of no benefit.

RETROCESSION.—There may be a retrocession of the eruption of measles, at any time after it has appeared. In the milder form of the disease, this increases the fever and the bronchial

14

irritation, and though unpleasant is not dangerous. But in other cases we will find the nervous system suffering severely from the retrocession, and if it continues the blood also becomes impaired. In these cases dullness, stupor, and coma follow one another rapidly; the skin is dusky, the temperature increased, and the tongue soon becomes brown, and sordes appear upon the teeth. These symptoms are of a grave character, and unless prompt means are employed to bring the eruption again upon the surface, it may terminate fatally in a short time.

THE SEQUELÆ OF MEASLES.—An irritation of the bronchial tubes and larynx, with a harassing cough, is very frequently left from measles. In some cases this seems to be the result of cold taken during the progress of the disease, or more frequently, shortly after the child has commenced getting about. The cough is very harassing, causes restlessness and broken sleep, affects the appetite and digestion, and indeed, all of the vital functions. Continuing, the bronchial irritation produces structural change, and after a time develops true phthisis.

A less common sequel of measles, is a subacute inflammation of the conjunctiva, principally confined to the lids, and finally terminating in granular lids, or ophthalmia tarsi.

In some other cases, otorrhœa results, and proves very stubborn. Deafness, partial or total, may also be produced by measles. The partial deafness is usually owing to disease of the Eustachian tubes, though it is sometimes of the middle ear.

DIAGNOSIS.—The marked catarrhal symptoms is the principal diagnostic feature in the early stage of this disease, as the severe pain is in small-pox, and the sore throat in scarlet fever. We distinguish the eruption by its irregular form, coalescing in blotches, not presenting the hardness of small-pox, or the vivid redness of scarlet fever.

PROGNOSIS.—The prognosis in measles is favorable; even in the malignant cases, if properly treated, there should be but a small mortality. It is true that in some epidemics in large cities, the disease is very severe and typhoid in its character, yet I think that the great mortality attending it is to be attributed to the medication as much as to the disease.

TREATMENT.—The treatment of measles is usually very simple. Bathe the patient's feet for half an hour in hot water, and give internally an infusion of two parts of Asclepias and one of Lobelia herb. It need not be given to produce nausea, though to the extent of slight nausea it favors the appearance of the eruption. After the eruption has made its appearance I frequently continue the same remedy in small doses, to relieve the bronchial irritation and check the cough.

In place of this we may put the patient upon the use of Veratrum or Aconite, in small doses, with gtt. x. to gtt. xx. of Lobelia to water ℥iv., using the general sponge-bath at first, and the hot foot-bath to favor their action. This is *not* associated with warm teas and stimulants as so commonly used, but on the contrary the patient is kept quiet in bed, not too warmly covered, and is allowed cool drinks.

If the cough is very troublesome, we may give an infusion of dried red clover, or from the pocket case, Tinct. Drosera gtt. x. to gtt. xx., water ℥iv., a teaspoonful every three hours. If there is anything specific in medicine it will be found in the use of Drosera in the cough of measles and in whooping cough.

In some cases we find the patient very restless, starting in sleep, with shrill cry, and we fear convulsions. The pulse is small and sharp, and the skin dry, contracted and burning. The remedy here will be Rhus, and its action is very prompt and kindly. Gelseminum, Bryonia, Apis, Phytolacca, Eupatorium, and other remedies, will find a place in the treatment of this disease, being guided by the special indications as heretofore given.

When bronchial irritation is great, or passed into the state of inflammation, we will find the use of an inhalation of the vapor of water, either alone, or medicated with Lobelia, or with some of the narcotics, will answer a better purpose than the usual internal remedies. A mush poultice applied to the chest in older children, and continued until the irritation is relieved, answers a good purpose. But with young children I prefer a cloth spread with lard, and sprinkled with the emetic powder of the Dispensatory.

Much care is required after the disease has subsided in order to confirm convalescence. The clothing should be warm, and the child not permitted to expose itself to draughts of air. If this is attended to and the secretions kept free, there is no more danger from measles than from any other disease.

In *malignant* rubeola we will have to give the case closer attention. The first evidence of the character of the disease in young
children will be manifested in the stupor and tendency to coma.
This will be met by the use of Belladonna,with Aconite, in the
doses heretofore named. Though the case seems grave to trust
to such small doses, yet a considerable experience has given me
great confidence in Belladonna as a specific against congestion,
especially of the nervous system.

With this I put the patient upon the use of the anti-zymotics,
to antagonize the septic condition of the blood; either of these
remedies may be indicated. Sulphite of soda, if the tongue is
broad, pallid and dirty; sulphurous acid, if it is deep-red and
moist, and covered with a glutinous nastiness; muriatic acid, if
it is deep-red, contracted, dry and covered with a brown coat;
Baptisia, if the patient shows that full livid face as of one exposed to severe cold; and chlorate of potash, if we have the
peculiar odor as from cynanche maligna.

When the eruption is tardy in making its appearance, the
patient being depressed, with stupor or coma, we may place great
dependence upon the action of an emetic. Or when a retrocession of the eruption takes place, with the same symptoms, it is
the most certain means of treatment. I prefer in these cases the
acetous emetic, and give it so as to get a gradual influence upon
the system, then carry it to thorough emesis.

This may be followed by the Aconite and Belladonna, and the
sulphite of soda, as named above.

To arrest the harassing cough that so frequently remains after
the eruption has disappeared, I prefer either the infusion of
clover hay, as before named, or the Tinct. of Drosera ʒss., to
water ʒiv., a teaspoonful every four hours. These remedies will
be found much better than the ordinary cough medicines in use.

SCARLATINA.

This is essentially a disease of childhood, and few persons will
take it after the age of twenty. Unlike measles, it is also milder,
as the patient is older. It is propagated by a specific contagion,
either by inhaling the exhalations, or contact with the clothes
of the patient. In some seasons it becomes epidemic; doubtless
because the poison is so intense as to be propagated readily and
at a considerable distance, and the condition of the atmosphere

is favorable to the ready propagation of a zymotic disease. Scarlatina has been divided into three forms: S. Simplex, S. Anginosa, and S. Maligna, differing in their intensity, severity of
symptoms and fatality. In some seasons the disease will present
the character of the first exclusively, in others it will be of the
anginose form, and again every case will be malignant.

SYMPTOMS.—From six to eight days elapse after exposure before the disease makes its appearance, and it is usually ushered
in with a chill. In *scarlatina simplex* the chill is not very well
marked, and lasts but a short time. The fever following presents
the common symptoms, increased heat of skin, arrest of secretion, frequent pulse, loss of appetite, etc. In the course of from
six to twenty-four hours, the eruption makes its appearance, in
the shape of patches of efflorescence upon the face and neck,
then extending to the body. If the eruption is minutely examined, it will he found to consist of an infinite number of small
red points, the rose-colored ground being simply the base upon
which they stand. Soreness of the throat, with slight difficulty
of deglutition, appears at the commencement, though not usually
severe, or accompanied with tumefaction. For eighteen to forty-
eight hours after the appearance of the eruption the fever continues as before, but then rapidly abates, and in from three
to five days the redness disappears, and is followed by branny
desquamation of the cuticle.

In *scarlatina anginosa*, the chill is usually marked, there is
nausea and vomiting, pain in the head and back, thirst, etc. The
fever which follows is intense, the skin is dry, husky and burning, the eyes dry and painful, the face congested and tumid,
bowels constipated, urine is scanty, frequently voided and high-
colored, with marked irritabillity of the nervous system. Soreness of the throat, with difficult deglutition, is complained of
from the first, and on examination we find the fauces tumid and
red, and the tonsils somewhat swollen. The nares are frequently
implicated with the angina, and there is consequent stuffing of
the nose, with difficult respiration and increased restlessness.

The eruption sometimes makes its appearance during the
latter part of the first day of the fever, but more frequently not
until the second or third day; about the third or fourth day it
has reached its hight. At the commencement there appears a
slight tumefaction of a portion of the surface, which gradually

assumes a scarlet color, and the minute red points are developed. These patches increase in size until the greater portion of the surface is involved. During the eruption there is an expression of anxiety and suffering; the child is restless and uneasy, and sleeplessness, which resists the usual means of rest, is caused by the heat and stinging of the surface and soreness of the throat.

The throat affection is here the most prominent feature; the soreness increases, the mucous membrane and subjacent tissue are engorged and tumid, and the secretion from the mucous follicles and salivary glands so viscid and tenacious as to cause great distress. In some cases, ulceration commences by the fifth or sixth day of the disease, and the secretion is difficult of removal and exceedingly offensive; occasionally the ulceration assumes a phagedenic form, and speedily terminates the life of the patient. Frequently enlargement of the cervical lymphatics commences from the third to the sixth day, and if not promptly treated, terminates in inflammation and suppuration. The fever, under appropriate treatment, commences to abate when the eruption has made its appearance, and disappears entirely by the fourth or sixth day, when desquamation commences. As this progresses the surface becomes paler, the epidermis exfoliating in whitish scales, or in large pieces when it is thick. Sometimes desquamation is retarded for two or three weeks.

Scarlatina maligna might be divided into two kinds, the distinctive symptoms being marked. In the one case there is marked evidence of prostration from the commencement. The chill is greatly prolonged, and the child seems dull and stupid, the countenance vacant or besotted. Febrile reaction comes up slowly, the body becomes hot, the heat being pungent, but the extremities are cold. The pulse is frequent, but soft and fluent, or else small and wiry. Frequently there is nausea and vomiting, sometimes diarrhœa. The tongue is broad, flabby, and covered with a foul, dirty mucus, and the patient has difficulty in controlling its movements. The eruption makes its appearance slowly, the redness being more or less dusky. The throat affection possesses the same characteristics; there is difficult deglutition and respiration, the mucous membrane presenting a dusky, tumid appearance. Ulceration is of frequent occurrence, their surfaces being foul, the edges ragged, and a strong tendency to phagedena. Enlargement of the cervical lymphatic glands is very common, with a tendency to the formation of a diffusive abscess, and also, if the patient lives, to the formation of secondary abscesses. As the disease progresses the symptoms are all of a typhoid character; there is the dark tongue, sordes on the teeth, feeble pulse, great oppression of the nervous system, tendency to diarrhœa, tympanitis, etc.

In the second case, the disease exhibits but few, if any, premonitory symptoms. The child is attacked suddenly; the chill lasting but a quarter or half an hour, is not well marked, and is succeeded by the most intense febrile reaction. The skin is intensely hot, dry, and husky; the mouth parched and dry; the eyes injected, dry, brilliant and painful. The patient is either delirious or suffers such intense pain as to be almost unconscious of what passes around him. There is nausea and vomiting, the irritation being sometimes so intense that nothing can be retained on the stomach. In these cases the patient is frequently exhausted by the intense reaction, and dies before the appearance of the eruption, or during the time that nature is trying to throw it on the surface.

In the last two forms of the disease, and sometimes in the simple form, we observe a want of power upon the part of the system to determine the eruption to the surface. In such case the skin appears tumid and dusky, there is tendency to coldness of the extremities, and marked oppression of the nervous system. In such case, prompt measures must be taken to bring the eruption to the surface, or the patient will die. Again, we observe cases in which the eruption comes out, but, from some cause, it retrocedes; in this case the same alarming symptoms manifest themselves. In other cases, the anginose affection is so severe that it seems that the patient has not sufficient power to carry on respiration; sometimes the difficulty depends upon the secretion of a viscid, tenacious mucus.

DIAGNOSIS.—Scarlet fever is diagnosed from other diseases of the skin by the rose-colored efflorescence, upon which are the innumerable small red points. A marked characteristic of the disease is the fact that the redness is effaced by pressure, and does not return for some little time. Thus, by taking a pencil or the finger-nail we can write our name, which remains for a moment, and then gradually fades out.

PROGNOSIS.—In the simple and anginose forms of the disease, the prognosis is favorable. In the malignant form, unless the treatment is prompt and effective, the prognosis is unfavorable. In all cases, if the eruption becomes dusky, if coma or typhomania ensue, or if the tongue becomes brown and foul, it is unfavorable.

SEQUELÆ.—Among the most frequent of the sequelæ of scarlet fever, are diseases of the kidneys, consisting of simple exhaustion and want of power to secrete, chronic inflammation and albuminuria. In the first we notice a marked dullness and hebetude, the appetite is poor, the bowels irregular, marked debility and tendency to cachectic diseases, the blood being greatly impaired. In the second, the pulse is hard and frequent, the dryness and huskiness of the skin continue, there is pain and soreness in the back and loins, the appetite is poor, the tongue dry, whitish, and fissured. In the third, dropsy makes its appearance when the child is supposed to be convalescing. Continued disease of the throat, with irritation and enlargement of the cervical lymphatic glands, is sometimes observed. Ozæna, with weakness and irritation of the eyes, and chronic disease of the ears, attended by purulent discharge and partial deafness, is not unfrequent.

TREATMENT.—In the treatment of this disease it is well to have some well-defined line of action—to determine exactly how we can benefit our patient. We know, by experience, that the higher the fever, and the longer it continues before the appearance of the eruption, the greater the danger, and that the case also becomes critical in proportion to the amount of eruption and arrest of secretion. Thus, in all cases, it is good practice to use such means as will control the primary fever, and favor the early appearance of the eruption. I have already mentioned, when speaking of small-pox, that keeping the secretions free during this period lessened the amount of eruption; this is the case here; hence depurants are advantageous. Care must be used, however, not to overstimulate either the skin, the kidneys, or the bowels. If the first, we will have too free eruption, and afterward inaction; if the kidneys, secretion may be arrested, or albuminuria induced ; if the bowels, it may cause retrocession of the eruption, or very great wrong of sympathetic innervation.

In scarlatina simplex we put the patient upon the use of Tinct. of Aconite and Belladonna—the usual doses repeated every hour. The alkaline bath is used sufficiently often to keep down the heat of the skin, and render it soft and pliable. If the skin is inactive the bath may be hot; sponging the child with water as hot as can be borne, stimulates the capillary circulation, and brings the eruption to the surface. This may be done before the fire, covering the patient with a blanket, or it may be done under the bed-covering. It is well to say to the nurse, in these cases, " if the child complains of any part, sponge it with hot water, and cover with dry flannel."

Generally we will find that these means relieve the irritation, the fever is lessened, and the eruption comes out early ; and continuing these for two or three days, there is but little fever after the eruption has made its appearance.

The treatment for the throat will consist of the use of occasional inhalations of one part of vinegar to three of water, and a flannel wrung out of equal parts of vinegar and water and applied around the throat, with a dry flannel over it; this may be changed every half hour or hour, and as the disease subsides may be replaced with a dry flannel. In some cases no application to the throat is necessary, patients doing better without it. In some Stillingia liniment may be used once a day.

In scarlatina anginosa we adopt pretty much the same treatment. If the febrile reaction is high we may add Veratrum to our sedative mixture, or we may substitute it for the Aconite. We do not expect immediate sedation, indeed it is not desirable, but as the remedies gradually influence the circulation, many of the unpleasant symptoms pass away.

In some cases there will be marked irritation of the nervous system, with determination of blood to the brain. In these we replace the Belladonna with Gelseminum, continuing it until these symptoms have disappeared.

In other cases there will be marked indications for Rhus, in the sharp stroke of the pulse, pain in the forehead, starting in the sleep, shrill cry, etc. This remedy is also indicated by burning of the surface, with extreme nervous irritation, and whilst it favors the appearance of the eruption, it promotes functional activity of the skin. Apis will be indicated by itching of the surface, and pinkish coloration of the face. Baptisia will be suggested by the full face, somewhat dusky, and by duskiness of the eruption.

The disease of the throat, which in this case is the most prominent feature, will demand much attention. The vinegar pack is the best external application here, as it was in the preceding case, and should be continuously employed. The inhalation also offers the best local application to the affected mucous surface. The simple inhalation of the vapor of water, or of water and vinegar, or an infusion of hops, of German chamomile or garden tansy, will give great relief, and repeated every two, three, or four hours, will be sufficient in many cases for the permanent cure. When additional remedies are deemed necessary, they are best prepared in powder with gum arabic and sugar, and allowed to slowly dissolve upon the tongue. We may thus use chlorate of potash, hydrochlorate of ammonia, sulphite of soda, borax, alum, etc. Used in this way, they relieve dryness of the throat, and the constant desire to swallow, which is so unpleasant.

If there is a marked tendency to enlargement of the lymphatic glands, I administer Phytolacca, alternated with the remedies that have been named. It has been my experience that if remedies are carefully selected, we will have but little glandular trouble. Phytolacca is an excellent external application, or in place of this, especially if suppuration is threatened, a solution of salicylic acid and borax, aa. ʒj. to water Oj. may be employed. Boracic acid in solution, ʒj. to Oj. of water, is also a good application.

In the first case of *scarlatina maligna*, I employ Aconite and Belladonna in the usual doses, using stimulant hot sponge-baths and the hot mustard foot-bath, to aid in restoring an equal and free circulation of blood. The indications for remedies are carefully noted. Baptisia is frequently indicated by duskiness of the skin and even of mucous membranes; Rhus by the sharp stroke of the pulse, starting in sleep, and contraction about the base of the brain; Phytolacca by the soreness of mouth and swelling of the cervical lymphatics; Ipecacuanha by irritation of the stomach and diarrhœa; Nux by the nausea and vomiting, associated with nervous prostration, etc.

As typhoid symptoms rapidly develop, we select our anti-zymotic early, sometimes commencing its use at our first visit. The pallid, *dirty* tongue (not a common case) is met by sulphite of soda; the red tongue, with glutinous dirty coat, calls for sulphurous acid. The latter remedy is indicated in a large number of cases, and is not only one of the best for internal administration, but also as a topical application to the throat. In severe cases it may be used with a spray apparatus. Muriatic acid may be added to the child's drink when the indication is strong—the deep red tongue. Baptisia has already been named, the indications being dusky coloration of face, surface, tongue, and throat. Chlorate of potash is indicated by the peculiar fetor, as from cynanche maligna, and may now be used as an internal remedy, as a gargle, or as a spray.

The throat receives the same treatment as the preceding case with the addition of an infusion of Baptisia tinctoria as a gargle, if it can be used in that way; if not, by a spray instrument, or by inhalation of the vapor. I would also give it internally, a teaspoonful every two or three hours. If the throat is tumid and dusky, or is ulcerated and sloughy, boracic acid or thymol in solution will be found an excellent application. The sulphurous acid, one part to four of water, is also very good when used with the spray apparatus. In very severe cases, when there is marked torpor of the nervous system, with rapidly approaching coma, I prefer to commence the treatment with an emetic. I use the acetous tincture of Lobelia and Sanguinaria, and give it so as to obtain a prompt and thorough action. It should relieve the depression of the nervous system, and produce a free and equal circulation of blood. If the eruption should not appear at the usual time, the symptoms becoming grave as named, I should also use the emetic, as I would if there was a retrocession of the eruption.

In this, as well as the other forms of scarlatina, in some cases, I prefer fatty inunction to the use of the bath. It relieves the irritation of the skin, keeps it soft, and thus favors some secretion. Many physicians direct their patients rubbed with a *bacon rind*. A prominent Homœopath of my acquaintance places much dependence upon it. Instead of this I use lard alone in the first forms of the disease, and an ammonia inunction (hyd. ammonia 3j, lard 3ij.) in the malignant form.

In congestive diseases it has been recommended to sponge the surface with a solution of carbonate of ammonia, or liquor ammonia. Several cases are recorded of malignant scarlet fever, purpura hemorrhagica, and typhus fever, in which it was employed with advantage. In using it, one ounce of muriate of ammonia is added to two quarts of hot water, and applied freely. The carbonate may be used in the same proportion in diseases of children.

It must not be forgotten that this, like other diseases, will vary in different years. In one season fatty inunction is one of our most important means, and water seems to exert an unpleasant influence upon the skin. But in another year, the fatty matter can not be used, for it seems to close the skin up and prevent excretion, and we are forced to use soap and water for its removal, and afterwards employ baths. I have had most admirable results from hot sponge baths, as I have had from the ammonia just named.

I may also note that tincture of muriate of iron has also been used in scarlet fever with advantage. It is given after the eruption has made its appearance; the combination with glycerine is the best form, and this may also be used as a gargle. It will also be of advantage during convalescence, and will tend to prevent dropsy and the other sequelæ.

The enlargement of the lymphatic glands is a source of much trouble in this form of the disease, appearing frequently when the patient is apparently convalescent. I prefer an application of the Salicylic Acid and Borax aa. 3ss., to water Oj., or sometimes double this strength. If it becomes evident that suppuration will occur, no benefit will follow from trying to keep it back; indeed we would thus endanger diffusive abscess. To hasten suppuration, we apply a poultice of a decoction of cornus, thickened with wheat bran or powdered ulmus, or of equal parts of hydrastis and elm.

It is well to continue the use of the salicylic acid solution with the poultices, as a preventive against diffusive abscesses and purulent absorption. And should there be symptoms indicating this result, put the patient upon the use of sulphite of soda, and give quinine and iron, with stimulants, if needed, and a nutritious diet.

If the lymphatic inflammation goes on to suppurotian, it is well to open the abscess early, and not let the pus burrow in the tissue. This not only relieves pain, but it lessens the danger of secondary abscesses.

The first intimation of dropsy should be noticed, and appropriate treatment adopted. Usually we will note its appearance in a swelling of the upper eyelids, and sometimes in a night the eyes will be closed. Œdema of the feet is also common, and in either case albumen will probably be found in the urine. In the majority of cases Apocynum, with the small dose of Aconite, will be the remedy. If not, we will probably choose between Aralia and sulphate of manganese.

PERTUSSIS—WHOOPING-COUGH.

This is eminently a contagious affection, though how it is propagated is more than is known. Usually it is contracted only when children are brought in such immediate proximity that the breath or exhalations of the diseased person are inhaled; this, however, is not always the case, as the poison seems to contaminate the atmosphere, so that persons take it when at considerable distance from those having it.

It is undoubtedly a disease of the nervous system, the parts implicated being the pneumogastric nerve and medulla oblongata at its origin. And yet, post-mortem examination has not shown any more serious lesion of the medulla than evidence of determination of blood, which we would be likely to find in any case of such prolonged irritation of the parts to which the nerve is distributed. Like other contagious diseases, it runs a very regular course, and gives immunity against a subsequent attack.

SYMPTOMS.—Whooping-cough manifests itself at first as a simple catarrh, the cough being gradually developed. Some days elapse before there is anything distinctive in it; and it is not usually well marked under from two to four weeks. The cough differs from others in that it seems to arise from an obstruction to respiration, and forcible inspiration is taken, and then there is a series of short expulsions until the air is all expelled. The tendency to cough still continuing, produces great distress, and more or less evidences of impaired respiration are noticed. The *whoop* is developed when the cough becomes intense, and is the shrill sound formed as the air is drawn through the yet contracted larynx in the forcible inspiration succeeding the cough. The cough is paroxysmal, the paroxysms recurring at longer or shorter intervals, in proportion to the severity of the disease.

There is a secretion of glairy mucus in most cases, which is raised at the latter part of the cough, and frequently seems to increase the suffering. If the disease is very severe, and sometimes when mild, there is a free yellowish expectoration. There is frequently some fever at the commencement of the disease, and it may occur during its progress.

Writers divide pertussis into three stages : the first, lasting from five to fifteen days, presents the symptoms of ordinary catarrh; the second, lasting from three to six weeks, presents the peculiar whoop, which gives name to the cough ; and the third, of variable duration, is the period of decline.

It is during the second stage of the disease that the symptoms become so aggravated as to demand relief. We sometimes see the paroxysms of cough so severe that the little patient will turn purple in the face, gasp for breath, and even for some time afterward exhibit marked evidences of imperfect respiration. Occasionally bronchitis sets in and is very troublesome ; sometimes there is marked congestion of the lungs; at others, the frequent and severe paroxysms of coughing prevent necessary rest, derange the functions of the body, and wear the patient out.

In some cases there is a tendency in the disease to recur for months after it has ceased, on exposure to cold, though almost always in a mild form. Instead of impairing the strength of the lungs in'feeble children, it seems rather to have increased it, and may sometimes be regarded as of marked advantage to the child.

DIAGNOSIS.—In the first stage it is with difficulty recognized, but in the second, the paroxysmal character of the cough, its long continuance without seeming cause, and the peculiar whoop, are sufficient for the diagnosis.

TREATMENT.—Except in some rare cases I discard all of the old methods in the treatment of this disease, and rely wholly on specific means. We get a clear idea of the variability of named diseases in this study, because it belongs to the class of contagious diseases, is produced by a specific poison, and has symptoms so characteristic that it cannot be mistaken, and to the casual observer all cases are alike except as to severity. Yet we find that it is not always the same disease, but varies from season to season, and the remedy that cures it now may be of no service in the next recurrence of the disease. If one will but think, it is just the same with other coughs.

We have a group of remedies for whooping cough, all of them good, all of them specific, if we have the right indication. They are Nitric Acid, Belladonna, Drosera, Bromide of Ammonium.

Nitric Acid is an old remedy, and a very good one, but will not benefit all cases of whooping cough—which are the cases? If the tongue has a violet color, showing over red, the remedy will cure whooping cough, as it would cure ague, or any other curable disease, with the same symptom.

Belladonna is also an old remedy, and a most excellent one, but it would not relieve the nitric acid cases. What cases will it cure? When the patient is dull and inclined to sleep, (the eyes are dull, the pupils dilated,) just as it is the remedy in other diseases showing these symptoms.

Drosera is the remedy for the cough of measles. There is something peculiar about this cough which will be readily recognized, and it is based upon a peculiar catarrhal condition of the mucous membranes. Whooping-cough may show these very conditions, and if so, Drosera is the remedy.

Bromide of Ammonium is peculiarly the epileptic remedy, and if our whooping-cough shows the peculiar convulsive condition, we expect to relieve with this agent.

What may we say then? Why certainly that we have four classes of whooping-coughs, distinct and separate, and which may be diagnosed, and the right remedy selected.

PAROTITIS.

Parotitis, or mumps, belongs to the class of diseases we have been studying, as it is caused by a specific contagion, generated during the progress of the disease, propagated by contact, and when a person has had the disease there is subsequent immunity.

In the case of small-pox the contagious material is something tangible; in scarlatina and rubeola there is a distinctive disease of the skin, and may be such change that it is propagated by the epidermis thrown off, or by the excreta in breath, from skin, kidneys and bowels. But in whooping-cough and in mumps we do not know the character of the contagion, or why in the one case it should involve the medulla oblongata, and in the other case the parotid glands.

PATHOLOGY.—Arising from a specific contagion, to which the patient has been exposed, usually by contact, we find the disease manifesting itself in a peculiar inflammation of the parotid gland. I say peculiar, because there is but little exudation of plastic material, and but little tendency to suppuration.   In some cases, indeed, it seems but little more than an irritation with marked determination of blood, with some watery or serous effusion into the tissues.   In other cases the inflammatory symptoms are all well developed, and the function of the gland entirely suspended. Other glandular tissues may be involved.   Thus we have a metastasis of the disease to the testes in the male, and to the mammary glands in the female, and I have noticed disease of the thyroid, and I think of the thymus bodies.

SYMPTOMS.—The period of incubation varies from five to twelve days, during which the person has no symptom of the coming trouble.   Frequently he will go to bed well, and awake with swelling about the jaws and ears, stiffness of the neck, and when he sits down to his breakfast will find both mastication and deglutition difficult.   Acids increase the pain, and a very common domestic means of diagnosis is to have the patient try to eat a pickle, when the sharp pain about the articulation of the jaw soon tells the story.

In some cases the disease will be ushered in with headache and pains in the back and extremities, a well marked chill, followed by febrile reaction.   In these cases the inflammatory action runs high and the pain in and about the parotid space and ear is exquisite, and the patient can hardly eat or speak because it increases it.   The fever will continue one or two days, and then pass away with increased excretion, but the parotial inflammation and swelling will continue for five or six days, and then slowly pass away.

But one side may be involved in the attack, and it is thus claimed that the person may contract the disease again on subsequent exposure, the other parotid being involved.

When there is a metastasis to the testes, the patient feels a deep aching dragging pain, with occasional lancinating pains through the organ.   The slightest touch or pressure occasions exquisite pain.   The testicle is found greatly swollen and reddened; very rarely both are affected at a time, but occasionally one is involved, and as the acute symptoms subside, the other testicle swells and becomes tender.   In these cases the fever is likely to recur, and the secretions are arrested.

Mastitis is not so common, yet I have seen cases in which the breast was very sensitive and painful.   The thyroid enlargement is associated with ovarian irritation in the woman after puberty, and may continue for some months.

TREATMENT.—The patient is put upon the use of Aconite or Veratrum, as indicated, with Phytolacca as the specific remedy in the larger number of cases. A very common prescription is, ℞ Tinct. Aconite gtt. v., Tinct. Phytolacca gtt. x., water ℥iv.; a teaspoonful every honr. Belladonna replaces the Phytolacca, if there is dullness and stupor; Rhus, if there is sharp pain, frontal headache, and sharp pulse; Gelseminum, if there is increased heat of the head, inability to sleep, with general headache. In some cases there is a clear indication for sulphurous acid in the red tongue, with nasty glutinous coating; and in others the indication is equally clear for sulphite of soda.

I do not think it best to make local applications to the swollen parotids—a flannel cloth tied around the head to insure warmth being sufficient. Still, if the pain is very severe, a lotion of—℞ Tinct. Aconite ℥j. Tinct. Phytolacca ℥j., water ℥j., may be applied every two or three hours. Sometimes a flannel cloth wrung out of hot water, and applied for a few hours, will give relief.

If there is a metastasis to the testes, they should be well supported with a flannel bandage, and in some cases the lotion of Aconite and Phytolacca may be used, and in others of Belladonna and Phytolacca. ·In older persons, if the swelling persists, we strap the swollen testicle firmly to the abdominal wall, using adhesive plaster.

SCROFULA.

Scrofula, or king's evil, is one of the most common diseases the physician has to treat; and manifesting itself in so many different forms, its symptoms are protean, and its treatment varied and difficult. It is undoubtedly a disease of the blood, though the secretions and nervous system are markedly affected. Copeland remarks that "The blood in scrofula and tubercles has long been considered popularly, and with much truth, to be of a poorer quality than in healthy constitutions." Simon states that the blood is deficient in solid constituents, especially in fibrin and in corpuscles. According to Duboise, the blood of scrofulous subjects coagulates slowly, the clot is small, soft, and diffluent; the serum is thin, and often of a reddish color. Under

15

the microscope, some of the corpuscles appear devoid of color at the edges only, some entirely colorless. Their size is not materially changed, but they appear flattened, spherical, or cylindrical. Hence he infers that there is a deficiency of the salts in the blood of scrofulous persons. Mr. Phillips remarks that, in every case in which he examined the blood of scrofulous subjects, the coagulum was relatively small, the serum large, the clot unusually soft, almost diffluent; in a few instances only, it was tolerably firm. In most cases the proportion of globules was considerably under the healthy standard. The fibrin had not generally undergone much change. The causes of scrofula, whether those acting on the parent, or the individual himself at a very early age, or even at later periods, whether external or internal, whether hereditary, congenital, or acquired, have all a similar tendency, namely, directly to depress, or to exhaust organic nervous, or vital power; and thereby to impair vital resistance, to prevent the processes of repair consequent upon morbid vascular action, and to arrest the formative or organizing tendency of the exudations produced by this action. Not only is there a disposition to a dyscrasia—to a solution of vital cohesion, observable in parts near the seat of scrofulosis, but there is also an absence of the formative effort in the fluids exuded by morbid actions in scrofulous constitutions. The state of vital power or endowment in the several tissues or organs of scrofulous persons, appears insufficient both for the healthy or sthenic actions or functions these parts should perform, and for the organization of the fluids or matters effused from their vessels. Hence the changes which the exuded matters undergo neither favor nor are followed by organization, even in its lowest grades; and most probably, the fluid itself is exuded from the capillaries of a kind and in a state which indisposes it to organization.

Scrofula is said to be hereditary, and so it is in this, that the child inherits a defective vitality, which manifests itself in imperfect elaboration of the blood, and enfeebled vitality of tissues and organs. Such persons may live for years without any manifestations of the disease, simply because there has been no cause acting to further depress vitality, or to determine scrofulous deposit. Finally, however, from arrest of secretion or other cause, the system is depressed, and an irritation of some part being set up at the same time, we have full manifestation of the disease.

If we have correctly stated the pathology of the disease, what measures may be adopted to remove this predisposition? Some contend that it can not be removed, but we have evidence sufficient to show that it can be entirely eradicated. To accomplish this we resort principally to hygienic measures, such as will stimulate healthy digestion, secretion, and innervation. Remove the child to the country, let it have plenty of out-door exercise with accompanying light and sunshine, give it nutritious food and eschew condiments, pastry, and sweetmeats, and the entire constitution of the child will undergo a change.

Scrofula manifests itself in various ways; very frequently the deposit commences in the lymphatic glands; sometimes in the viscera, as of the lungs, liver, brain, etc.; again in the bones, in the muscles, in the skin, in fact in all the tissues of the body. The determining cause of the deposit is undoubtedly an irritatation of the part, causing determination of blood.

SYMPTOMS.—The symptoms of a scrofulous constitution are not well marked, though it has been frequently described as if they were. It is true that it occurs most frequently in children of fair skin, blue eyes, light hair, and regular features; but it is so often met with in persons of dark skin, hair and eyes, irregular features, and rough development, that it is impossible to say, by a child's appearance, whether it is scrofulous or not. There is, however, in very many cases, such manifest imperfections in assimilation, circulation, and nutrition, and feeble vitality, that we are enabled to recognize the scrofulous constitution. Usually, the previous history of the family will throw some light on the matter; but, as Prof. Powell has well demonstrated, the scrofulous constitution may be and often is developed in children by incompatibility of the parents.

Scrofula manifests itself when, from any cause, the vitality of the system is so depressed that the blood is not properly elaborated, or the detritus of the system is not removed, either by an imperfection in the process of retrograde metamorphosis, or by failure of the excretory organs. The situation is determined in all cases by the existence of a local irritation or inflammation in or adjacent to the parts affected. Thus, we observe scrofulous deposit and disease of the cervical lymphatic glands, from disease or irritation of the mouth or throat; involvement of the axillary glands, from disease of the arm or breast; of the inguinal glands from disease of the lower extremities or genital organs; of the mesenteric glands, from disease of the bowels; of the lungs, from irritation produced by cold; and in the muscles and bones, from the same cause. It might be divided into two forms, as it occurs in the lymphatic glands, or as a deposit in the form of tubercles in the structure of a part, but no practical benefit would grow out of such distinction. As we have, in other places, described scrofulous or tubercular affections of the principal organs, we will confine ourselves here to a description of it as it affects the lymphatic glands.

In many cases the irritation giving rise to the development of scrofula is very manifest, and occasionally demands treatment, but in others it is very slight. The superficial lymphatic glands are then observed to become slightly enlarged and hard, so as to be very perceptible when the finger is passed over them. This occurs frequently in scrofulous children in the superficial cervical glands, without further development, and is considered by many as the best indication of a scrofulous constitution. When the disease has fully commenced, one or more of the glands continue to enlarge, a low form of inflammation sets in, and deposit takes place in the adjacent tissues, which become swollen and hard. Now the inflammation becomes more or less acute, the part is reddened, painful, hot, tender on pressure, and the swelling increases rapidly. Continuing in this way for a longer or shorter time, suppuration commences, and the deposit is gradually changed into pus, which in time makes its way to the surface, and is discharged. This occupies a variable period of time, sometimes passing through all its stages in eight or ten days, and at others occupying as many weeks. In some cases the inflammation is acute and the pain severe, but in others it progresses without much redness, heat, or pain.

The pus forms slowly in many cases, and there is but little tendency to its discharge, and in others weeks pass over, the part still continuing hard; and at last, when our patience is nearly exhausted, suppuration occurs rapidly. Sometimes the pus is well formed and healthy, and when discharged the part heals readily; but at others it is watery, of a greenish-brown color, or clear, with more or less flocculent material mixed with it. Occasionally the abscess exhibits no tendency to point, but the pus burrows in the tissues for a long time, unless it is opened. In other cases, when the pus is discharged the abscess does not heal, but continues to discharge a dirty, flocculent pus; and if we examine it, we will find the walls ragged, and often a chain of lymphatic glands dissected out, and lying at the bottom.

The constitutional disturbance varies greatly. Sometimes there is quite brisk febrile action when inflammation first comes up, with loss of appetite, arrest of secretion, and much prostration. In these cases, suppuration is frequently marked with a chill or rigor, and occasionally attended with hectic fever and night-sweats. In other cases, there is no constitutional disturbance further than loss of strength, and some derangement of secretion, languor, and a peculiar pallid appearance of the surface.

DIAGNOSIS.—Scrofulous enlargement is readily recognized from its situation, and from the attendant symptoms above named.

PROGNOSIS.—In very many cases the prognosis will be favorable, as the tendency to the disease is not so strong but that it may be removed by appropriate treatment, and measures calculated to improve the general health. There is no doubt but that by proper care the constitution of a child can be so entirely changed, in the course of time, that the tendency to this disease will be entirely removed. There are other cases, however, in which, though we may get the patients safely through the present attack, they will inevitably die, sooner or later, of this or some analogous affection.

TREATMENT.—When children are predisposed to scrofula, a judicious hygienic plan should be adopted to strengthen the constitution, by improving the functions of digestion, assimilation and nutrition. Such children are said to be tender, and hence they are kept in the house a considerable part of the time for fear of colds and sickness, and being weakly they are petted, and their appetites pampered; and not spending their time in play, as they should do, their minds are precociously developed at the expense of their bodies. Instead of this, such children should be accustomed to the open air from an early age. As with plants, the human species can not be robust and stout without fresh air and sunshine. As soon as they commence walking they should play in the open air whenever the weather is suitable. In this way, the constitution is strengthened, and the liability to colds by alternations of temperature much reduced. Sleeping-rooms should in all cases be large, well-ventilated, and exposed to the direct rays of the sun during some portions of the day.

Up to the age of eight or ten years, the child's occupation should be out of doors, and whether it was play or work, it should be of such a character as to bring into action all the muscles of the body. Before this age the child should not be required to study, neither should it be sent to school, there being sufficient time after this for all laudable educational purposes. Regular meals of good, hearty food, with fruits in their season, with a sedulous avoidance of all cakes, sweetmeats, etc., are of the highest importance. An observance of these rules, the children being raised in the country, will almost invariably result in a complete change of constitution, and such increased vitality that not only is the predisposition to this disease removed, but the child becomes a vigorous, hearty man or woman, instead of dropping into a premature grave from phthisis or some kindred affection.

In the treatment of the disease, the indications are to, first, improve the quality of the blood, and raise it above the point at which scrofulous material is effused, and, second, to promote the absorption and elimination of such material as may have been deposited. To accomplish these indications various means are resorted to, according to the condition of the patient. Alteratives are relied upon to a very great extent, and various agents of this class are employed. By some the compound syrup of Stillingia and iodide of potassium are considered the preferable agents, and are used to a very great extent. My experience has not been favorable to these remedies, and I have been compelled to select others. I now use the Phytolacca, Rumex Crispus, Alnus Serrulata, Iris, Scrophularia, Podophyllum, Corydalis, and some two or three other agents, sometimes singly, or two or three combined to suit the indications of the case. Acetate of potash or other saline diuretic is my main dependence to promote absorption and elimination by the kidneys. I believe it to be as much more efficient than iodide of potassium, as this is over epsom salts; at the same time employing the bitter tonics, iron, the hypophosphites, and cod-liver oil.

Very much depends upon getting proper action of the three principal emunctories—the skin, kidneys, and bowels. Great care is necessary, however, in the severer cases, not to overstimulate and exhaust these organs. To restore the secretion of the skin, I employ—if it is dry and husky—oleaginous frictions, followed by thorough washing with castile soap and water; if soft, relaxed, and flabby, I use the bitter tonic baths; if there is deficient capillary circulation, with coldness of the extremities, a sponge-bath of dilute Tinct. of Capsicum.

As a local application to promote resolution, I have used equal parts of tinctures of Belladonna and Stramonium, and glycerine, or if there is much fever, an equal part of tincture of Aconite. In other cases, a wash of equal parts of tincture of muriate of iron and glycerine may be used, or the part may be painted with the iron, and then followed by the lotion named. In some cases we obtain good results from the use of the Mayer's ointment or the black salve. Finely pulverized Indian turnip, made into a poultice, is an excellent application. If there is much heat and redness, we may use fomentations of Stramonium leaves, or a poultice of a decoction of Cornus and wheat-bran. If it is seen that resolution can not be effected, we will employ poultices to facilitate suppuration, and if pus has formed to any extent, instead of permitting it to burrow, we will immediately open the abscess. The poultice may be continued for a few days longer, until the inflammation has passed off, when it may be dressed with Mayer's ointment, or other stimulant application, until it heals. If it does not discharge well, and looks ragged, it will be best to use a solution of sesquicarbonate of potash until suppuration becomes free. And in those cases in which the healing process is slow, and the discharge thin and watery, it may also be employed with advantage.

In some cases the healing process progresses until the abscess is nearly closed, but a red, ugly cicatrix is left, from which there is more or less oozing; or, if it closes, it breaks out frequently, and, after running for a few days, again closes, with a thin, bluish cicatrix. These cases are remarkably tedious, and are very difficult to cure. I have treated them by employing the zinc paste to entirely destroy the morbid cicatrix, and then healing with some mild stimulating ointment; or, instead of this, we may sometimes dissect the cicatrix out, and draw the parts together with adhesive straps. In other cases we will find that a decoction of equal parts of Cornus, Rumex, and Alnus, continually applied, and taken internally, will in time overcome the disease.

DYSCRASIAS.

The definition of *dyscrasia* by Dunglison, " *a bad habit of body,*" would answer our purpose very well as describing a bad blood and an impairment of nutrition, and from this enfeebled tissues. The older pathologists used the term to express " an ill habit or state of the humors," *i. e.,* of the fluids of the body.

It is used to describe a condition of life, and not a special form of diseased action, though whatever form this may assume, it possesses the characteristics of the entire group. In scrofula the impairment of the blood manifests itself in the deposit of imperfect albumen, most frequently in the neighborhood of lymphatic glands. In this the nutritive fluids are impaired, and the tissues formed from them are also imperfect. In addition to this the excretory organs being insufficient for its removal, we have it thrown off in the skin, producing skin diseases ; in cellular tissue, producing low grades of inflammation, and in other tissues giving rise to degenerations and inflammatory affections.

Causes.—The causes giving rise to this condition are numerous ; indeed whatever depresses the vital powers, either an impairment of digestion and assimilation, or retrograde metamorphosis and excretion, will produce it. Hereditary feebleness of vitality or formative power is very frequently the cause. Add to this imperfect food, deficient ventilation, impure air, want of sunshine and exercise, and we have the common causes.

We have also to take into consideration the fact that bad blood or bad tissue manifests a constant tendency to reproduce itself—indeed, that whenever a fluid or tissue has had its vitality thus impaired, it perpetuates the impairment ; also, that every point where the disease manifests itself becomes, to a greater or less extent, a depot of supply or depravation. Thus the fluids are being constantly impaired by materials taken into them from these sources.

In some of these cases, the lymphatics suffer more than other parts, and the lymph being impaired, the blood which is formed from it is impaired to the same extent.

Symptoms.—The evidence of bad blood and bad tissue may be found in the general impairment of function, as well as in the many local diseases arising from it. Nutrition being imperfect, the tissues are soft, and have lost their tone and elasticity; the circulation is feeble and unequal; the appetite is variable, and the digestive act imperfect; the tongue being pale, broad, and frequently covered with a pasty white coat.

There is a want of activity of the excretory organs. The skin is dry, rough, and harsh, or soft and flabby, in neither case performing its function well. The urine is changed, containing the triple phosphates or urates, or at times of low specific gravity and deficient in urea; while the bowels are irregular, neither acting well as a digestive or an excretory apparatus.

The local diseases vary in character, but they are alike in giving rise to deterioration of structure, low grades of inflammation, a poor, purulent product, and deficient power of repair.

TREATMENT.—The indications for treatment are very plain in these cases. We have to get rid of the imperfect blood and imperfect tissues, and replace them with good blood and good tissue. We get rid of the bad material by increasing the process of retrograde metamorphosis, and stimulating the excretory organs—the skin, the kidneys, and the bowels. We obtain better blood and better tissue by the use of means that improve the appetite and digestion, and that restore to the blood the materials in which it is deficient, and which stimulate the nutritive processes.

The selection of remedies to accomplish these objects is not always easy. The processes that we desire to act upon are vital processes, and remedies that increase their activity, if properly used, may depress them, or even arrest them, if used without care.

Excretion, or the removal of the bad blood or tissue by the skin, kidneys, and bowels, occupies the first place. Usually we will have no trouble in obtaining this influence, if we are willing to give time enough, and employ simple agents in small doses. I prefer the vegetable alteratives in infusion, singly, or two or three in combination. The Alnus, Rumex, and Scrophularia are favorites of mine, and I think they will give satisfaction.

The skin is reached by the use of baths and frictions. If dry and harsh I prefer fatty inunction, with brisk friction, occasionally using a small portion of quinine in this way, if there seems need for its tonic influence upon the nervous system. If relaxed and flabby, stimulant, tonic, or astringent baths are the best.

If the bowels are inactive, minute doses of Podophyllin,

thoroughly triturated with sugar, answer a very good purpose, but the doses should be so small as not to produce purgation.

The kidneys are called into action by the use of acetate of potash, better than by other remedies, the solids of the urine being especially increased. I have regarded this as our most powerful alterative with children, being much better than the iodides in such common use.

While employing these means the patient is put upon the use of iron, the hypophosphites, cod oil, and the bitter tonics. In many cases the tincture of muriate of iron with glycerine, as heretofore named, will answer the purpose ; or if a stomachic is needed, a small portion of tincture of Nux vomica or solution of strychnine may be added.

Of course a nutritious diet, carefully adapted to the condition of the patient is indispensable, and the selection of this and its preparation, will require the advice of the physician. Add to this good ventilation of and sunlight in the sleeping apartments, and out-door exercise, attention being paid to warmth and cleanliness, and we have an excellent treatment.

INFANTILE SYPHILIS.

A child may receive the primary disease from a chancre of the breast, while nursing, or from a chancre of the lips in the act of kissing, or by the use of cloths or clothing soiled by the discharges from a chancre. Though examples of each of these are upon record, they are of very rare occurrence. In my practice I have seen one chancre of the breast, which, had the woman been nursing, would have communicated the disease to the child, and I have seen three cases of chancre of the lip, in one of which such transmission occurred.

In such cases, the presence of the specific sore will be sufficient for the diagnosis, if the physician gives it a close examination. All the symptoms of the primary disease will be there, and after a period varying from two months to a year, constitutional symptoms will be observed, if the system becomes infected.

But we have a more common source of the syphilitic lesion, in the transmission of the disease from one or the other of the parents to the child in utero. This is hereditary syphilis, and is the form we desire to study particularly.

Secondary syphilis may be transmitted by either parent to the child, which will be born impregnated by the poison. The following are the propositions laid down by M. Ricord regarding its transmission, and they may be received as facts well established:

" 1. The father and mother may transmit the disease to their child indifferently, if either or both of them be affected.

" 2. Transmission may occur from the parents to the child, when they are affected with constitutional symptoms, or when a concealed syphilitic diathesis exists in them.

" 3. The absence or existence of constitutional symptoms in parents at the moment of impregnation and conception exerts no influence on the form of the disease, which may afterward appear in the child. The distinction established by M. Cazenave between congenital and hereditary syphilis, and which is based on the absence of constitutional symptoms in the parents at the moment of generation, or which have been developed in the mother during gestation, is totally erroneous; and indeed M. Cazenave confesses that his opportunities of observing have not been ample.

" 4. The character and period of the manifestation of the symptom in the child are governed by the stage to which the disease had advanced in the parents at the moment of generation. The treatment to which the parents were subjected may also retard, prevent, or modify its appearance in the child.

" 5. If the parents are both healthy at the time of generation, and the mother contracts syphilis during gestation, she may transmit the disease to her child. Of this I have seen several examples at various periods of pregnancy, even to the seventh month inclusive.

" 6. When the venereal poison is transmitted from the mother to the child during pregnancy, infection takes place through the medium of the placenta, and in this case appears to occur after the fourth month of utero-gestation.

" If the father alone be diseased at the moment of generation, an abortion may occur at any period of pregnancy. If the mother alone be diseased at the time of conception, the abortion will not take place until after the fourth month.

" 7. Children born of a father and mother affected with syphilis may escape infection; for a certain disposition to receive constitutional disease is necessary for the child as well as the adult, and this may be absent.

" 8. Observations made as accurately as possible seem to prove that constitutional syphilis may be transmitted from the child to the mother during utero-gestation."

SYMPTOMS.—The most common, as well as the most reliable, evidence of hereditary syphilis, after the child has attained the age of ten years, is the peculiar appearance presented by the permanent teeth, especially the upper incisors. These are usually short and narrow, with a broad vertical notch in their edges, and their corners rounded off. Horizontal notches or furrows are often seen, but they as a rule have nothing to do with syphilis.

Associated with this condition of the teeth, and earlier as to time, being noticed by the end of the first year, are the following symptoms : " The skin is almost always thick, pasty, and opaque. It often shows little pits or scars, the relics of a former eruption, and at the angles of the mouth are radiating linear scars, running out into the cheeks. The bridge of the nose is almost always broader than usual, and low ; often it is remarkably sunk and expanded. The forehead is usually large and protuberant in the regions of the frontal eminence ; often there is a well marked, broad depression a little above the eye-brows. The hair is usually thin and dry, and now and then (but only rarely) the nails are broken and splitting into layers. If the eyes have already suffered, a hazy state of the corneæ, and a peculiar leaden, lusterless condition of the irides, with or without synechiæ, may be expected. If, however, the eyes have not yet been attacked by a syphilitic inflammation, they will present no deviation from the state of perfect health and brilliancy.

" The occurrence of well characterized intestinal keratitis is now considered by several high authorities as pathognomonic of inherited taint. It is also invariably coincident with the syphilitic type of teeth; and when the two conditions are found together in the same individual, I should certainly feel that the diagnosis was beyond doubt."—Hutchinson.

The symptoms present at birth or occurring during infancy vary in different cases. In some cases the child is born shriveled and emaciated, the skin hanging in folds in different parts of the body. The throat is sore, the voice rough and unnatural, and an unnatural discharge is noticed from the nose. Associated with this may be a pustular or squamous skin disease, and copper-colored discolorations. Occasionally there are unpleasant cicatrices or marks of an intra-uterine lesion.

In other cases, and more frequently, the disease first manifests itself two or three weeks after birth in the appearance of the syphilitic exanthemata or vesicula. Following this, there is the deterioration of the general health, above named, with squamous, papular, and pustular eruptions, and the associate ulcerations of the throat and mouth. At a still further advanced stage, there is disease of the nasal cavities, mucous tubercles about the anus, and finally disease of the bones.

Dr. Golding Bird described a characteristic snuffling as one of the most marked symptoms of infantile syphilis. With this, the "puckered mouth, the position of the very characteristic eruption round the lips and anus, in addition to the peculiar and fissured appearance of the surface from which the scales have faded, will seldom, if ever, fail to convert a suspicion of the disease into positive certainty."

DIAGNOSIS.—While it is sometimes easy to make the diagnosis, the symptoms being characteristic, as above described, at others it is extremely difficult. These may be but the symptoms of bad blood and impaired nutrition, as described in the quotation from Hutchinson. Yet whenever one form of cutaneous eruption follows another, after a while associated with disease of the throat and nasal cavities, and attended by impairment of the general health, we will have cause to make strict inquiry whether there is the peculiar discolorations of the skin or not.

PROGNOSIS.—When the symptoms of constitutional syphilis are present at birth, the prognosis is unfavorable; indeed, as a general rule, it is better that such children should die early, for it is almost an impossibility for it to be removed and good health restored. But when it appears some weeks after birth, in the form of *syphilida*, in a well-developed child, we may expect to effect a permanent cure, if the mother's health is not much impaired.

But in most of these cases, the evidence of the syphilitic lesion will still remain as named. The disease makes a permanent impress on the nutrition of the individual, and may be seen for several generations, though all its active symptoms are arrested.

TREATMENT.—The objects of treatment in hereditary syphilis are: to increase the process of retrograde matamorphosis, and the removal of the broken-down material by way of the skin, kidneys, and bowels; and to improve digestion, assimilation, blood-making, and nutrition, thus renewing the blood and tissues with healthy material.

This must be carefully done, for the influence of the syphilitic poison is to depress vitality, and it is much easier to break down tissue than to replace it. The two processes—waste and renewal —should go on together, the one being the exact complement of the other. Thus, as time passes, the syphilized blood and tissues are replaced with new material not contaminated with the poison, and in the course of some months the disease is wholly removed.

There is no *specific* against the syphilitic virus; if there was its administration would suffice. The use of mercury and arsenic temporarily arrests its activity in some cases, but they sometimes render it difficult of cure. When the tongue is small and red, I make the following prescription: ℞ Donovan's Solution of Arsenic gtt. x. to gtt. xx., Tinct. Phytolacca gtt. x., water ℥iv., a teaspoonful every four hours. When indicated it sometimes exerts a remarkable influence, clearing the skin, arresting the ulcerative process, and improving digestion and nutrition. In same cases the mother may be put upon the same remedies in larger doses, as: ℞ Donovan's Solution ʒj., Tinct. Phytolacca ʒij., water ℥iv., a teaspoonful three or four times a day.

Iodide of Potassium is an admirable remedy when the tongue is broad and has a leaden pallor. Sometimes the gums and mucous membranes show the same pallor. For the child I order: ℞ Iodide of Potassium ʒj., water ℥iv., from half to one teaspoonful every three or four hours. For the mother, if the child is nursing, the proportions of: ℞ Iodide of Potassium ℥ss., water ℥iv., a teaspoonful three or four times a day. Sometimes much smaller doses may answer for both child and mother.

I prefer to treat the young child partly through the mother, if it is nursed. She may be placed upon the use of the vegetable alteratives—Corydalis, Alnus, Rumex, Stillingia, etc.—using them in weak infusion, but quite freely. They can be gotten in pleasant form by the use of some agreeable aromatic, as in the following formula: ℞ Corydalis ℥ss., Alnus and Sassafras, aa. ʒj., boiling water Oiv., put in a covered vessel and keep hot for four hours, then strain and sweeten to the taste. Any aromatic that the patient might prefer can be substituted for the sassafras. In using the vegetable alteratives, I think it best to change them every week, as we may in this way obtain a stronger influence, and the remedies are less distasteful.

With this the physician may give acetate of potash to the extent of three drachms, or iodide of potassium to the amount of from one-half to one drachm daily.

. At the same time the patient, if past the second summer, is put upon the use of iron and the bitter tonics, and has a full diet in which animal food preponderates. Occasionally we will find that the tonic and stomachic is most efficient when combined with small doses of Podophyllin, as in the following formula: ℞ Podophyllin gr. j., Hydrastia, Quinine, aa. gr. xx., divide in twenty powders, and give one three times a day.

The compound tonic mixture,* the formula for which will be found in the foot note, is an excellent remedy, and may be employed in this case with advantage; or in place of it, we may sometimes use the following: ℞ Tincture of Muriate of Iron ℥ss., Solution of Strychnia ℨj., Glycerine ℨiijss., a teaspoonful four times a day.

The treatment of the child may consist in part of the administration of the alterative infusions already named. They should be made pleasant, so that the child will take them freely and without disgust.

Iron with glycerine, as so frequently recommended, will prove the best stomachic and restorative. If the child is feeble, especially if there is occasional febrile action, I like the inunction of quinine, indeed, I am not certain but what it will be found of advantage in most cases.

I am partial to the use of alterative baths in infantile syphilis, and have been able to obtain a very decided benefit from them.

---

* FORMULA FOR COMPOUND TONIC MIXTURE.—℞ Ferri Sulph. ℨj., Soda Phos. ℨvj., Quinia Sul. grs. 192, Sul. Acid dil. q. s., Aqua Ammonia q. s., Strychnia grs. vj., Acid Phosph. dil. f℥xiv., Sacch. Alba ℥xiv. Dissolve the Iron in ℥j of boiling water, and the Phosphate of Soda in ℥ij., of boiling water. Mix the solutions and wash the precipitated phosphate of iron until the washing is tasteless. With q. s. of dil. sul. acid dissolve the quinine in two ounces of water, precipitate the quinine with ammonia, and carefully wash. Dissolve the phosphate of iron and the quinine thus obtained, as also the strychnia, in the dilute phosphoric acid, then add the sugar and dissolve the whole without heat.

I generally employ, externally, the same remedies that are used internally. For instance, if Corydalis and Rumex are given. a strong infusion of the same is used as a bath.

POISONOUS BITES AND STINGS.

Occasionally a physician is called to treat a child who has been bitten by a poisonous serpent; but more frequently one who is suffering the effects of the sting of the bee, wasp, or others of like species.

The history of the accident in these cases is usually plain, and their symptoms very marked ; so that there is little danger of mistaking the character of the injury.

SYMPTOMS.—The symptoms from the bite of a poisonous serpent are manifest in a short time. The patient is prostrated, the countenance pale and listless, body bedewed with a cold perspiration, the pulse small, rapid and fluttering, with drowsiness and disinclination to speak or answer questions.

The part bitten usually swells rapidly, and becomes bluish discolored. In some cases the swelling extends to adjoining parts, and finally the whole body is more or less swollen and sometimes discolored.

All the symptoms are those of prostration, and we may regard the poison as a depressant, having a somewhat similar action to hydrocyanic acid, and at the same time a blood-poison, setting up a process of decomposition.

The poison of the sting of the bee, wasp, hornet, and others of like species, is somewhat similar in kind, though in much less degree. Usually we will find the disturbance principally local, except the irritation of the nervous system from the extreme pain, which sometimes goes so far as to produce convulsions. The part is much swollen, pale at the part stung, but with a red areola, and is exquisitely painful.

In some persons, extremely susceptible to the influence of the poison, we will have marked prostration from a single sting. Usually, however, the constitutional symptoms are seen when the person has been stung in many places. In these cases the pulse is small and feeble, the extremities cold, the face pallid, a sense of weight and oppression in the præcordia, difficulty of breathing, and sense of general prostration. In one case that came to my knowledge the patient was unconscious for some hours, and seemingly lifeless for a time. As a general rule, these symptoms pass off in the course of twenty-four hours.

TREATMENT.—The treatment for the bite of a serpent will be of a stimulant character. Let the wound be freely incised, and cupped, or drawn by the mouth, and afterward a strong solution of ammonia applied. Place the patient in bed, covering warmly, and applying dry heat freely.

Give internally the aromatic spirit of ammonia with tincture of asafœtida, in full doses, repeated frequently, with as much strong coffee as the patient can drink.

The common treatment in the South-west is to give whisky freely, to the extent of a pint or more in a short time, for an adult; but I think it doubtful whether this is as good as the plan proposed.

In the case of a *sting* I have slices of raw onion applied to the part, and changed frequently. It is very certain and speedy in its action, relieving the pain, and dispersing the swelling, usually in the course of half an hour. If there are a number of stings, it should be applied to every one of them, and bound on firmly.

In case the general symptoms are developed, I should recommend the treatment given for the bite of a serpent. If convulsions ensue, chloroform will probably prove the best remedy.

16

CHAPTER VI.

DISEASES OF THE RESPIRATORY APPARATUS.

The respiratory apparatus consists of the cavities of the nose, the pharynx, larynx, trachea, bronchial tubes, parenchyma of the lung, and the investing membrane—the pleura. Each of these parts may be the seat of diseased action separately, or two or more may be engaged at one time. The diseases are mostly of an inflammatory character, but in addition we have them dependent upon morbid innervation, and change of structure dependent upon causes other than inflammatory.

We diagnose these diseases in part by the general symptoms, as we do affections of other parts of the body. But to confirm such diagnosis and render it more exact, we are guided by certain *physical* signs, manifest to the senses of sight, hearing, and touch. This is called physical diagnosis, to distinguish it from the ordinary means by general symptoms.

We purpose giving but a brief space to the subject of physical diagnosis here, merely pointing out the differences between the child and the adult. For further information the reader is referred to special monographs on the subject, or to treatises on the practice of medicine.

The information for physical diagnosis is obtained from an examination of *the conformation of the thorax, respiration, cough, sputa,* and from *auscultation and percussion.*

CONFORMATION OF THE THORAX.—But little information is obtained from the form of the thorax in acute disease. In some cases of chronic disease, the narrow and flattened chest, with feeble respiratory power, will be additional evidence of feeble viability. But it is of more importance if it impresses upon us the necessity of passive and active exercise of the thoracic muscles, and increased inspiratory effort to obtain a better development. There is no doubt but that the movement cure may be applied in this case with great advantage.

RESPIRATION.—The respiratory movements of the child are more open and free than in the adult, and at the same time more superficial. Changes in the respiratory movements have less diagnostic value than in the adult, because they may be influenced by slight lesions, or by derangements of the nervous system. Considerable irregularity of the respiratory movement is permitted, without seriously interfering with the function.

Respiration normally is both thoracic and abdominal; if it is changed in this respect, it is evidence of disease. Thoracic respiration is caused by acute inflammation of the liver, stomach, or spleen, or from peritonitis; but the most frequent cause is from an irritable state of the bowels, sometimes met with in severe cases of colic. It is abdominal in pleurisy, pericarditis, and in extreme debility.

Respiration is increased in frequency from two causes: First, in consequence of an increased frequency of the circulation, to which it bears the normal relation of one to four; and second, from disease of the respiratory apparatus, the capacity of the lungs being lessened, the respiratory movement is proportionally frequent.

A slow and free respiration indicates an easy circulation of the blood, sound lungs, and an unimpaired distension of them. If the respiration is large and attended with difficulty, much exertion being necessary, it indicates loss of nervous power and approaching coma or stupor. The short respiration, when unattended with pain, is a very certain symptom of obstruction of the lungs.

Difficult respiration is manifested by labored breathing, the inclination being to assume a sitting position, and in the child by the clutching movements of the arms and hands. There are also the additional evidences from imperfect aeration of the blood.

Cough.—Cough is an indication of irritation of the respiratory mucous membranes, and may be produced from disease of any part. The purpose fulfilled by the normal act of coughing is the removal of irritating matters which may be in the air-passages. The irritation of disease gives the sensation of something irritant within the bronchial tubes, hence the cough. The act of coughing removes increased secretion of mucous, and keeps the air-passages free for the performance of the respiratory function.

Cough may be sympathetic—depending upon disease of some part other than the lungs, as when it arises from disease of the stomach, liver, or other abdominal viscera. Or it may be dependent upon the irritation of the nervous centers, especially of the base of the brain, as we have a marked example in whooping-cough.

The short irritative cough is generally met with in the first stages of inflammation of the parenchyma of the lungs. The hoarse and stridulous cough results from laryngeal disease. The hollow, rattling cough is found in bronchitis, with increased secretion. A dry cough indicates want of secretion; a moist or *mucous* cough, increased secretion.

SPUTA.—While we obtain considerable information in diseases of the adult from an examination of the sputa, we learn but little in diseases of childhood. The adult raises the secretion by an act of coughing, and spits it out; the child raises it to the larynx, and swallows it. If the sputa is ejected, and can be examined, it will give the same information as in the adult.

PERCUSSION.—The practice of percussion does not give the same information as in the adult. The walls of the chest are not so elastic, and the comparative resonance or dullness is affected by minor circumstances in a greater degree. Neither do we have solidification of the lungs even in pneumonia, as in the adult.

Still, *resonance* on percussion gives the information that the lungs are permeable for air, and there is no structural reason for impaired respiration. And *dullness* on percussion is evidence of congestion, or of effusion into the structure of the lungs.

AUSCULTATION.—We obtain the most certain information in regard to the condition of the respiratory organs from auscultation; yet should we be guided by the statements of writers on auscultation, we would be led into frequent errors. We have no *sibilant, sonorous, dry crackling, crepitant, subcrepitant, mucous, or cavernous* rhonchi in children, and those who auscultate with the expectation of hearing such sounds will be mistaken.

The sounds heard may be divided into two classes, *dry and moist blowing sounds*; they are all *blowing* sounds, and I use the word in its literal signification. They are dry in various degrees, and they are moist in various degrees, which the ear soon learns to appreciate. Dryness indicates an arrest of secretion, and contraction of the bronchial tubes; as is the dryness, so is this condition. Moist blowing sounds indicate the establishment of secretion, and in proportion to the moisture, or rattling, is the abundance of the secretion.

The sounds also give evidence of the condition of the bronchial tubes, as regards tone. If the sound is well defined or acute, the tubes are contracted, and in proportion as the sound is wavering and hollow, they are relaxed. Getting the *timbre* of the sound is the most important part of the education of the ear.

The respiratory murmur is very distinctly heard in the child, being once or twice as loud as in the adult, so that beginners in auscultation are recommended to commence the training of the ear, and to get a better knowledge of this natural sound, by auscultation of the child. When this sound is present we know the parenchyma of the lung is not involved. It is changed in inflammation of the lungs, becoming coarser and less even, and broken by mucous cracklings. It is masked by the moist blowing sounds of bronchitis, and these sometimes mask the change in its character—in the latter stages of an inflammation of the lungs.

CORYZA.

A mild form of coryza is of very common occurrence in the child, and while it is mild and demands but little treatment, it is a source of great annoyance to the child, and from the restlessness that attends it, to the family.

The severe form of the disease is a true inflammation of the nasal cavities, and causes considerable constitutional disturbance in addition to the local uneasiness.

SYMPTOMS.—In the first form, the nose becomes stopped by accumulations of mucus, and the child draws its breath through the nose with difficulty, and finally is compelled to breathe wholly through the mouth. The adult finds such a state of things sufficiently annoying, but the child can see no reason for it, and no cause why it shall have patience under the infliction, and consequently makes known its objections in violent crying.

The nose may remain stopped in this way for several hours, and it may be repeated day after day for a week or more. Usually it is associated with slight cold, and when this passes away the trouble ceases.

In the second form there is a similar condition of the nose, and the same restlessness and suffering in consequence. In addition to this the child manifests symptoms of febrile action and arrest of secretion; the face is flushed, the eyes injected, and the head is hot.

The closing of the nose in this case, during the first two or three days, is dependent upon swelling of the nasal mucous membrane. After this there is free secretion and discharge, but when this is retained it closes the passages.

With the subsidence of the fever and the establishment of secretion, the disease passes away.

TREATMENT.—In the first and simple form, thoroughly rubbing the nose and over the frontal sinus with hot lard, or any fatty matter, will give present relief. If the child is suffering from cold, the use of an infusion of Asclepias will generally be all that is required.

In the second form of the disease the patient is put upon the use of the sedative, Aconite or Veratrum, in the usual doses, and the action is aided by the hot foot-bath. In some cases the discharge is acrid, and though the child may not complain of burning, we can see by the expression of its face that there is much irritation of the mucous membrane: we use Rhus with the Aconite. If the child is dull, and wants to sleep a great deal, Belladonna is used. If the head is hot, and the face flushed, the remedy is Gelseminum. If the tongue is pale, the throat sore, or any enlargement of the lymphatic glands, give Phytolacca. Some cases will present themselves in which marked benefit will be derived from sulphite of soda or sulphurous acid. The first, if the tongue is pallid and dirty, or if there is eczema; the second when the tongue is red and dirty, and the tissues somewhat puffy.

The use of inhalations of the vapor of water and vinegar give great relief, and may be employed several times a day in bad cases. In place of this we may use a friction to the nose of tincture of Aconite gtt. x. to gtt. xx., lard 3j.; mix.

CHRONIC CATARRH.

Chronic catarrh is not of as frequent occurrence in childhood as in the adult, yet we meet with cases from the age of three years upwards. In some cases it seems to be dependent upon an enlargement of the mucous follicles, thickening of the mucous membrane, and increased circulation. In this case the discharge from the nose is of mucus only. In other cases there is a true chronic inflammation, producing thickening of the mucous membrane and superficial ulceration, and the discharge is muco-purulent.

SYMPTOMS.—In the first case there is too free secretion from the nose, the constant discharge becoming disagreeable. In addition to this the voice is changed, and has a *nasal* tone, which is sometimes disagreeable. In some cases the nasal cavities become closed at times, from accumulation of the secretion, and this is very unpleasant to the patient.

In the second case there is the free discharge from the nose, but now it is frequently fetid, and is otherwise unpleasant. The patient complains of pain over the nasal bones, and frequently a frontal headache, which may be located in the frontal and orbital sinuses. In addition to this, there is the unpleasant nasal tone to the voice, and the occasional stopping up of the nose, compelling the child to breathe through the mouth.

In some cases the disease is confined principally to the superior portion of the cavity, extending backward to the posterior nares. In others the posterior nares and the upper part of the pharynx back of the soft palate, are principally involved. This gives rise to a peculiar hollow voice, and occasionally to some difficulty in articulation, and an unpleasant gurgling noise in swallowing. In both of these cases the secretion pours backward, and dropping into the pharynx, is ejected by an act of hawking and expectoration.

The general health suffers more or less as the disease continues, especially in the second form. The appetite is impaired, digestion deranged, and poor blood is formed, and from this we have imperfect nutrition. I do not think such cases are necessarily scrofulous, but as the disease progresses, a condition very similar to scrofula is developed.

DIAGNOSIS.—The diagnosis is not difficult, as the continued abundant discharge from the nose, its occasional closure, the nasal tone of voice, and the pain and uneasy sensations in the nose and frontal region, are very prominent.

PROGNOSIS.—The disease will rarely if ever get well without treatment; but an appropriate general and local treatment, if continued sufficiently long, will effect a radical cure.

TREATMENT.—In many cases a general treatment is necessary to success; in some cases, however, nothing but local means will be indicated. The general treatment will vary according to the indications, and much care will be necessary to select the right remedy. The possible remedies might be named, as Arsenic, Iron, Lime, Sulphite of Soda, Sulphurous Acid, Hamamelis, Alnus, Scrophularia, Phytolacca, Donovan's Solution, Iodide of Potassium. The relaxed, atonic skin, with feeble circulation, will be benefited by arsenic; pallid skin, with blue veins, iron; dusky redness of mucous membranes of nose and throat, with occasional erysipelatous flushings of skin, tincture of muriate of iron. Lime is used when there is acid dyspepsia, or a tendency to inflammation of cellular tissue—lime-water or hypophosphite of lime; sulphite of soda, if there is eczema of the face; sulphurous acid, if there is deep redness of mucous membranes, with offensive discharge; Hamamelis, if the tissues are full and atonic; Alnus and Scrophularia, if there is purulent secretion, with the formation of unpleasant crusts or scabs; Phytolacca, · if the throat is sore, or the cervical glands enlarged; Donovan's solution, if the bones are being involved, the skin dirty, and especially if a syphilitic taint is suspected; iodide of potassium, if there is a broad atonic tongue, with leaden pallor.

In some cases I have employed Rhus with Aconite or Veratrum, with most marked benefit. The Rhus has the usual indications, frontal headache, persistent, with burning in the eyes or nose. The Aconite is suggested by the irritable tonsils and throat, and the Veratrum by cough and mucous rattling in bronchia.

At the same time the child should have its regular bath every one or two days, using the alkalies when indicated, the tonic bath if the child is anæmic, or the fatty inunction if the skin is dry and harsh.

With young children it is somewhat difficult to make proper use of local remedies. With such I prefer to use them by inhalation, or with the steam or air atomizer. In this way they can be brought into direct contact with the diseased structures. When the child is old enough (six years) to use it, the nasal douche may be employed (Thudicum's method), the fluid being passed into one nostril and flowing through the nasal cavity and out at the other.

The Thudicum apparatus consists of a glass vessel, holding a pint or quart, to the bottom of which is attached a piece of rubber tubing six or eight feet long, furnished with a stop-cock to turn on or off the fluid, and a nose-piece to fit in one nostril. The method of using it is as follows : Place the vessel containing the fluid on a shelf or stand three or four feet higher than the patient's head, and place the nose piece in one nostril; now let the patient open his mouth, breathing through it entirely, turn on the fluid, and it will pass through the nose, and out of the other nostril, as described. A vessel may be placed before the patient to receive it.

If such apparatus is not at hand, one may be improvised by taking an ordinary bottle to contain the fluid, and a piece of half-inch rubber tubing six or eight feet long to conduct it to the nose. Place one end of the tube in the bottle so that it may reach to the bottom, apply the mouth to the other end, and by suction draw the fluid through, then compressing it with the fingers, introduce it into the nostril, and proceed as before. It is the ordinary syphon, and will continue to discharge until all the fluid in the vessel is removed ; its force will depend on the elevation of the vessel. Another method is to use the ordinary rubber-pump syringe, the fluid passing into one side, going through the nasal cavities and out at the other nostril, so long as the patient breathes through the mouth. With either of these, so soon as the patient shuts the mouth the fluid will pass down the throat, or into the larynx if the patient makes an inspiration. Of course the sense of strangulation stops its use until the patient gets breath and starts fairly again.

The remedies used by the atomizer will differ somewhat from those used by the hydrostatic method. I have obtained good results from the use of a solution of chlorate of potash (gr. x. to water ʒiv.), alternated with Pond's extract of Hamamelis; or in place of the preparation of Hamamelis named, we may use one part of the tincture to three of water. When the discharge is purulent and offensive we may use the solution of permanganate of potash grs. x. to water ʒiv to ʒviij., depending upon the condition of the mucous membrane. In place of this, a solution of carbolic acid, one part to twenty, may be used with good results.

Within the past two or three years I have used a solution of Salicylic Acid with Borax or with chlorate of potash with excellent effect. The strength would be, ℞ Salicylic Acid gr. x., Borax or Chlorate of Potash gr. x., water ℥iv. The apparatus for atomizing is now so good and so cheap, (the Essex selling for $1.50 and $2.00, the Adams for $2.00,) and the method is so easily employed even with young children, that I give it the preference.

It will not do to forget the old means in looking after the new, and there are cases in which a filtered infusion of equal parts of Alnus, Rumex, and Quercus Rubra, will be preferable to any other remedy. These are cases of profuse secretion of muco-pus, with probable disease of bone, and marked development of the nasal voice.

With the hydrostatic apparatus, we use a solution of common salt ℥ss. to the pint of water, to cleanse the nose and remove the mucus or muco-pus. Not unfrequently considerable quantities of offensive material, with large unpleasant looking crusts, will be brought away. Following this, use a solution of chlorate of potash, varying in strength from gr. x. to ℥j., to the pint of water. In a large number of cases we find that these simple means are sufficient for a cure. Occasionally it may be well to alternate with this an infusion of Hydrastis or Hamamelis, or the distilled extract of Hamamelis diluted with one part of water.

If the discharge is very offensive, we employ the solution of salicylic acid and chlorate of potash ; or instead of this, a weak solution of carbolic acid may be used.

To relieve the pain and unpleasant sensations in the frontal region, I always direct the local application of tincture of Aconite root over the part. It is applied with the finger, and of course is used in small quantity. The use of the voltaic battery, the current being passed from the nape of the neck through to the part in front where the pain points or reversed, relieves the unpleasant sensation, and it is claimed aids in the cure.

CHRONIC PHARYNGITIS.

Chronic disease of the pharynx, fauces, soft palate, and tonsils is occasionally met with in the child, and is generally of an inflammatory character. All of these parts may be involved in the disease at once, or one or two may be affected separately. It is frequently associated with disease of the posterior nares, and occsionally with disease of the larynx and bronchia.

CAUSES.—Occurring in persons of feeble vital power, and consequently poor blood and impaired nutrition, we can easily see how any irritation may progress and run into this disease. With such conditions of the system it is most frequently caused by cold, and follows ordinary sore throat. Repeated attacks of tonsillitis, or of sore throat, may give rise to it in persons otherwise of good constitution. It may also be a sequel of diphtheria, which has prevailed so extensively during the past ten years.

PATHOLOGY.—The mucous membrane covering these parts is thickened, the blood-vessels, especially the veins, enlarged, and the mucous follicles increased in size and activity. With this changed and enfeebled condition the transformation of epithelial cells into pus cells, and the reparative material into pus, is easy ; and at a further advanced stage erosion, superficial ulceration, and at last deep ulceration, would naturally follow this condition of the tissues.

SYMPTOMS.—It is noticed that the child is easily affected by cold, frequently complaining of sore throat ; that it frequently clears the throat by an act of hawking, and expectorates a mucus or muco-purulent material in considerable quantity. There is also a change in the voice, readily recognized by one who has not become accustomed to it like the family. Quite often the child will make an unnatural gurgling noise when it sleeps, or it may be, it will snore like an adult.

The constitutional lesions vary in different cases, in some marked, and in others slight, but as they are only incidentally related to the disease we are describing, it is not necessary to enter into a detailed description.

DIAGNOSIS.—The symptoms above named having drawn our attention to the throat as the seat of a disease, we make an examination of it, by placing the patient in a good light and depressing the tongue. I prefer the use of a hand-mirror to reflect the light, concentrating it in the throat. In using this the patient is placed with his back to the window, in such position that we can catch the direct rays of the sun upon the mirror, and throw them in the throat. Or, using a lamp, which is more convenient, it is placed by the side of the patient on a level with the head, when the light may be readily thrown into the pharynx with the mirror.

Making such examination we find the mucous membrane thickened, relaxed, changed in color, and the tonsils enlarged and spongy, and covered by the mucus or muco-purulent secretion.

TREATMENT.—The general treatment named for chronic catarrh will be applicable to this case. If called in consequence of a recent cold, (which always increases the throat trouble,) we will administer Aconite with Phytolacca or with Rhus as may be indicated. If the mucous tissues are deep-red and relaxed, sulphurous acid can be given with advantage, especially if the tongue is dirty and the breath bad. With the dull purplish color of mucous membrane, and sometimes also of face, the patient should have Baptisia.

Restoratives are very generally indicated, and we select them as heretofore named. Minute doses of Veratrum and Arsenic, Compound Syrup of the Hypophosphites, Cod-liver Oil, the diffent preparations of malt, Hypophosphate of Lime, inunctions with Quinine, etc.

If the child is old enough to use a gargle we may employ the Hamamelis, alternated with chlorate of potash. The Hamamelis may be used in the form of the tincture, ℥ss. to water ℥iv., the chlorate of potash in solution, ℨj. to water Oj. In place of the Hamamelis we may use the Hydrastis, Cornus, Alnus, or Marsh Rosemary.

If the child can not use a gargle, the same remedies may be employed with the *spray* apparatus. Occasionally good results may be obtained by using a remedy in powder, allowing it to dissolve on the tongue, swallowing slowly. In this way we may employ the following: ℞ Chlorate of Potash ℨj., (or Alum gr. xx.,) Tannic Acid gr. v., Gum Arabic powdered, White Sugar, aa. ℨij.; triturate thoroughly and divide into powders of five grains.

In place of this we might use a pastile or lozenge in the form of Dr. Anton's, which has proven quite available. The formula is given in the foot-note.*

---

* ℞ Cubebs, fresh and finely pulverized, two ounces; Chlorate Potash half ounce; Gum Tragacanth one and a quarter ounces; Refined Sugar

A flannel cloth wrung out of cold vinegar and applied around the throat, with a dry flannel over it, on going to bed, is a very important aid to the treatment. If the vinegar produces irritation of the skin it should be diluted. When the cloth is removed in the morning, the neck and shoulders should be washed in cold water, using it freely, but drying with brisk friction. This is the most certain means to prevent the frequent taking cold, which in some cases seems to be the greatest obstacle to permanent recovery.

TONSILLITIS.

Inflammation of the tonsils is a very peculiar disease, in that the tendency to it is hereditary in some families, and that, having once occurred, there is a continued predisposition to it, and it continues to recur, sometimes during the entire life-time. It is also peculiar, in that an inflammation so active in form should be confined to a small gland, and not extend to adjacent structures.

CAUSES.—Tonsillitis occurs most frequently at the commencement or breaking up of winter, when the weather is very changeable. A slight cold, contracted at such times, will be followed by an attack.

PATHOLOGY.—The tonsils are composed of an association of follicles, terminating on the free surface by twelve or fifteen ducts, through which the secretion is passed for the lubrication of the fauces. These follicles are bound together with a rather loose areolar tissue, and the whole is invested by a reticulated fibrous capsule, and covered externally by mucous membrane.

---

fourteen ounces; Tincture Collinsonia Canadensis three fluid ounces; Tincture Stillingia half fluid ounce; Tincture Capsicum thirty drops; Essence of Peppermint thirty drops.

Add the tinctures to a few drachms of the sugar and evaporate, then mix all the ingredients and triturate thoroughly; add of boiling water two and a half ounces; mix and work well in the mortar, and set aside (bottom upward to prevent evaporation) for twenty-four hours, to allow the water to soften and incorporate with the gum and sugar. Cut out slices, flatten out with a roller on a pill tile to a uniform thickness, using sugar of milk or starch powder to prevent the mass from adhering to the tile and the roller, and the lozenges to each other. Cut out the lozenges with a flaring tin punch, a full half inch in diameter, so the lozenge will weigh eight grains. Let them dry until hard before using. Avoirdupois weight is used.

The structure is such as to permit very great variations in size. Thus, in simple congestion, they may attain a size three or four times as large as in the normal state, and under inflammatory action with exudation, their bulk is still further increased. The looseness of the tissue likewise permits organized exudative material to once or twice the usual size of the organ, without materially interfering with their function, as we see in protracted cases of tonsillitis.

SYMPTOMS.—Quinsy usually manifests itself first, by soreness and stiffness of the throat, with difficult deglutition, and more or less derangement of the digestive functions; occasionally it is ushered in with a marked chill, followed by febrile reaction. There is always some fever, dryness and constriction of the skin, and general arrest of secretion. In a few hours the patient complains of pain, and a sensation as if some foreign body were present in the throat, with heat and constant desire to swallow. When fully developed, deglutition becomes so difficult and painful as to occasion extreme suffering, and in some cases it is impossible. A guttural cough with frequent desire to remove the secretion from the throat; a hoarse and difficult respiration, and confused whispering and guttural articulation, or sometimes entire loss of voice, is observed. In the severer cases it becomes impossible for the patient to lie down, and in many, but little rest is obtained in consequence of the difficult respiration when the will is in abeyance. If we examine the throat in this disease, we will find the tonsils enlarged and reddened; sometimes so large as to entirely close the opening of the fauces.

An attack of quinsy continues for a variable length of time; usually from four to twenty days, and terminates sometimes by resolution, at others by suppuration. When it terminates the latter way, the gland rapidly enlarges; there is a dull throbbing pain or aching, and a yellowish color near where the pus points; usually it readily comes to the surface and discharges without assistance, but sometimes it is very slow and requires the bistoury.

A condition of chronic inflammation and enlargement frequently continues, in those predisposed to the disease. The glands appear prominent on examination; the mucous follicles enlarged; the color a dusky-red, and considerable tenderness. Associated with this, we frequently have a chronic irritation with determination of blood to the entire isthmus of the fauces, and elongation of the uvula, giving rise to a continuous disagreeable cough, derangement of the general health, finally inducing serious disease of the respiratory apparatus.

DIAGNOSIS.—The diagnosis is very readily effected, as the symptoms pointing to disease of the throat are so prominent as to lead to its examination at once. Upon depressing the tongue one or both tonsils will be seen enlarged and reddened. Day by day we find the swelling increasing, until, if both tonsils are engaged in the disease, they will have quite closed up the isthmus of the fauces. The deep, throbbing pain in the part, greater difficulty of respiration and deglutition, with yellowish discoloration, give information of the establishment of suppuration.

PROGNOSIS.—Though these symptoms are sometimes very urgent, and the patient suffers extremely from a sense of impending suffocation, not one in a thousand will die of the disease. Yet it has been one peculiarly difficult to influence with remedies; and the radical cure of the disease, where there has been a predisposition to it, has been considered impossible, except by total ablation.

TREATMENT.—The use of Aconite with the spray instrument is almost specific in the early stage of the disease. I have employed the steam atomizer, but the Bergson tubes operated with the rubber bellows, the Richardson apparatus in its many modified forms, will answer the purpose, or the small Essex spray is very good. With the steam atomizer I use it in the proportion of gtt. xx. to water ℥ij., with the air spray gtt. x., to water ℥iv. The spray is used as often as every four hours, for five minutes at a time, until relief is obtained. In many cases I have succeeded in arresting the disease with the one application. It is well to have the patient spit out the Aconite that accumulates in the mouth, as there will be too much to swallow.

When these instruments are not at hand, let the patient inhale the vapor of vinegar and water, and apply to the throat the Linamentum Stillingia on flannel. The internal remedy in this case will be Aconite alone, using it in the usual proportion, and repeating it every hour. There is no doubt about the specific action of the remedy, even when taken in these small doses, though it is not so certain as when used with the spray. Sometimes Phytolacca, Belladonna or Rhus may be added to the sedative. If the tongue is red and dirty, and the tissues of the throat red and relaxed, sulphurous acid is a very good remedy.

Penciling the tonsils with the strong tincture of Veratrum will also exercise a marked influence on the inflammation, and will sometimes arrest it at once.

These means should be persisted in, and if they do not arrest the inflammation they will most frequently prevent suppuration. When they prove ineffectual, I am satisfied that there are no means which would have given better results, and we wait the result of suppurative action with patience. Much relief is now given by the use of inhalations, and sometimes by hot fomentations applied to the throat. As a general rule the abscess will open itself, and this we would always prefer in children. If it does not, and the symptoms of obstruction in the throat become alarming, we will have to lance the tonsils. This is not very easily done, but by guarding the bistoury with the finger, it may be accomplished without danger.

The treatment for the radical cure of the disease will vary in different cases. If the tonsils alone are affected, the general health being good, I think we may hope for a cure in the child. For this I rely principally upon the local application of persulphate of iron ; at first one part to three of glycerine, but increasing its strength as the treatment progresses, until it is used of full strength if necessary. The continued use of the tincture of Hamamelis, applied to the tonsils once or twice daily, will also give good results. If there is disease of adjacent parts, the treatment advised in chronic pharyngitis will be used in addition.

When these means fail, we will have to take into consideration the propriety of excision. It is claimed by some, that the removal of the tonsils leads to tuberculosis of the lungs, and this claim is based upon considerable experience in sections of country where tonsillitis prevails. Why such a result should follow I can not see.

The tonsils are removed with a *tonsillitome,* or guillotine, and is easily effected, and without risk. The important part of the operation is to include the whole of tonsil in the ring of the instrument, so as to remove it when the knife is thrown forward. If not wholly removede, the disease may be reproduced, just as if nothing had been done. If there should be hemorrhage following the operation, pencil the part with persulphate of iron, or a saturated solution of alum.

CROUP.

Croup is laryngitis of the child, and yet it differs very materially from that disease in the adult. This difference is owing in part to the imperfect development of the larynx in the child, as compared with the adult, and to a limited extent, the difference in the progress and results of the inflammatory action. Thus it is well to retain the name of *croup* for the laryngeal affection of the child, to prevent confounding the disease with the *laryngitis* of the adult.

We recognize three forms of croup, *mucous*, *pseudo-membranous*, and *spasmodic*. Though alike in some of their features and symptoms, it is well to study them separately.

MUCOUS CROUP.

This is the most common form of the disease in our western country, comprising, probably, two-thirds of all the cases met with. It varies in intensity in different localities, at different seasons, and in different persons. Thus in some places it is a very common disease, in the late autumn and early winter, and in the spring, when winter is breaking up ; as on the shores of Lake Erie. While in some other sections croup is rarely seen, as in Cincinnati away from the river.

CAUSES.—The cause of this, as well as other forms of croup is cold, with its arrested secretion and derangement of the circulation. Why it should particularly affect the small surface of mucous membranes lining the larynx, we are unable to say, as we are unable to account for many things we meet with in the study of medicine.

PATHOLOGY.—This is a true inflammation of the mucous membrane of the larynx, though not of an active character. It is sufficient, however, to cause an increased circulation of blood to it, some impairment of the capillary circulation, and an increased activity of the mucous follicles. It might properly be termed *follicular laryngitis*.

There is but little thickening of the mucous membrane ; indeed, all there is depends upon the increased distension of the blood-vessels. We have, therefore, to look somewhere else for

17

the cause of the difficult breathing, which is so prominent a symptom. This is partly owing to accumulations of mucus in the larynx, but principally to contraction of the *intrinsic* muscles of the organ, from the irritation of the mucous membrane.

Thus post-mortem examination does not show the larynx occluded by structural change, as many would suppose. In some cases, it is true, there are considerable accumulations of mucus, but in most the larynx is sufficiently free for respiration, and we must conclude that the child has been asphyxiated by the spastic contraction of the laryngeal muscles.

SYMPTOMS.—Frequently for a day or two before the attack the child will have had symptoms of cold, with a slight cough. Both the cough and voice are frequently a little hoarse and rough, and would be recognized by a person acquainted with the disease as *croupal.*

The attack of croup occurs most frequently at night, though it may be in the day time. The child seems to be suffering more with its cold during the evening, but still it is not sick, and it is put to bed without, probably, a thought of danger. But along about the middle of the night the parents are aroused by the child starting out of sleep with difficult respiration, a hoarse voice, and a croupal cough.

The respiration is rough and whistling, the cry hoarse and feeble, except when a great effort is made, when it becomes shrill and piping. At first the difficulty of respiration is intermittent, but after an hour or two it becomes permanent, and there is a peculiar whistling or gurgling sound as the air passes into and out of the larynx.

As the disease progresses the difficulty of respiration becomes more marked, and the cough is hoarser, has a peculiar metallic tone, and the voice sinks to a whisper. If the child sleeps mucus accumulates in the throat, the breathing becomes more and more difficult, until at last the child wakes with symptoms of asphyxia. At first the skin is dry, its temperature increased and the pulse full and hard. But as the respiration becomes more difficult, a cold clammy perspiration breaks out, the extremities become cold, and the pulse frequent and feeble. The disease runs its course in from twelve to twenty-four hours, terminating in a subsidence of the disease or death.

DIAGNOSIS.—The hoarse metallic (croupal) cough, with hoarseness and change of voice is sufficient evidence of croup, but it does not inform us which of the three varieties it is. In mucous croup, there is the slight febrile action to distinguish it from the spasmodic variety, and the evident presence of mucus in the larynx manifested by the rattling sound heard on auscultation, and when the patient coughs, which distinguishes it from the pseudo-membranous form. The evidence of increased secretion of mucus in the throat is the diagnostic feature of this disease, though other points of distinction will be named when we describe the two other forms.

PROGNOSIS—Though some cases of mucous croup are very severe, and require careful and close attention, yet we may regard the prognosis as favorable. I do not think the mortality should exceed from two to five per cent.

TREATMENT.—Three methods of treatment may be pursued, and either of them will give success if properly carried out; or, if the practitioner chooses, he may take a part of each and form his own treatment out of it.

I may premise by saying that the treatment generally adopted, of giving emetics for the purpose of emesis, is the worst that could possibly be done, if we except the insane idea of the local application of nitrate of silver, by the followers of Dr. Green. It makes very little difference what emetic agents are employed, if the object is speedy vomiting; for such action is not attended with relaxation or sedation in one-half the cases, while the efforts at vomiting throw the larynx into action, cause increased determination of blood, and increased irritation of the intrinsic muscles. There is no more certain way to destroy the life of the child than this.

Emetic agents, however, may be used in the treatment of croup with success, and we will consider this use, as the first plan. The objects we wish to accomplish are: first, relaxation of the larynx to give better respiration; second, to produce sedation, and thus lessen inflammatory action; third, to increase secretion and thereby get a material less tenacious and more easily removed, and at the same time deplete the engorged vessels of the part. For this purpose they should be used in small doses frequently repeated, so as to produce nausea and its effects—relaxation and

sedation—without emesis. I prefer preparations of Lobelia, but
other nauseant emetics may be used with good results. The
acetous emetic tincture of our Dispensatory will be found to fulfill all the indications.

Aconite alone is the principal internal remedy in the second
plan of treatment. I have no doubt of its specific action in this
case, even in small doses. I would prescribe it in the proportion of: ℞ Tincture of Aconite gtts. j. to iij., water ℥iv., a
teaspoonful every fifteen minutes. Veratrum may be used in
some cases in the proportion of gtt. v. to gtt. x., to water ℥iv.,
and alternated with the Aconite.

The external remedy in this case, as well as in the others, is
the application of the Stillingia liniment to the throat over the
larynx. It may be applied with the finger or a piece of soft
flannel or cotton, and the application repeated every half-hour
or hour. In some cases this remedy locally applied is all-sufficient for a cure, and relief will be noticed within an hour.
Occasionally we administer it internally in doses of half a drop
to one drop on sugar; it may be repeated every half-hour or
hour. In families where croup is of frequent occurence, it is
well that the mother should have a bottle of Stillingia liniment,
and be instructed to use it on the first appearance of hoarseness.
The formula for this preparation will be found in the first part.
I greatly prefer the second plan of treatment.

When the breathing is very difficult, hot water may be applied
to the throat with flannel cloths assiduously, until relief is
obtained. This part of the treatment is of much importance,
and requires that the application of heat be constant, and that
the surface is not allowed to be chilled by exposure as the cloths
are changed, or by wetting the clothing.

The third plan of treatment is by the use of *inhalations*, the
vapor of water or vinegar being the basis, and medicated as the
case seems to demand. The steam spray apparatus is an excellent instrument for using inhalations in this case, though if it
is not at hand we may improvise our means from a tin vessel
containing the fluid, and a hot iron to raise the vapor. With
the spray apparatus I use the preparation of Aconite, named
above for internal use, or equal parts of vinegar and water in
the cup, or if the sounds are whistling, lime water. The ordinary inhalation will be of vinegar and water, or of an infusion
of hops or tansy.

With either of these means the hot application to the throat is important. Or, when it can not be conveniently employed, or we are afraid to trust the nurse, we will apply the compound Stillingia liniment, or in place of this, a cloth sufficiently large to cover the throat and upper part of the chest is spread with lard and freely sprinkled with the compound powder of Lobelia and Capsicum, and applied. The old-fashioned snuff plaster, made in the same way, answered a good purpose.

PSEUDO-MEMBRANOUS CROUP.

This form of croup is fortunately of rare occurrence, as it is much more severe than the others, and, if not treated promptly and skillfully, will terminate fatally.

CAUSES.—This variety is produced by the same cause, cold, as the other forms, though there is some difference in the general health of the patient, which, in one case, gives increased secretion of mucus, and in the other a plastic exudation.

PATHOLOGY.—This is a true inflammation of the larynx, coming on gradually and progressing slowly in the majority of cases, and attended with plastic exudation upon the mucous membrane. In this case we find the mucous membrane no more thickened than could be accounted for by its injection with blood ; and even where loosely attached there is no submucous infiltration.

Post-mortem examination shows the formation of a false membrane from the one-twelfth to the one-sixth of an inch in thickness. This of a grayish-white or yellowish-white color, opaque, and of considerable tenacity. Microscopic examination shows it to be composed of mucus, epithelial cells, and an obscure fibrous structure, the result of organization of the plastic exudation. It varies greatly in its tenacity of adhesion to the mucous membrane; in some cases it seems almost to form a part of it, and is detached with great difficulty ; in others it is very loosely attached, and may be loosened and removed by simple pressure on the larynx externally. In a large majority of cases this false membrane is not of sufficient thickness to account for the arrest of the respiratory function, and we must regard the *spastic* contraction of the muscles of the larynx, from irritation, as one of the causes of death.

SYMPTOMS.—The coming on of the attack of pseudo-membranous croup may sometimes be recognized for three or four days, or even a week. The child does not seem sick, and is playing about the house as usual, but has some cold, and the parents notice a slight hoarseness of voice and cough. We will notice, however, a peculiar metallic resonance to the voice, cry, and cough, but more especially that there is a dry and whistling respiration. This is so marked that the breathing may be heard across the room.

The attack of croup most frequently comes on at night, as in the other cases. In the evening it is noticed that there is more hoarseness of the voice and the cough is somewhat croupal, but as the child breathes pretty well and does not seem sick, the parents flatter themselves that it is but a cold, and will give no trouble. The mother has told me of going to the child's bed or crib, attracted by the peculiar *whistling* respiration, impressed that there was something wrong, but fearing ridicule if she sent for the physician.

As time passes the child becomes restless from difficult breathing, has slight attacks of cough in his sleep, which are clearly croupal. In another hour or two he awakes with a start and assumes a sitting position, evidently suffering much from difficult respiration, which is increased by the attacks of coughing.

The symptoms are now very marked, the respiration is sibilant or whistling, and difficult; the cough hoarse and metallic; the voice roughened or sunk to a whisper, and the cry shrill and piping; the skin is dry, the pulse hard and increased in frequency; urine scanty, and the patient restless and uneasy.

As the disease progresses there is a gradual increase of all these symptoms, but especially of difficult respiration, which is constant. The cough is spasmodic in its character, and when it comes on the patient suffers very greatly from want of air. After a time evidences of asphyxia appear in the bluish lips, distended veins, leaden appearance of the surface, cold extremities, dullness of the nervous system, and finally coma and death.

The entire duration of the final attack will be from six to forty-eight hours.

DIAGNOSIS..—That it is a case of croup is evidenced by the peculiar cough and the change of the voice and cry; that it is pseudo-membranous croup by the constantly increasing difficulty

of respiration, the marked dryness and sibilance in the sound of
the air passing through the larynx, and in the peculiarly dry and
metallic cough. *Dryness* and metallic resonance, in addition to
the croupal cough and voice, are the diagnostic points.

PROGNOSIS.—The prognosis is not as favorable as in the other
two varieties, and with the best treatment some cases will prove
fatal. Still I should not be willing to indorse the old statement—"a large majority *must* die"—but think, with proper
treatment, the majority will recover.

TREATMENT.—The indications of treatment in this case are:
To produce relaxation of the intrinsic muscles of the larynx,
and thus give freedom to the respiration, while we pursue the
main treatment; to lessen inflammatory action, and obtain free
secretion of mucus, for the purpose of effecting the detachment
of the false membrane; and finally to effect the removal of this.

To fulfill the first indication, we employ inhalations of the
vapor of water, or water and vinegar or lime water, as will be
hereafter named. With this we direct the continuous application to the throat of flannel cloths wrung out of hot water, in
the meanwhile bathing the throat with the Compound Stillingia
Liniment. These are important means, and should never be
neglected.

There are two plans of accomplishing the second indication.
The one is, by the use of tincture of Veratrum viride or Aconite,
aided by inhalations of lime water, and is a very good treatment
and much pleasanter than the use of nauseants. I prescribe the
Veratrum in the proportion of gtt. x. to water ℥iv., a teaspoonful every fifteen minutes, until it produces a marked influence
upon the pulse, then in smaller doses to continue its effect.

Aconite is preferred when the pulse is small and frequent, and
is administered in the usual small doses: ℞ Tinct. Aconite gtt.
ij. water ℥iv., a teaspoonful every fifteen minutes. If the child
is very sensitive to the action of the remedy the dose should be
still further reduced, and if we find the lips dry and contracted,
and the child grasping at the mouth with its hands, it should be
suspended and Veratrum substituted.

If the tongue is pallid and shows small spots of red, Phytolacca may be combined with the medicine. If the little patient
is dull and stupid and wants to sleep, give Belladonna. If
it has a sharp stroke of pulse, and moves its head restlessly back-

ward and forward, throwing it backwards as if it would bury
the occiput in the pillow, give it Rhus. This remedy is also
indicated by the shrill cry, as if frightened, and sudden starting
from sleep. Gelseminum is indicated by the flushed face, bright
eyes, and contracted pupils, with restlessness and great irritation.
These remedies are secondary, it is true, but it is a case that requires all that we can do, and if by one of these we strengthen
the Aconite and Veratrum, we give our patient an additional
chance.

What the physican needs, most of all, is a steady hand. The
treatment requires time, and we must not get excited. If the
patient is growing no worse, we should feel satisfied for a time;
if there is but a slow improvement, as marked by more ease of
respiration, a better circulation, warmth and moisture of feet,
legs and forehead, we will feel encouraged and hold fast to the
treatment.

The use of lime water as an inhalation is a very important
part of this treatment. It is claimed that it alone is sufficient
to arrest the inflammatory action and cause the detachment of
the membrane; and I have employed it with success when other
means have failed. The Veratrum and Aconite also have
proven very successful alone, and they will fulfill the first two
indications.

The other and older plan of treatment is by the use of the
nauseant emetics, and if properly used will give excellent results. I may add, that if improperly used, i. e., so as to irritate
the stomach with retching and ineffectual efforts to vomit, they
will hasten the fatal termination.

Of these remedies I prefer: ℞ Acetous Tincture of Lobelia,
Acetous Tincture of Sanguinaria, aa. ʒj., Molasses ʒj., Chlorate
of Potash, finely powdered ʒss., let them be combined with heat,
and add the potash. We give this in doses of a teaspoonful
every ten or fifteen minutes until nausea is induced, then in
smaller doses so as to continue the nausea without vomiting.
The greater and more constant the nausea without efforts at
vomiting, the greater the success of the treatment.

Using the hot applications to the throat, and the inhalations
of vinegar and water, we continue the nausea for some hours, at
least until we have evidence of secretion, and the commencing
detachment of the false membrane. This will readily be detected by the moist sound of respiration, and a gurgling flapping

sound in the act of coughing. If the child is breathing pretty freely, we may wait for the removal of the membrane by the cough, as it will be brought away by shreds.

But if, with the loosening of it, it seems to be drawn upward in expiration, and downward in inspiration, tending to block up the passages, and producing evident symptoms of asphyxia, we carry our remedies to thorough and prompt emesis. Generally it will be well enough to prepare an infusion of the compound powder of Lobelia and Capsicum for use at this time, as we will have established a degree of tolerance for the other preparation. Occasionally we will meet with a case requiring prompt relief. Here the child may be turned on its abdomen, and a finger introduced into the mouth, drawing the tongue forward and exciting the fauces, will be followed by a forcible expulsive effort, and the membrane will be detached. A case of this kind occurred in my practice—the membrane became detached and entirely stopped the larynx, the child was asphyxiated and would have died in five minutes. I snatched it from the mother, turned it on its face, inserted my finger as far down as the larynx; a forcible effort at vomiting ensued, and the whole membrane was removed at once, being a perfect cast of the larynx. The child recovered.

Success in the treatment of pseudo-membranous croup, whatever means may be pursued, depends upon keeping the larynx relaxed to permit aeration of the blood, until in the course of time we get the detachment of the false membrane. It demands patience and perseverance in the use of the means named, and success will follow.

The convalescence of such a case demands much care, as the patient will have suffered from imperfect respiration, from the effects of the remedies employed, and the larynx and other mucous surfaces will be in a debilitated condition. Put the patient upon the use of bitter tonics, as: ℞ Quinia, Hydrastine, aa. gr. vj., make twelve powders, give one every three hours. The child should be kept warm and quiet, in the recumbent position, and guarded from draughts of air, and in addition it should be kept free from excitement.

SPASMODIC CROUP.

In some sections of country this is the common form of croup, though in others the mucous form is of most frequent occurrence. It is not so severe, and ordinarily patients recover without difficulty. Very rarely a case is met with running to a fatal termination.

CAUSES.—The cause of this, as of the other forms, is cold, sudden change of temperature, exposure to an east or north wind, sitting in a draught, irritation of the digestive apparatus, etc.

PATHOLOGY.—In spasmodic croup there is irritation of the mucous membrane of the larynx with determination of blood, or in some cases a slight superficial inflammation. The irritation is sufficient to excite spasmodic contraction of the muscles of the larynx, hence the symptoms of croup. In many cases the irritation extends to the bronchial tubes as well, and they are more or less contracted. Thus, while the difficulty of breathing is principally laryngeal, it may be to a limited extent asthmatic.

SYMPTOMS.—In this case there is usually but slight symptoms of cold before the attack, though occasionally the child will have a severe cold. Frequently there is slight hoarseness in the evening and a little cough, though not sufficient to attract attention. The child is put to bed and sleeps for an hour or two, then becomes restless, and finally wakes with a start, suffering severely from difficult breathing.

Now the breathing is stridulous, the cough hoarse and croupal, the voice hoarse or whispering, and the cry shrill and piping. The skin is soft and moist, the pulse soft and regular, and the nervous system shows no traces of excitement. In a few minutes the child breathes easier, and may fall asleep, but the period of ease is short, a paroxysm of cough occurs, and the breathing is as difficult as before. Thus the disease will continue for hours, broken up into exacerbations and remissions, until finally, the paroxysms becoming lighter and lighter, the breathing is wholly relieved, and nothing is left but a slightly hoarse cough.

DIAGNOSIS.—The diagnosis of spasmodic croup may be easily made if we notice, first, that there is an entire absence of febrile symptoms; second, that it is remittent in character and broken up into exacerbations and remissions. There is an absence of the mucous rattling, as in mucous croup, and the extreme dryness of respiration and cough, as in the pseudo-membranous; and neither in the respiration nor cough do we detect any evidences of change in the condition of the mucous membrane, as is so distinct in the other two forms. On the contrary, all the symptoms point to irritation of the intrinsic muscles of the larynx, and the consequent diminution of its calliber, as the true condition.

PROGNOSIS.—The prognosis is favorable. For, while in exceptional cases, the impairment of respiration may be so great as to destroy life, in the majority recovery would occur without aid from medicine.

TREATMENT.—The treatment of mucous croup may be employed in this case, especially the treatment by the use of nauseants. I prefer preparations of Lobelia to any other, as this is our most powerful anti-spasmodic. The compound tincture of Lobelia and Capsicum (King's Anti-spasmodic Tincture) is a very good form, but almost any preparation of the remedy will answer. It should be given in doses sufficiently large to produce nausea, but always short of emesis.

In the milder cases the administration of the compound tincture of oils of Lobelia and Stillingia (Stillingia liniment), in doses of one drop, repeated every quarter or half hour, with its external application to the throat, will be sufficient to arrest the disease. In giving this, it is dropped on sugar, which is allowed to dissolve in the mouth and swallowed without water. This will also be found a good treatment in the milder cases of mucous croup, or in its earlier stages. In families where the children are subject to croup, it will save much trouble to provide the mother with the remedy, with directions for its use. It will save many unpleasant trips at night, and be a very great satisfaction to the parents.

Spasmodic croup is frequently sympathetic, and repeated attacks occur much to the surprise of the physician. There is an irritation of the stomach requiring treatment, or the patient wants sulphite of soda or sulphurous acid to remove an unpleasant coat from the tongue, (and a similar unpleasantness from the stomach), or it may be troubled with worms and require Santonine with Podophyllin.

In some of these cases the disease is distinctly periodic, and quinine should be given in full antiperiodic doses. In other cases the recurrence of the disease is prevented by the administration of Bromide of Ammonium ℨij. water ℨiv., a teaspoonful every four hours.

In still other cases the croup is the result of a suppressed eruption, frequently urticaria or hives, or it may arise from retrocession of an eruption, and erythema, roseola, or even that which is popularly known as heat. These cases will be reached by the administration of small doses of Aconite with Belladonna, and sometimes by sponging the surface with hot water.

ŒDEMA GLOTTIDIS.

Œdema glottidis, or *asthenic* laryngitis, may occur at any age, but is most frequent in childhood. It is not met with very frequently ; indeed; many physicians will practice a life-time without seeing a case.

CAUSES.—This disease seems to be dependent upon cold, but why it should invade this particular part we are unable to tell. As it occurs in feeble children, and therein manifesting a tendency to disease of the throat, it is probably owing to debility of the tissues.

PATHOLOGY.—The disease is undoubtedly inflammatory in its character, but the inflammation is subacute, and involves the submucous tissue. In its progress there is effusion into this, and it becomes swollen, and where this tissue is loose, as the inward surface of the epiglottis, and the upper portion of the larynx, it produces such engorgement as to obstruct the passage of air. As this distension is greatest in the epiglottis, the difficulty is much greater in inspiration than in expiration.

SYMPTOMS.—The disease commences with a continually increasing impediment to respiration, and a feeling of fullness and constriction, and continuous desire to clear the throat, as if the irritation were caused by some foreign body. The voice becomes hoarse, then croupal, and afterward sharp, stridulous, whispering, and is then lost entirely. There is a hoarse, convulsive cough, with fits of suffocation, causing great agony.

The most marked feature of the disease is, that while inspiration is prolonged, stridulous, and exceedingly difficult, expiration is comparatively easy. This feature is so constant as to be pathognomonic.

There is no fever, but as the disease progresses the pulse becomes more frequent, small, and irregular. The difficulty of breathing increases, the paroxysms of coughing and suffocation are more frequent, symptoms of asphyxia appear, the cerebral functions are disturbed, and at last death ensues from inability to inflate the lungs.

DIAGNOSIS.—The diagnosis in this disease will be readily made, if it is recollected that the difficulty is in inspiration, while expiration is comparatively free. In the later stages of the disease, when it is likely to prove fatal, the diagnosis will be more difficult.

PROGNOSIS.—In the early stages of the disease, the prognosis is favorable; but when symptoms of asphyxia begin to appear, it is doubtful.

TREATMENT.—The treatment of this case will be wholly different from that adopted in croup, being stimulant instead of relaxing. We would dry cup the throat and upper part of the back, and repeat it if the case was serious; applying to the throat and breast a cloth spread with lard and sprinkled with the compound powder of Lobelia and Capsicum, changing it two or three times daily. A mustard foot-bath may be used with good effect, and repeated two or three times, following it by the application of dry heat. We guard against coldness of the extremities, which so frequently follows the ordinary use of the foot-bath, and which almost invariably increases the disease.

If the pulse is frequent and small, we will give Aconite, combining with it Apocynum, as: ℞ Tinct. Aconite gtt. ij. to gtt. v., Tinct. Apocynum gtt. iij. to gtt. xv., water ℥iv., a teaspoonful every hour. Apocynum will act upon the bowels in quite small doses, but if the bowels are torpid the proportion of this remedy may be increased. Phytolacca is sometimes indicated by the soreness of the mouth and throat, and engorgement of the lymphatic glands. Belladonna is a good remedy in some cases, the indication being the dullness and stupor, and inclination to sleep.

In some cases stimulant doses of Lobelia, with some pleasant

aromatic to prevent nausea, may be prescribed. The following
formula answers well: ℞ Tincture of Lobelia ℨij., Compound
Tincture of Lavender f℥ss., Simple Syrup to ℥ij.; a teaspoonful
every hour, or, better, half teaspoonful every half hour. If at
any time it should produce nausea the dose must be lessened.

This may be aided by the use of stimulant inhalations with the
spray apparatus. A very good inhalation may be formed by
adding Carbolic Acid, grs. v., to water ℥iv., or, Hydrochlorate
of Ammonia, grs. x., to water ℥iv.; or in place of them, lime
water, of full strength, may be employed. ·

Associated with these means, in a malarial country we would
give quinine in stimulant doses, say half to one grain every three
hours. It may be given alone, or, what is better, associated with
hydrastine.

Sometimes we find a very troublesome cough, that greatly increases the patient's danger, by its interference with the respiratory movement. Without the cough the child would get a sufficient supply of air, and its blood would be decarbonized; with
the cough it suffers from defective oxygenation and retained carbonic acid. In this case the life of the patient may depend upon
quieting the cough. For this purpose I would make the following prescription: ℞ Sulphate of Morphia gr. j., Alum ℨss.,
Gum Arabic, White Sugar, aa. ℨi.; mix and divide in twenty
parts. One of these may be placed upon the tongue, and allowed
to dissolve, and be swallowed without taking fluid, and repeated
as often as may be necessary to quiet the irritation.

BRONCHIAL CATARRH.

This is one form of cold—that in which it principally affects
the bronchial tubes—as in nasal catarrh it is principally confined
to the cavities of the nose. It is of frequent occurrence in the
fall and early winter, and in the changeable weather of spring.
Sudden changes of temperature, or the barometric condition of
the atmosphere, is a very common cause. Exposure to north or
east winds, or too long or great exposure in cold weather, will
also cause it.

PATHOLOGY.—The disease varies from simple irritation with
determination of blood, to a slight grade of inflammatory action.
In all cases there is a stage in which secretion is arrested. This,

however, is short, and when the patient is seen by the physician, increased secretion is a present condition.

SYMPTOMS.—The child manifests the usual symptoms of cold, and in addition has a cough, with some difficulty of respiration. It is noticed that it does not feel so well .as common, that it does not move freely, or if eating, that its appetite is impaired. There is always more or less febrile action for two or three days, and sometimes the fever is so prominent for a day or two that it obscures the symptoms of. local disease.

The physician notices in addition to all this, that there is a want of free respiratory movement, and an unnatural rattling sound in respiration. On auscultation we find the evidence of increased mucous secretion over all parts of the chest, in the moist blowing sound (mucous rhonchus).

After the third or fourth day, the febrile action passes away, the difficulty of breathing goes with it, leaving but the cough, and increased secretion, which gradually declines until in from ten days to two weeks, it has wholly disappeared.

TREATMENT.—In the earlier stage of the disease we give Aconite or Veratrum, according as the pulse is small or full, with such other remedies as may be indicated. If the patient seems to have pain in the chest, on coughing or in respiration, Bryonia is suggested. If the pulse is sharp, and there is restlessness with sudden starting in sleep, the remedy will be Rhus. If there is much mucous secretion, and rattling in bronchial tubes, with suppressed breathing, Lobelia is added to the sedative. In some cases the old prescription—℞ Tinct. Lobelia ℥j., Comp. Tinct. Lavender ℥iij., Simple Syrup ℥jss.—is a most excellent remedy ; it can be given in small doses, frequently repeated, just short of nausea, until the patient is relieved.

If there is much oppression of the chest with difficult respiration, use the cloth spread with lard, and sprinkled with the compound powder of Lobelia and Capsicum, applying it to the entire anterior and lateral walls of the chest, and changing two or three times a day.

ASTHENIC BRONCHITIS.

A form of bronchial disease is met with in young children, which we have been accustomed to describe under this head, though it does not express the conditions as well as we would wish.

CAUSES.—The exciting cause of the disease is cold, from exposure, from sudden changes of temperature, or of the hygrometric condition of the atmosphere. The predisposing causes are such as impair the general health of the patient: bad air, insufficient diet, want of cleanliness and the condition heretofore described under the head of *dyscrasias*—bad blood and imperfect nutrition. Though the disease most frequently occurs in such persons, it will occasionally be met with in children who have been healthy, and who have all the care and comforts that can be given them.

PATHOLOGY.—The inflammation of the bronchial mucous membrane is of a low grade, and is followed by relaxation and a sluggish circulation with increased secretion. It is this loss of irritability of the bronchial muscular fiber, with the greatly increased secretion, that is the principal source of danger.

SYMPTOMS.—This is the *peripneumonia notha* of authors, and generally occurs in very young or old persons, or in those of exhausted constitution, or who have been liable to coughs with profuse watery expectoration. It usually commences with symptoms of cold and oppression in the chest, with slight febrile reaction. The cough is severe, occurring in paroxysms; the breathing is oppressed, laborious and wheezing; the expectoration, scanty at first, soon becomes abundant, thin and frothy; the pulse is quick, small, and irregular, the heat of the surface but little if any increased, the extremities generally being cool; the tongue is loaded with a foul, dirty mucus, the appetite is gone, and the bowels constipated at first, become irregular as the disease advances. As the disease becomes more intense, the countenance is pale and anxious, there are exacerbating fits of dyspnœa, in which it seems almost impossible for the patient to breathe, and if the patient attempts to take a full breath to relieve this, or changes his position, a severe fit of coughing is brought on, sometimes terminating in vomiting which gives temporary relief. If the case terminates fatally, the tongue becomes livid, the face dusky, the patient can not lie down, and if he sleeps it is but for a few.moments, and wakes threatened with impending suffocation; delirium sets in, with cold and clammy perspiration, and the system is soon exhausted.

In weak and poorly nourished children, this disease is of frequent occurrence. At first it is noticed that the little patient has a protracted chill, followed by febrile exacerbation. The fever is higher in the afternoon, but becomes less and less marked as the disease advances. Respiration is quick and wheezing, the pulse frequent and full, though soft and easily compressed. The cough is persistent, deep, and hollow; the expectoration, at first a viscid mucus, becomes, as the disease advances, yellowish, greenish, and opaque. Dyspnœa is marked when the disease is fully developed, and coming on in paroxysms it is followed by a long harassing cough, which frequently terminates in vomiting, giving relief for the time being. The disease sometimes continues for days, or even weeks, terminating favorably; or the dyspnœa becoming more intense, we observe symptoms of asphyxia rapidly increasing, and the child dies of apnœa.

DIAGNOSIS..—We form our diagnosis in this affection, by the low grade of febrile reaction, marked derangement of function, and prostration, that the inflammation is asthenic; by the cough and difficulty of respiration, that the respiratory organs are the seat of the disease; and by the presence of the mucous rhoncus and resonance on percussion, that the bronchial tubes are the parts involved.

PROGNOSIS.—When the disease is mild, a favorable prognosis may be given, but when severe, it is an exceedingly dangerous affection, and our prognosis must be guarded.

TREATMENT.—In young children I invariably pursue the one course, and thus far with success. Taking a soft cotton cloth, sufficiently large to cover the anterior and lateral walls of the chest, it is spread evenly with lard, and dusted with the compound powder of Lobelia and Capsicum (emetic powder of the Dispensatory); this is applied to the chest and changed two or three times in the twenty-four hours. Internally I prescribe:

18

℞ Tincture of Lobelia ʒj., Compound Tincture of Lavender ʒij., Simple Syrup ʒjss., M.; half a teaspoonful every thirty minutes until the breathing becomes easier, then every hour.

In all of these cases it is necessary that the child be kept quiet and warm; and I always prefer that it be kept in the cradle or crib, rather than in the mother's or nurse's arms. Keep the extremities warm—close attention to this, will in many cases be the difference between a successful or fatal termination. Hot flannel to the lower part of the body, with dry heat in the form of bottles of hot water, will fulfill this indication.

In older children when the pulse is frequent, I give small doses of Aconite to give strength to the circulation. It will be observed that as the pulse becomes less frequent under its influence, it attains strength and volume. In some cases the dullness and hebetude, tending to coma, will be an unfavorable symptom. Here I would add Belladonna, using it in doses heretofore named.

Other remedies may be indicated, for the disease presents many phases. I have-seen cases where the *hot* sponge bath, stimulating the skin, or even where· the addition of a small portion of hydrochlorate of ammonia to the *hot* water was an important means. Others are markedly benefited by the quinine inunction. We always have the therapeutic maxim clearly before us, " that without reference to the·name of the disease, whichever remedy is indicated is to be employed."

ACUTE BRONCHITIS.

Acute bronchitis is of frequent occurrence in childhood, and forms the larger part of those cases which are known as *lung fever*, and *winter fever*. In general it does not present those acute features that are noticed in the sthenic bronchitis of the adult, but many cases will be quite severe.

CAUSES.—Inflammation of the bronchial tubes arises from the same causes that produce inflammation in other tissues. Exposure to cold, especially to the north or east wind, sudden changes of temperature, etc., are the common exciting causes. The predisposing causes are all such as impair the general health; and as a rule, it may be stated that such causes will not produce inflammation unless the vitality or resisting power of the system is lowered from other causes.

PATHOLOGY.—The inflammation is confined to the bronchial mucous membrane, and in some cases involves but a portion of it, but in severe cases it extends to all the bronchial tubes. At first there is determination of blood, and the tissue is injected, swollen, and dry, but afterward there is impairment of the circulation and increased secretion. When children die of the disease, post-mortem examination shows the mucous membrane thickened and softened, and the bronchial tubes more or less clogged with accumulated mucus and muco-pus.

SYMPTOMS.—For a day or two the child has symptoms of cold, with some cough. This is succeeded by a chill, sometimes scarcely noticed by the mother, but lasting from one to four hours. Following this, febrile reaction comes up briskly; the surface becomes hot and dry, the pulse is hard and frequent, and the child irritable and restless; the urine is scanty and the bowels constipated, the tongue having a pretty uniform white coat. The fever gradually increases to the third or fourth day, remains stationary for from one to four days, and then declines as the inflammation passes off.

With the first coming on of febrile reaction the attention will be attracted by the irritative cough, that repeats itself every few minutes. It is not yet the full bronchial cough, that is so marked a feature after the third day, but is half way between this and the hacking cough of pneumonia.

The cough is dry and the respiration dry and whistling, and when the ear is applied over the chest, there is a marked sibilance or dry blowing sound.

By the end of the second day we notice that there is slight secretion of a transparent, tenacious white mucus, streaked with blood. This is increased the third day, and by the fourth it commences to assume a yellowish, opaque appearance. Up to this time the secretion of mucus seems rather to increase the cough, as it is a source of irritation and is raised with difficulty.

When secretion is fully established we have a moist, blowing sound, or mucous rhoncus, which is very marked. After the sixth day, the mucus becoming yellow and opaque, it is raised with less effort, and the cough is not so hard or so frequent, and respiration is much easier. From this time there is a gradual decline in all the symptoms, and the patient is convalescent from the seventh to the fourteenth day of the disease.

PROGNOSIS.—Though a severe disease, we do not look upon it as a fatal one, though occasionally from its intensity it becomes difficult to manage. If secretion commences, becomes opaque, easily expectorated, with an abatement of the fever, the case is progressing well; but if symptoms of imperfect depuration of the blood are developed, with delirium, the case is a grave one. During the disease, if the sputa changes from an opaque to a glairy white mucus, we may be satisfied that the inflammation is redeveloped in its original intensity.

TREATMENT.—The treatment will be directed to lessening the frequency of the pulse and temperature, relieving the irritation and stopping determination of blood to the bronchiæ and lungs, relieving irritation or other wrongs of the nerve centers, and establishing secretion. It is well to have the subject thus clearly in mind that we may be able to make our diagnosis and select our remedies.

Quite as essential is it that we know what *not* to do, and what to keep friends from doing. We do not propose that the stomach or bowels of our patient shall be irritated by cathartics or nauseants, and we object to the teas and potions that friends are inclined to give or insist upon our giving.

To control the pulse we administer Aconite, if the pulse is small and frequent (Aconite is the child's sedative), Veratrum if it has volume and is frequent. They are both good remedies, and relieve irritation of the lungs and bronchiæ as well as the general irritation which gives the frequency of pulse.

If the patient suffers pain, (this may be known by the expression of the face and the cry,) Bryonia should be added to the sedative. If the patient has a persistent cough with whistling in the smaller bronchiæ, and further on, if there are blowing sounds with free secretion, Ipecac should be employed. If there is an oppressed respiration, with abundant mucous secretion after some days, Lobelia is the remedy. A very hot skin, but seeming almost on the point of breaking out into a perspiration, and a full, oppressed pulse, calls for Eupatorium perf.

The remedies that relieve wrongs of the nerve centers are

suggested by symptoms that force themselves upon our notice. The extreme restlessness, sensitiveness to impressions, evidences of pain, sleeplessness, with flushed face, bright eyes, contracted pupils, and increased heat of the head, call for Gelseminum. The same restlessness and irritation, but with contraction of the frontal muscles, sharp pulse, and reddened papillæ at tip of the tongue, call for Rhus. The dullness and inclination to sleep, want Belladonna.

In addition to the remedies which influence the pulse, and relieve irritation of the nerve centers, which also lessen the temperature, we employ baths to lessen the heat of the body and improve the functional activity of the skin. The bath may be soap and water, the alkaline bath (bicarbonate of potash is the best), the acid bath, or fatty inunction. The temperature of the bath is to be determined by the condition of the patient, as heretofore named.

If the temperature is high, and inflammation active, a cold pack may occasionally be used with great advantage—even an entire wet-sheet pack is a good thing. In the majority of cases, however, a hot pack is best, and may consist of three or four thicknesses of flannel wrung out of hot water, and applied to the chest over a single thickness of flannel which is allowed to remain. Sometimes sponging the chest with hot water for a few minutes, and then covering it with flannel, exerts a good influence. My favorite local application, however, and one that is always safe, is the soft cloth spread with lard, and sprinkled with the Compound Powder of Capsicum. This may be renewed twice a day, scraping the old off with a knife and respreading.

If the irritation of the bronchiæ is so great as to cause much difficulty of respiration, or if the cough should be so harassing as to prevent sleep and aggravate the fever, we may relieve the patient by the use of inhalations. The steam atomizer is an admirable apparatus in this case, but if not at hand we will improvise an inhaling apparatus out of a half gallon cup and a hot iron. The vapor of water alone is sufficient in most cases; or in some we may use an infusion of hops, of tansy, of poppy heads, or a small portion of tincture of opium may be added. The inhalation may be repeated as often as every one, two or three hours, until the patient has relief.

The plan laid down I think much better than the old method by the use of nauseant expectorants, and while a case might occasionally arise in the adult, where I would use the old method, I do not think I would employ it in the child.

CHRONIC BRONCHITIS.

We occasionally meet with a case that may be classified as chronic bronchitis, though it does not present all the symptoms of that disease in the adult. Like the disease in the adult, it is associated with an impairment of the general health, and may terminate in infantile phthisis.

CAUSES.—The cause of this bronchial disease in the child is the frequent repetition of bronchial catarrh from cold. At first the patient recovers from the attack well, but as it continues to be repeated the recovery is tardy, and less and less perfect, until finally we have the condition of the respiratory organs under consideration. In other cases the disease is the sequel of measles.

PATHOLOGY.—The disease is not strictly a chronic inflammation, but rather a condition of the mucous membrane resembling that produced by such inflammation. The nutrition of the mucous tissue is impaired, the circulation is sluggish, the tissue thickened, and there is an increased mucous secretion which clogs up the bronchial tubes. In the severer cases, the secretion is muco-pus, and may be in considerable quantity and quite offensive.

SYMPTOMS.—The child has lost flesh and strength, and has a pale or sallow appearance. The pulse is feeble, the extremities become cold easily, the appetite and digestion are impaired, the bowels irregular, and the child is fretful and peevish, as we should expect under the circumstances.

Our attention is called to a bad cough, from which the child has suffered some weeks, and which has been very intractable. We notice that it is loose and somewhat hollow, and that more or less mucous is brought up by the act of coughing, but generally swallowed. Applying the ear to the chest, we hear a very marked moist blowing sound—indeed, when secretion is free, it had better be called a *gurgling* or rattling—as the air passes into and out of the lungs.

As the disease progresses, the child becomes feebler, its appetite and digestion more impaired, and the cough worse. It has chills, followed by hectic fever, during which it is irritable and restless, and finally the strength is so exhausted that it dies of inanition.

TREATMENT.—The treatment of this case requires considerable care, and all harsh and unpleasant remedies should be avoided. There are two principal objects in view—to modify the irritation and lessen the cough; and to improve the appetite and digestion, and thus get better nutrition and restore the structure of the part affected.

If both can be accomplished by one class of remedies it will be best, as the child's stomach is easily irritated, and will not tolerate too much medicine. Frequently we will attain the object by the use of Colliusonia alternated with tincture of muriate of iron and glycerine; it may be prescribed in the following form: ℞ Tincture of Collinsonia ℥ij., Simple Syrup ℥ij.; a half teaspoonful three times a day. The preparation of iron may sometimes be changed for the compound syrup of the hypophosphites.

Occasionally we may require something more for the relief of the cough, in the earlier part of the treatment. Here I like the action of an infusion of clover hay, sweetened and given as required. Or in place ·of this the Stillingia Liniment may be given in doses of half to one drop on sugar, every three or four hours. When the disease has arisen from measles, the Drosera will be found a valuable remedy, in the proportion of gtt. x. to gtt. xv. of the tincture to water ℥iv.; a teaspoonful four times a day.

In some cases Veratrum and Bryonia seem to exert an excellent influence in quieting bronchial irritation, checking the cough, and improving the general health. In other cases we employ Arsenic, as—℞ Tinct. Veratrum, Fowler's Solution of Arsenic, āā. gtt. v., water ℥iv.; a teaspoonful every two or three hours. Rumex is sometimes an excellent remedy for cough, gtt. v. to gtt. x. being added to water ℥iv. and given in teaspoonful doses. If there is a very free secretion of mucus, with much rattling in the chest, Lobelia with Compound Tincture of Lavender may be employed, as heretofore named. For further information, the reader is referred to my *Practice of Medicine*, article "Cough."

Inunction with quinine is an excellent measure in those cases in which there is feeble innervation or impaired circulation. Generally fatty inunction serves a better purpose than the ordinary use of baths; the surface being rubbed freely with the hand or with flannel.

Exercise in the open air, the free admission of light and sun-shine to the room, and a nutritious diet, are important aids to the cure.

PNEUMONIA.

Inflammation of the lungs is met with most frequently during the winter months, and like acute bronchitis it forms a portion of those cases known as lung or winter fever. In some localities it is a very common disease, but in others it is seldom met with. From what I have noticed I should judge it to be most common in localities exposed to north and east winds, or indeed to any winds, and rare in sheltered places, as by woods, hills, etc. Thus in Cincinnati, away from the river, it is rarely met with, but on high grounds we occasionally see cases. On the prairies of Illinois, and in Northern Indiana, Ohio, and New York, it is of frequent occurrence.

CAUSES.—The common cause of inflammation of the lungs in children is cold, either from direct exposure, or from sudden alternations af temperature.

PATHOLOGY.—In this case the inflammation is of the parenchyma of the lung. In some it will progress as in the adult, embracing all of that part of the lung which is the seat of inflammation. But in the majority of cases it is confined to the lobules, and is thus scattered through a large portion of one or both lungs, sound lobules being interspersed between the inflamed ones. From this character of the disease it has received the name of *lobular pneumonia*.

For the first two or three days the lung presents the appearance of determination of blood, the capillaries being full, and the tissue consequently reddened; the lung still floats on water, though not so freely as the healthy tissue. From this time there is a continued increase in its density, the air-cells being effaced by the engorgement of the capillary vessels, and by exudation into the intercapillary spaces, and into the air-cells. This is the stage of *red hepatization*, and is fully completed by the sixth day. Now, in the majority of cases, resolution commences, and the circulation is gradually restored from the circumference to the center, the effused material being taken up as soon as the blood begins to flow freely. Thus the lung is free of both the arrested circulation and the effusion, and becomes again permeable to air. Resolution is usually complete by the ninth day.

But should the inflammation run so high as to impair the vitality of the tissues, then the condition of *gray hepatization* may ensue. In this the effused material is formed into pus, and in a still further advanced stage, the tissues losing their vitality, soften and are likewise transformed. If we examine a lung in this condition, we will find it soft and friable, and readily lacerated by the fingers.

The cut surface presents a yellow mottled appearance, and when pressure is made upon it an imperfect purulent fluid exudes. In a still further advanced stage small abscesses are formed in the lungs at the site of the diseased lobules.

Symptoms.—Usually the child shows symptoms of cold for two or three days before the attack, and is restless and fretful. In many cases the commencement of the inflammation is announced by a persistent hacking cough, which seems to arise from irritation of the throat or larynx. What is peculiar about this cough is, that it seems to be made by the child, and frequently it will be scolded for persisting to cough when it seems to be unnecessary. Such a cough is always annoying to the persons in the room for its persistency. Of course there are many cases in which an ordinary cough follows the invasion of the inflammation, and does not precede it as described.

In all cases there is a chill, but in most it is so mild that it is scarcely noticed by the parents. In the majority, febrile reaction comes up slowly, and for the first twenty-four hours the child does not seem to have more fever than would attend an ordinary *bad* cold, but even in those cases it will be noticed that the cheeks are flushed, that there is too great dryness of skin, and sharpness and hardness of the pulse, for a slight disease. In other cases the fever comes up rapidly, and runs very high.

By the third day of the disease, a casual observer would notice that the child was quite sick. The skin is hot and dry, the pulse frequent and hard, urine scanty, bowels constipated, no appetite, a coated tongue, one or both cheeks flushed, and some difficulty in respiration. The cough is usually paroxysmal, and varies from the hacking cough first named to a deep bronchial cough. In some cases the child is irritable and restless, but in most lies quite still, even though it suffers, and sleeps with its eyes partly open.

With these symptoms but little changed, the disease continues
to the sixth day, when, in a majority of cases, we find an amendment. The skin is less dry and hot, the pulse softens and is
less frequent, secretion is gradually established, the difficulty of
respiration and cough passes away, and by the ninth day the
child is free from fever and convalescent.

When it passes into the stage of gray hepatization, the prostration becomes alarming. The mind wanders, and in a short
time coma comes on, and, gradually deepening, is attended with
evidences of imperfect aeration of the blood, from which the
child dies.

Occasionally a very severe case is met with, in which evidences of asphyxia are manifested as red hepatization advances,
and in which the disease will terminate fatally before the sixth
day, on account of the extreme engorgement. And in very
feeble children there is occasionally a case in which the determination of blood is followed by congestion, and such engorgement of the lungs is produced in a few hours, that death results.
This resembles the pulmonary apoplexy of the adult.

DIAGNOSIS.—The symptoms above named, cough and difficult
respiration, point to the lungs as the seat of disease, while the
febrile reaction evidences its inflammatory character. Auscultation during the first three days detects a marked change in the
respiratory murmur, which is roughened and louder, and associated with a marked *blowing* sound, at first dry, afterward moist.
From the third to the sixth day the respiratory murmur becomes
less and less marked, until at last it is scarcely heard over the
seat of inflammation, but is replaced by a moist blowing sound
(mucous rhoncus). With this change there is increasing dullness
on percussion, though it is rarely so marked as in the adult.

We have less evidence of diseased action from the sputa than
in the adult, for children can rarely be induced to spit out the
secretion removed by coughing. At first there is dryness of the
mucous membrane; in two or three days the secretion is white,
tenacious, and has the globular form so characteristic of the disease in the adult ; about the fifth day it acquires a reddish-brown
tinge, but rarely has the rusty hue that is seen in the adult.
After this time, if convalescence progresses, it becomes yellowish,
and resembles the sputa of bronchitis.

Prognosis.—The prognosis in the pneumonia of children is favorable, the mortality being but little increased over remittent fever. This, of course, is supposing that the patient has proper treatment and good nursing. Those cases that manifest symptoms of asphyxia early, are to be regarded as unfavorable, as are those in which the disease progresses unabated beyond the seventh to the ninth day.

Treatment.—The treatment of pneumonia in the child should be quite simple, as it is not a case that will be benefited by or bear much medication. The indications are plain—to lessen the frequency and obtain a uniform circulation of the blood, to allay irritation of the lungs, and to establish secretion, and this is done in the order in which they are named.

To accomplish the first object we prescribe the special sedatives, Veratrum and Aconite, in the usual doses, repeated every hour. If the fever runs high, and the pulse is full and hard, the dose of the Veratrum may be doubled until it influences the pulse. All we expect of the remedy is that it will hold the febrile action and the inflammation in check, and after twenty-four or forty-eight hours will cause a gradual decline in both. That action of Veratrum which brings the pulse down in six or eight hours from one hundred and thirty beats per minute to eighty or seventy beats per minute, is not desirable, because it can not be maintained, and because it is opposed to nature's methods of cure.

Again, we note the indications for other remedies which may be added to the sedative solution or alternated with it. Bryonia is given if the child shows evidences of pain in respiration or on coughing; Rhus, if there is contraction of the facial muscles, and a sharp stroke to pulse, which is small and frequent; Lobelia, if there is marked oppression in the chest, and difficult breathing; Gelseminum, if there are evidences of determination of blood to the brain; Belladonna, if the child is dull and inclined to sleep a great deal; Eupatorium, if the pulse is full and oppressed, and perspiration is caused by coughing, notwithstanding the heat of skin; Phytolacca, if patient complains of sore mouth and throat; Baptisia, if the face is full and purplish, the tongue and lips showing the same color.

When the tongue is dirty—a yellowish, glutinous coat—sulphurous acid exerts an excellent influence, and will sometimes be

nearly all that the patient will require. Occasionally the sulphite of soda is indicated, and more rarely the chlorate of potash. A trituration of podophyllin (second decimal or first centessimal), in doses short of an action upon the bowels, will give
good results in those cases showing fullness of face and tissues
generally, and especially fullness of veins, which should be noted.

Cathartics as a rule are contra-indicated, and if the bowels are
locked up, it is much better to use an enema to open them. Two,
three or four days without an evacuation of the bowels, is not
commonly injurious, at least it is much better than the irritated
stomach and intestinal canal from physic.

Ipecacuanha exerts a specific action upon the respiratory apparatus, removing irritation and stopping determination of blood.
It has been used alone with marked success, being given triturated with sugar in doses just short of nausea, or with slight
nausea for a few hours and just short of vomiting. I employ
it in tincture and combine it with Aconite gtt. v. to gtt. x., to
water ʒiv.

As the disease advances, if there be a very free secretion, with
much rattling in the chest, and difficulty of breathing from
obstructed bronchial tubes, the preparation of Lobelia with
Comp. Tinct. of Lavender, as named under the head of asthenic
laryngitis, will answer a most excellent purpose. In this case
the cloth spread with lard and sprinkled with Compound Powder
of Lobelia should not be neglected.

To fulfill the second indication we apply a hot fomentation as
heretofore named, or the " mush jacket ; " or in young children,
where this would be inconvenient, the cotton cloth spread with
lard and sprinkled lightly with emetic powder. The mush jacket
is made of corn mush of ordinary consistence, spread an inch
thick on cloth large enough to cover the anterior and lateral
portions of the chest, and covered with musquito-bar, netting,
or other thin goods, to retain the poultice. It is applied hot,
a thinkness of flannel being first laid over the chest, and changed
twice daily ; the heat is retained by the thickness of the poultice,
and by keeping the body warmly covered. The hot application
should not be continued longer than the first day, and then if
needed replaced by the Lobelia and Capsicum.

In all cases the air of the room should be kept moist, and if
the irritation of the lungs is great, with harassing cough, we
may obtain much advantage from the use of inhalations of the

vapor of water, of an infusion of hops, or tansy, or poppy heads, or of one part of vinegar to three of water. This not only gives present relief, but it aids the cure; indeed, I am not certain but that the disease could be very successfully treated by the use of inhalations alone, adding the sedative to the water employed.

As a general rule, we will find the third indications of cure accomplished by the means already named. As soon as the pulse is brought under the influence of the sedatives there is a natural tendency to a re-establishment of secretion. At the commencement of the treatment the patient has a general bath with soap and water to cleanse the skin, and a hot foot-bath for its quieting influence upon the nervous system. Further than this, I do not think bathing is beneficial, except sponging the surface under the cover, drying speedily and with friction.

The child should be kept in the recumbent position in its cradle, crib, or bed, and should be kept pleasantly warm and free from all draughts and changes of temperature. If at any time during the progress of the disease, there is tendency to coldness of the extremities, apply dry heat, wrapping the feet and legs in flannel.

If the child needs a stimulant and tonic, I think we may obtain the best results from the inunction of quinine, as heretofore named.

When the disease passes into the stage of gray hepatization, we can do nothing more than support the strength by the use of the bitter tonics, as they can be given, and such food as we can prevail upon the child to take. In some cases advantage might result from the use of sulphite of soda.

PHTHISIS PULMONALIS.

Infantile phthisis is generally acute, resembling the hasty consumption of the adult. In this form we meet with it at all ages from the first weeks of life to the age of five or six years. After this time, phthisis pulmonalis is of very rare occurrence, until the age of puberty is past.

CAUSES.—The predisposing cause is a feeble vitality, manifested especially in the formation of blood and nutrition of tissue. In this sense infantile phthisis is always hereditary, as it is hardly possible that a child possessing normal vital or formative power should have this disease, even under the most adverse circumstances.

The exciting causes are all such as still further lower the vitality of the child. Diseases affecting the stomach and bowels, thus impairing digestion, or those which impair secretion, lower the vitality of the blood and its adaptation as a nutritive fluid. If now the child is exposed to the ordinary causes of cold, and irritation of the respiratory organs is set up with determination of blood, the imperfectly formed albumen will be thrown out in the parenchyma of the lungs as tubercle.

PATHOLOGY.—As above stated, a congenital defect of viability lies at the foundation of the disease, and as has been previously stated, it is impossible for some children to live to adult years for this reason. This feeble viability is manifested in impairment of all the functions of the body, but especially in the formation of poor blood. This imperfect elaboration furnishes the material for tubercles, which only requires an arrest of secretion by which it is ordinarily removed, and a determination of blood to the lungs for its deposit in that tissue.

Post-mortem examination shows very extensive disease. In the early stage the lung is infiltrated with a yellow albuminoid material of cheesy consistence, and showing but very slight, if any, traces of organization. It resembles the lower form of yellow tubercle, as seen in the adult. This exudation does not wholly prevent the entrance of air into the part diseased, for even when the deposit is extensive, some lobules will be found free.

At a further advanced period we will find this material in every stage of breaking down, from that in which it has softened to the ordinary consistence of pus, to that in which the tissue of the lung has yielded, giving rise to large and ragged abscesses, containing the debris of the lung and the tubercular material.

SYMPTOMS.—The disease frequently commences as a simple cold, manifested by coryza, slight cough, seemingly from irritation of the throat, impaired appetite, chilly sensations in the morning, with febrile exacerbations in the after part of the day. The child is restless and fretful, and does not sleep well during the day, at its usual times. These symptoms continue for a week or ten days, the coryza disappearing, but the cough increases, the chills are more marked, and the febrile reaction higher.

Prostration becomes marked after a few days, and the appetite is almost wholly lost. The patient is thirsty, and drinks often, especially at night, waking every hour or two. Its sleep is also disturbed, and it changes its position often, throwing off the cover, and frequently expressing its uneasiness in moaning and short cries.

Thus week after week we observe an increasing loss of flesh and strength, until the child is as much emaciated as in cholera infantum. And yet the cough is not very severe, neither is there much difficulty of respiration, or imperfection in the aeration of the blood. Still, these symptoms are sufficient to call our attention to the lungs as the seat of the disease. Quite frequently the *blowing* sound will be so marked that it is readily heard when we sit down by the side of the child. Upon auscultation it is found to extend to every part of the lung tissues.

DIAGNOSIS.—The marked emaciation and debility is evidence that the disease, whatever it may be, is one affecting the formation of blood and nutrition of tissue. If in the summer, we would suspect it to be one of the intestinal canal, but a careful examination determines that this is not the source of the lesion. The cough, some difficulty in respiration, and the blowing or rattling sounds heard when near the child, cause us to make an examination of the chest.

On percussion we usually find a dullness over the superior lobes of the lungs, sometimes quite marked. But over the middle and lower lobes the resonance is clear, except in cases with congestion, when there is some dullness even here.

On auscultation we find a uniform blowing sound over the entire chest, with more or less mucous rattling, as there is more or less secretion. Though it seems louder in the sound lung, it does not mask the respiratory murmur which is heard distinctly. In those parts of the lung infiltrated with tubercle, the blowing sound is dull, as if deadened by passing through an infiltrated tissue, or the sounds are small, whistling, creeping in and out of the tissues, sometimes like a nest of mice. Not unfrequently the *dry-crackling* that so generally attends the same disease of the adult will be heard, though but for short periods.

PROGNOSIS.—The prognosis is unfavorable. The disease occurs in children of feeble viability, and who have little power to resist disease or of repair when this is arrested. Still, in some cases, an infantile phthisis may be arrested, and by a proper restorative treatment, aided by good hygiene, a moderately good constitution may be developed in time.

TREATMENT.—The treatment of this case can only be outlined, as the indications for remedies will vary, and here, as elsewhere, we wish the right remedy for the conditions present, rather than a treatment for the name of the disease.

With a frequent pulse and increased temperature, we think of Veratrum as a good basis of treatment. If there is pain in the chest, Bryonia is given for a few days, and is then replaced with Arsenic if the skin shows atony; continuing in this way for a time, these remedies may be replaced with hypophosphite of lime alone, or with some preparation of malt given after meals.

We can not, at first, use bitter tonics and restoratives by mouth with advantage, and we will therefore employ them endermically. The inunction of quinine, heretofore spoken of, will answer a very excellent purpose. Even simple fatty inunction will be found beneficial. If the parents can afford it as well as not, let the inunction be of good cod-liver oil, with quinine in the proportion of ʒij. to the pint. Let the entire surface be thoroughly rubbed with this once or twice daily, using considerable friction. No exposure of the surface should be permitted at these times, but the child kept carefully covered. The fatty matter may be removed from the skin with a soft flannel.

Washing with water is absolutely prohibited, except the face and hands.

The same directions for diet should be given as in a case of cholera infantum, and if there is irritation of the stomach and bowels, the same remedies may be used to relieve it.

As the child improves, we may put it upon the use of tincture of muriate of iron and glycerine, as heretofore named, and sometimes we may add a small portion of the tincture of Nux Vomica to it.

Probably a better restorative in this case will be found in the compound syrup of the hypophosphites, which is not only very pleasant to the taste, but is easily appropriated. The dose for a child two years old will be one-third of a teaspoonful four times a day.

Cod-liver oil answers a very good purpose when it can be taken. But with the majority of children under six, we will find it impossible to give it.

As the child becomes stronger, it should be taken out of doors when the weather is warm, and the air calm; and it should also be persuaded to use its limbs, and take moderate exercise. When it can not be taken out of doors, its chair or crib should be so placed that it may receive abundance of light, and even sunshine, if not too bright.

CHAPTER VII.

DISEASES OF THE DIGESTIVE APPARATUS.

THE digestive apparatus embraces the mouth, the throat, the stomach, and small and large intestines. The diseases we have to study are those of the mucous membrane of these several parts, and of their functions. We take it for granted that diseases of mucous structure will demand a somewhat similar treatment, no matter what its location. If this view is correct, we will have simplified the therapeutics of this class of diseases, and will also have made it more definite.

DENTITION, AND ITS DERANGEMENTS.

Dentition is usually regarded as the great trial of the health, if not of the life, of the child, and we find parents and even physicians speaking of it as if it were a morbid process, instead of a physiological development. It is true that during the cutting of the teeth there is occasional disturbance, sometimes of one function, sometimes of another, but more frequently of the digestive tract. But the majority of these are but indirectly influenced by dentition, and the cause should be sought elsewhere.

The error is an important one, because it obscures real causes of disease, and prevents the direction of remedial agents for their removal. And with some persons it leads to interference with the natural eruption of the teeth, which proves injurious.

As a general rule, the child exhibits evidences of buccal excitement about the fourth or fifth month. There is an increase of the salivary secretion, and it has a strong desire to bite or press its gums strongly against whatever is given it.

19

This excitement continues to the eruption of the first teeth--
the front incisors—which occurs from the sixth to the eighth
month.   In some cases, the excitement becomes general, and
there is restlessness and irritability of the nervous system, fever,
derangement of digestion, and sometimes diarrhœa.

The treatment of such a case is very simple.   The child has a
bath, and once or twice a day a hot foot-bath.   We prescribe
Aconite in the usual doses, and if there is much derangement of
the stomach and bowels, small doses of an infusion of compound
powder of rhubarb and potash, to produce a gentle laxative influence upon the bowels.

Usually when dentition is attended with such derangements,
the subsequent cutting of the teeth will be attended with more
irritation than the first.   Especially is this the case in feeble
children, at the eruption of the molars and canine teeth, during
the second summer.   But these cases need but the treatment
just named, unless there is a distinct complication, which will
be treated as noticed hereafter.

LANCING THE GUMS.—Lancing the gums is an old practice,
and is now somewhat out of vogue.   When the lancet was the
remedy for adult fevers and inflammation, lancing the gums was
the remedy for all irritations supposed to arise from teething.
Its use, therefore, while sometimes beneficial, was frequently
injurious.

The eruption of the teeth through the gums is not a sudden
nor a forcible process.   As the teeth attain increased size they
press upon the gum, which is as gradually absorbed before them,
and in the true physiological process there is no disturbance
whatever.   The lesions we notice result from the deterioration
of the child, induced by our abnormal civilization.

The cases in which the gums may be cut to advantage are
those in which there being the lesions of the nervous system and
fever, that have been already noticed, there is marked swelling
and lividity or duskiness of the gums, over the point where the
tooth is to make its appearance.   In such cases, occasionally,
prompt relief follows the incision of the gum, and the relief of
its tension by the discharge of a small quantity of blood.   It
seems evident that it is the discharge of the blood which relieves
the tension upon the parts, and in the same degree the irritation,
and not the coming through of the tooth, as many persons think.

The tooth is cut none the quicker for the lancing of the gum. It has to come through by a process of growth, and this is a matter of time, so that days, and sometimes weeks, will elapse before it has fairly made its appearance.

In lancing the gums a sharp knife, bistoury, or gum lancet, should be used, and the incision should be directly down upon the crown of the tooth. The lancet should not be cut against the tooth, however, so as to injure it, as it is yet delicate.

Hemorrhage after lancing a tooth sometimes occurs, and I have seen two cases in which it proved fatal. Of course, these were exceptional cases, the children being of a hemorrhagic diathesis. Still, free and somewhat persistent bleeding is not uncommon. Occasionally the application of common salt or wood soot will arrest it, or a portion of spider's web held on the incision for a few minutes with the fingers will answer the purpose. The most certain means of arresting hemorrhage in any such operation about the mouth, is the application of tanner's shavings to the part. It seems something like the crude medication of centuries ago to recommend tanner's shavings or a cobweb, when our materia medica is so full of astringents; but a rather troublesome experience has taught me that such means are not to be rejected, when found beneficial. In my own person I had hemorrhage for four days, from the extraction of a molar tooth, until, from loss of blood, I could not maintain the erect position. All the ordinary means had been tried, without any permanent effect, including the persistent use of persulphate of iron. The part was cleansed of clots, and while bleeding freely, the tanner's shavings were applied, and in fifteen minutes it was wholly arrested.

TAKING CARE OF THE TEETH.—There are but few parents who everthink of caring for the teeth of the child. They know they have to serve but a short time, and then be replaced with permanent ones. These milk teeth, however, serve just as important a purpose with the child as with the adult. Their loss, I am satisfied, tends to injure the health of the child, as it impairs mastication, insalivation and digestion. Many a case of dyspepsia of the child three or four years old may be traced to this cause.

As a means of preserving the teeth, I would recommend that the child be taught to wash the mouth with cold water, or water to which a small quantity of salt has been added. The decay of the teeth is owing in part to the decomposition of food between and adherent to the teeth, and if the mouth is thus cleansed after eating, this cause, at least, will be removed.

The first permanent molars make their appearance about the third or four year, and are very commonly mistaken for temporary teeth. These not unfrequently become carious, from the cause above named, from the sixth to the eighth year. When this is the case, advise that the child be taken to a dentist, and have the teeth properly filled.

SECOND DENTITION.—The loss of the deciduous or milk teeth commences between the fifth and the sixth year—or should commence at that time. There is a gradual absorption of the fangs of the teeth, as the permanent ones are developed beneath them. After a time they become loosened, so that sometimes they can readily be pulled out with the fingers. If not removed in time, the permanent teeth press them to one side, or forward, or backward. This gives rise to present deformity, disease of the gum, and may so change the position of the permanent teeth as to occasion lasting deformity. When it is noticed that they have become loosened, or that the permanent teeth are coming through, they should be extracted. This is but a little matter, as the teeth have mostly lost their fangs.

I do not like the extraction of the deciduous teeth before this time, not only because the child loses the use of them, but because the jaw-bone is sometimes imperfectly developed, in consequence, and we have irregularity of the permanent set.

It is of importance that the permanent set of teeth be well-formed and properly placed, so as to form a regular and uniform arch. It is not only a matter of good looks, but also of good teeth.

It is my impression that if the general health of the patient is good, and it is furnished with the material for the formation of bone, in such form as it can appropriate it, the teeth will be regular and good.

If a patient is presented to me showing an irregular or deformed dentition, I prescribe for the general disease, as if manifested by other symptoms of imperfect nutrition, the usual proportion of iron for children : R Tincture of Muriate of Iron 3j., Glycerine 3iv., a teaspoonful three times a day. The hypophosphite of lime, or more frequently phosphate of soda, in small quantities, furnishes the phosphorus; while a nutritious diet, properly selected, (oat-meal is very good, so is cracked wheat,) and such hygienic means as seem to be necessary in the case, complete the treatment.

When the teeth are developed irregularly, as regards the arch, or overlap one another, or are angular, as it were twisted upon their axis, much may sometimes be done by simple means. When a tooth is outside or inside of the line of the arch, pressure with the thumbs or fingers toward the proper place two or three times a day, will usually accomplish the object. Of course, in these as well as other irregularities, the deciduous teeth should be removed. Occasionally we will find a fang of one of these teeth overgrown by the gum so that it is not noticed. A careful examination discovers it, and being removed, the cause of the deformity being gone, we find the teeth speedily assuming their proper position.

The physician may instruct the parents as regards such manipulation, who will pursue it for some months, as the change is a very slow one. When the deformity is greater than we can expect to overcome by such means, we recommend that a dentist be consulted, who may apply such artificial support or pressure as will accomplish the object.

STOMATITIS.

Inflammation of the mouth in childhood assumes a number of different forms, which really differ in pathology and treatment. In some of them the sore mouth is the primary disease; in others it is secondary and symptomatic of some other affection. They range from a simple erythematous inflammation, which disappears of itself in a day or two, to that profound lesion in which the tissues soften and break down as it progresses.

We may divide these diseases into the following classes: Stomatitis Simplex, Aphthæ, Stomatitis Ulcerata, Gangrenous Stomatitis.

STOMATITIS SIMPLEX.

Simple sore mouth may be caused by cold, or by gastric irritation, in some cases being the result of the two. The child first complains of its food burning or smarting the mouth, or if too young to make its sufferings known, it will be noticed that it cries when it nurses. On examination the mouth will be found reddened and hot.

Frequently the inflammation is so slight that it will pass away of itself in two or three days; but in other cases, especially those which are dependent upon gastric disease, it will be remarkably persistent.

TREATMENT.—In a large number of cases, Phytolacca is a very certain and speedy remedy, and I prefer its internal use to "mouth washes." As there is usually slight fever, the prescription will be—℞ Tinct. Aconite gtt. ij. to gtt. v., Phytolacca gtt. v. to gtt. x., water ℥iv.; a teaspoonful every hour. Occasionally the indications will point distinctly to Rhus, when it may replace the Aconite.

The local treatment is quite simple. An infusion of sage, sweetened with honey or loaf sugar, and borax added in the proportion of one drachm to four ounces answers very well. An infusion of Hydrastis, of Coptis, or of Hamamelis, will also answer a good purpose. Or, instead of these, a solution of chlorate of potash, gr. x. to water ℥iv., may be employed.

In those cases which depend upon gastric irritation, ipecac may be added to internal remedies, or an infusion of compound powder of rhubarb and potash may be given until it has a slight laxative effect. If the case is a persistent one, this may be followed by the subnitrate of bismuth, or liquor bismuth. In other cases benefit will result from the use of small doses of tincture of muriate of iron with glycerine.

APHTHÆ.

Aphthæ occurs in three conditions of the system : as a simple sore mouth with exudation from gastric irritation, in cachectic conditions of the system, of which it may be an evidence, and as an attendant upon severe forms of disease, being met with especially in their later stages when the vital power is nearly exhausted.

In the first case the aphthous sore mouth yields readily to treatment, and is never regarded as presenting any danger. In the other two it is a very grave symptom, indicating a depression of the power to live, which renders the prognosis unfavorable.

SYMPTOMS.—The symptoms of aphthæ are well marked. The child exhibits evidences of a sore mouth, and complains of tenderness, smarting or burning, when eating ; or, if nursing, it cries when it takes the breast, and sometimes refuses it except when pressed by hunger. As it occurs in very young children, this will be the principal evidence of the seat of the disease. The child is fretful, uneasy, does not sleep well, has considerable fever, and cries persistently when it is given the breast, and frequently lets it go, bursting out into a prolonged and painful cry.

Our attention being thus attracted to the mouth as the seat of a disease, we find the mucous membrane somewhat swollen, reddened, especially in patches, upon which there are numerous small white points of exudation. The mouth is quite tender to the touch, and the child cries when it is examined. As the disease progresses these points of exudation become more numerous, until finally the greater part of the buccal surface is thus covered. When it occurs in cachectic children, or in low forms of disease, the exudation occasionally extends to the tongue, covers the roof of the mouth, and sometimes involves the throat and nasal passages.

The evidence of gastric disturbance will frequently be obscured by the sore mouth, but in almost every case there will be found some lesion of this kind. There is also a real febrile action, and not, as some might think, simply a sympathetic disturbance of the circulation.

TREATMENT.—In many cases the small dose of Aconite with Phytolacca will be a very much better treatment than the old plan. If there is no fever the Phytolacca may be given alone, gtt. v. to gtt. x. to water ℥iv.; a teaspoonful every one to three hours. · If the face is full and purplish, and the tongue and mucous membrane of the mouth have a purplish-red hue, Baptisia should be substituted for the Phytolacca. If the tongue is red and covered with a glutinous coating, sulphurous acid may be given in small doses, and also used as a wash for the mouth. Indeed I think that in many cases it gives us the best local application. If there are erythematous spots of eruption on the surface, or small vesicles appear about the mouth, or there are startings in sleep, with contraction of the facial muscles, Rhus should be alternated with Phytolacca. If there has been a retrocession of an eruption, or we think there is an eruption under the skin, the patient will usually be dull and inclined to sleep, and we give Belladonna.

These means I regard as of more importance than the local treatment, for with the removal of the internal disease the aphthæ disappears. The use of Aconite and small doses of sulphite of soda, are the most valuable of these in a majority of cases.

As a local application, an infusion of Baptisia, a solution of chlorate of potash, or sulphite of soda, or sulphurous acid, may be employed. It may be applied with a soft cloth or sponge, care being taken not to rub or irritate the mouth. If the directions are not explicitly given, we will find the mother or nurse trying to rub the white spots off with the wash, or, as she expresses it, she is trying to *clean* the mouth. You do not want the mouth cleaned, but the medicine applied.

When the aphthæ is associated with a cachectic condition, the treatment should be decidedly tonic and restorative, and this should be aided by a nutritious diet and such hygienic means as will improve the general health. When it can be taken, cod-liver oil answers a very good purpose. Alternated with this we may give small doses of tincture of muriate of iron with glycerine. The hypophosphite of soda, twice a day, will be a good addition to the treatment in some cases; or in place of the iron and hypophosphite as named, the compound syrup of the hypophosphites may be used.

The quinia inunction will prove beneficial in these cases. It should be applied with brisk friction, so as to stimulate the skin to absorption.

The food should be principally animal; the lean of beef, mutton, game, milk, eggs, etc., will give a proper diet. Not unfrequently changing the diet from farinaceous slops to well cooked meats, and a plentiful supply of milk and eggs, is better than any medication.

No treatment can be laid down for those cases of aphthæ, which appear in the later stages of acute disease. It evidences the need of restorative medication, and the appropriation of food, but this has been clearly seen before the aphthæ made its appearance. All that can be said is, let every effort be directed to placing the stomach in condition for the appropriation of some food; and by the use of restoratives increase the power of digestion and assimilation, and by the use of that class called antiseptic, lessen the rapidity of decomposition.

STOMATITIS ULCERATA.

This is not a very common form of sore mouth with children, but is sometimes quite severe. It is difficult to determine the cause, and why an ulcerative inflammation should be set up in a person seemingly otherwise healthy. We may surmise some depravation of the blood, but we can give no reason why it should manifest itself in disease of the mucous membrane of the mouth, rather than other mucous membranes, or even other tissues.

SYMPTOMS.—We have in this case the same evidences that the mouth is sore, that have been named before. The child manifests pain when it nurses or when it eats. If old enough to express its sufferings, it complains of burning and heat in the mouth, and a feeling of stiffness. Sometimes its voice will be changed, and it can not articulate distinctly.

On examination the mucous membrane will be found presenting a uniform red blush, but patches will be markedly reddened, usually somewhat discolored, dusky, livid, or blanched, and somewhat swollen. At a later stage of the disease, these swellings will have increased, and a vesicle having formed and ruptured, a superficial ulcer will present. These ulcers will vary in size from the head of a pin to a dime, and in depth from a simple excoriation or removal of epithelium, to a deep excavation which passes through the mucous membrane. The ulcers are covered with a grayish or yellowish exudation, which when removed, leaves a red base.

In some exceptional cases, the disease continues to increase for several days, and the ulcers are numerous and deep. There is free secretion of saliva of a sticky and tenacious character, which, with the secretion from the ulcers, and an increased mucous secretion, is removed with some difficulty, and keep the mouth in a very unpleasant condition.

TREATMENT.—The constitutional treatment in this case is of more importance than the local means. True, in some mild cases, the use of chlorate of potash as a wash, and internally, will be sufficient. When the tongue is broad, pallid and seems swollen, I like the action of sulphite of soda, in doses of two to five grains, every three hours. Where it is red and covered with a glutinous nasty coat, sulphurous acid is the remedy. In other cases the use of tincture of muriate of iron with glycerine may be alternated with it.

The remedies named under the head of aphthæ may find a place here, according to the indications, Phytolacca being especially a good remedy. ·Following this, quinia with hydrastine may be given, or when there are objections to the use of quinia by mouth, it may be employed by inunction. Occasionally, small doses of triturated Podophyllin can be used with advantage, but it must never be carried beyond a slight laxative effect. The cases in which I would deem it necessary, would be those where the tongue was covered uniformly with a yellowish coat, some tumidity of bowels, with papescent and clay-colored stools.

The local applications in this form of sore mouth are, an infusion of the baptisia, sulphurous acid, or sulphite of soda. As already named, the chlorate of potash answers well in the mild cases, but can not be depended upon in those more severe. An infusion of Hamamelis is a good mouth wash in some cases. And in very stubborn ones a decoction of equal parts of Rumex crispus, Alnus serrulata, and Quercus rubra.

Occasionally one or two ulcers will be very persistent, and not yield to the ordinary treatment. Such may be lightly touched with the stick nitrate of silver.

GANGRENOUS STOMATITIS—CANCRUM ORIS.

Cancrum oris may follow stomatitis ulcerata, but in the majority of cases is a distinct disease. There seems to be a depravation of the fluids, and an impairment of the vitality of the solids. It bears a very close relationship to *cynanche maligna,* and presents many of the same symptoms.

SYMPTOMS.—Gangrenous stomatitis commences with a hardness and swelling of the cheek and lips; when this appears externally, it presents a blanched glossy appearance. On examining the mouth we find but little tenderness, the part swollen being but slightly redder than usual, more frequently dusky, livid, or blanched, and having in its center an ash-colored eschar. The tongue is pale and somewhat coated, the stomach and bowels deranged, with marked exhaustion and cachexia, with languor and restlessness.

The eschar soon spreads, sometimes extending to the lips and gums, and is attended with a copious discharge of saliva, which soon becomes fetid; the secretions of the mouth and the breath are quite offensive. As the ulcer progresses it extends both in depth and circumference. Passing through the mucous membrane, it involves the cellular tissue, and frequently dissects the parts for some distance beyond the circle of ulceration. Thus the ulcer is larger, and, in every respect worse, deeper in between the tissues, than it shows in the mouth.

Still progressing, it destroys part after part, until near the surface. Now a small vesicle or pale ashy-colored spot is formed upon the skin, which soon becomes livid and sloughs. The ulceration now spreads rapidly, destroying the muscles, integument, and bones, until death terminates the child's sufferings.

When the ulceration has progressed in this way, recovery occasionally takes place, leaving the child fearfully deformed. In olden times, when mercury was constantly used for all ailments, such cases of stomatitis were not unfrequent. As mercury was the direct cause, these cases received the name of mercurial stomatitis.

In some of these cases of mercurial sore mouth the gums will be swollen, spongy and pale, sometimes bleeding upon slight pressure. A nasty white sordes accumulates about the teeth, which are loosened, and presently we observe a white eschar forming on the edge of the gums, which separating leaves a foul ulcer. The child suffers constantly, can take but little food, and the fetor is so great that it is perceptible in all parts of the room. I have seen just such results from the Homœopathic use of mercury.

DIAGNOSIS.—This disease will usually be readily determined. The child complains of sore mouth, and presents a markedly cachectic appearance. The breath is fetid, the secretions unpleasant, the mucous membranes pallid or livid. Then the circumscribed swelling and hardness as described with the commencing ulceration, will be sufficient.

TREATMENT.—The treatment of this case will be varied to meet the indications, yet we obtain the best results from the use of chlorate of potash and chlorine water, sulphurous acid, sulphite of soda, Phytolacca or Baptisia. In some cases quinine (in malarial regions) is administered internally, but in many the inunction of quinine once a day will answer the best purpose.

Especial attention must be given to the food of the child, which, of course, is to be fluid. Hot milk with a small portion of salt, is a very common and good food; beef-tea, if there is a feeble circulation and muscular feebleness; rice nicely prepared with milk, tapioca, farina, etc., must be selected as occasion requires.

If there is some febrile action, small doses of Aconite can be used with benefit. Occasionally the bowels are sluggish and the patient suffers from retention of feces; in place of giving a cathartic, the bowels may be moved by a stimulant enema.

The points of ulceration I touch with nitric acid, being careful to reach all the parts involved. I use the strong acid with a piece of pine wood, so shaped as to be most easily applied. There is no trouble or danger in the use of this, if the precaution is used of holding it for a moment until any free acid has disappeared. The application should be thorough, both for the purpose of arresting the spread of ulceration at once, and to prevent the pain of frequent applications. Sometimes one will be sufficient, at other times it will have to be repeated for three or four days. In addition to this cauterization, the mouth should be washed with an infusion of Baptisia, or of the Rumex, Alnus, and Quercus, heretofore named. These will sometimes be found better than the chlorate of potash, so commonly used.

SORE THROAT.

We have no technical term that will answer to designate the different lesions that are grouped together under the name of *sore throat*. The situation of the disease varies in different cases. In some the inflammation is of the fauces, tonsils, and base of the tongue, in others it is of the pharynx proper, and in others still, it involves the posterior nares, the velum, and to some extent, the larynx. The character of the disease also varies in different cases. In some, it seems scarcely more than an irritation. In others, it is an acute erythmatous inflammation. In others, the deeper tissues are involved, and there is considerable swelling. And in still others, the inflammation progresses to change of structure and ulceration.

CAUSES.—The most common cause of sore throat is cold, though we can give no reason why it should so frequently affect the throat. In some cases, it seems to be associated with gastrointestinal irritation.

SYMPTOMS.—We determine the existence of sore throat in the nursing child, by the uneasiness and evident pain as it nurses and swallows. Sometimes it will refuse the breast on this account. Usually the child is more or less feverish, is irritable, sleeps poorly, and wants to drink frequently, though to swallow the water occasions pain. The throat becomes dry, and is the principal reason for the craving for drink.

On examination we find the throat presents evidences of inflammation, in its redness, dryness, and heat. It is not always that we can make a satisfactory examination of the throat, and we will have to be satisfied from the redness observed about the base of the tongue and fauces.

TREATMENT.—The simple sore throat of childhood yields readily to simple treatment. The patient is placed upon the use of Aconite, with Phytolacca, in the usual doses, which are continued every hour, until all evidence of the disease has disappeared. The child has a hot mustard foot-bath once or twice daily, and occasionally a general bath. The local application to the throat externally is, where one is necessary, a flannel cloth wrung out of cold vinegar, with a dry one over it. If there is some hoarseness the Stillingia liniment is advisable.

The young child can not use a gargle, indeed, it is almost useless until they reach the age of ten or twelve years. In its stead, if anything is necessary order the remedy in this form: ℞ Powdered Gum Arabic, Loaf Sugar, Chlorate of Potash, aa. make powders of five grains, and direct that they be put upon the child's tongue dry, and dissolved by the secretions of the mouth, slowly swallowed. These may be repeated as often as every hour, and will generally answer the purpose well. In older persons, a gargle of chlorate of potash, of sulphite of soda, of Baptisia, or of Hydrastis, may be used with success.

When the disease is severe, and requires more prompt relief, I direct that an inhalation of the vapor of water, or water and vinegar, be used. If no apparatus for inhalation is convenient, we may employ the means found in every house; sometimes sprinkling the fluid on a heated shovel or iron will answer, the child being placed in such position that the vapor may be inhaled. A better plan is to heat the fluid in a tin or other vessel, and place it near the child, throwing a light blanket or shawl over it and the child's head ; a hot iron is gradually put in the fluid, raising the necessary quantity of steam. Occasionally the inhalation may be medicated with advantage ; an infusion of Baptisia, of Hamamelis, or of tansy from the garden, will answer a good purpose

CYNANCHE MALIGNA.

Malignant sore throat usually prevails as an endemic, sometimes as an epidemic, and occurs most frequently in the winter and spring. The cause is somewhat obscure. For some reason the tissues of the throat are weakened, and a low grade of inflammation is set up. There is, doubtless, blood poisoning, as we observe in analogous cases.

PATHOLOGY.—There is a low grade of inflammation, of which malignant sore throat and epidemic dysentery are examples, that most frequently has its origin in some local miasm, animal or vegetable, which affects the atmosphere of limited portions of country. All the symptoms point to a poisoning of the blood, and depravation of this fluid, and consequently of the secretions, and of nutrition, as one of the principal elements of the disease. The local disease, however severe, would not occasion uneasiness, if the general health was good, but with depravation of the fluids and solids, and the attendant prostration, the disease becomes one of the severest we are called to treat.

SYMPTOMS.—For two or three days, sometimes for a week, it is noticed that the child looks pallid, its skin waxy or pasty, and that there is a want of expression in the countenance and of energy in play. The breath is also bad, the tongue broad and pale and somewhat loaded. The child's appetite is variable, it rejects its breakfast, but eats dinner or supper. Frequently it refuses its usual food, and wants sweets and fruit. For a day or two, in some cases, it complains some of its throat.

In some cases the disease is fully announced by a chill, of longer or shorter duration. But in others there is such a gradual increase in the symptoms that it is difficult to separate the forming stage from the fully developed disease.

When the physician is called he finds evidences of a general and a severe local disease. The pulse is soft, easily compressed, and increased in frequency from ten to thirty beats per minute. The extremities are kept warm with difficulty, the skin is pallid or sallow, and presents a peculiar waxy appearance, looking many times as if it was œdematous, and would pit on pressure. The face is pallid and expressionless, with a dark line under the eyes, which also are dull, with dilated pupils. The bowels are irregular, the feces clay-colored and papescent; the urine free, pale, and of low specific gravity. There is no appetite; indeed, from the condition of the mouth and throat, there is disgust for food.

On examining the mouth and throat we find the mucous membranes pallid, the tongue broad, pitting where it comes in contact with the teeth, and covered with a pasty, white coat. The mucous membrane of the throat is swollen and discolored; in some cases it is livid, in others of a dusky-red, and in some few it presents a peculiar blanched appearance. The tissue seems relaxed and flaccid, and the circulation sluggish.

In a couple of days small points of ulceration will be seen, sometimes superficial, at others with a tendency to extend in depth. These ulcers increase in size more or less rapidly, according to the severity of the disease, and the throat will present a remarkably ragged and foul appearance. In very severe cases the ulcers pass through the mucous membrane and invade the cellular tissue, so that in fatal cases the structures are destroyed to a greater extent than we would deem compatible with life, for some hours before death ensues.

A distinctive symptom of malignant sore throat is the change in the tone of the voice; it is not so much hoarse as hollow and sepulchral—as a musician would say, " it has lost its timbre."

DIAGNOSIS.—This disease is readily recognized by the fetid breath, the abundant secretion from the throat and mouth, and by the peculiar relaxed condition of the structures. Add to this the general cachexia, which is peculiar to this, and, to some extent, to cancrum oris, and we have a grouping of symptoms that can not be mistaken.

PROGNOSIS.—Though the disease is a very unpleasant one, and attended with such depravation of the fluids and solids, the prognosis is not unfavorable. A large majority of cases will recover, probably as much as ninety or ninety-five per cent.

TREATMENT.—The treatment of cynanche maligna will be both constitutional and local. We want to antagonize the septic influence, improve the circulation of the blood, increase the tone of the system, and place the stomach in condition to receive and appropriate food, and re-establish secretion.

Aconite and Belladonna may be given in small doses, to improve the circulation, or if the cervical glands are enlarged Phytolacca takes the place of the Belladonna. Under their influence we find the pulse becoming stronger and more full, the capillary circulation better, and the temperature of the body more uniform.

Of the antiseptics I prefer sulphite of soda in the majority of cases, giving it in the usual doses, every three hours. In some cases chlorate of potash may be used instead, or alternated with the sulphite. Triturated with gum arabic and sugar, as named for diphtheria, will probably be the best form of administration. The Baptisia in infusion is an excellent antiseptic, and may be associated with either of them.

In addition to this, I prescribe quinine in stimulant doses, sometimes alone, at others in combination with Hydrastine. The dose will be about one-half grain, three or four times a day. Tincture of muriate of iron can also be used with advantage in some cases. It may be specially named as an important remedy in those cases which manifest an erysipelatous tendency.

The local means will vary in different cases. In the milder ones a decoction of Baptisia, used as a gargle, will be sufficient. In others we may alternate this with a gargle of chlorate of potash, and in others the sulphite of soda will answer a good purpose. In those cases where the tissues are relaxed, and the ulceration progressing rapidly, the sulphurous acid will be the most powerful, as well as the most certain local remedy we can use. We would make the solution of the strength of one ounce of the acid to three ounces of water. When it is used with a pencil or probang it may be applied of full strength.

But gargles can only be used with children who have attained to six or ten years, and in younger children we will have to depend upon other means. Of course, the sulphite of soda, chlorate of potash, and Baptisia, can be so employed as to get their topical action, as they are taken. I do not like the use of the probang, or the *swab*, to make local applications to the throat of children. Instead of this I use inhalation, preferring the *spray apparatus*, either air or steam, to any other apparatus. But it does not require an instrument, for, as we have already shown, an inhalation can be given with nothing but a vessel to hold the fluid and a heated iron to raise a vapor. The vapor of vinegar and water answers an excellent purpose, as does an infusion of tansy, or of Baptisia. In using the spray apparatus, I use the same remedies named for gargles.

The external application in this, as in many other diseases of the throat, is a flannel wrung out of cold vinegar, with a dry flannel over it. We call it the vinegar pack, but a cold water pack to the throat will answer the purpose.

CHRONIC SORE THROAT.

We are not accustomed to think of chronic disease in childhood, and yet we have many that are as strictly so as any disease of the adult. Among these is a chronic sore throat, which is very troublesome, and, like other chronic diseases, manifests no tendency to recovery.

CAUSES.—It is sometimes difficult to determine the cause of this disease. In all cases, I believe, it will be found associated with a *scrofulous cachexia*, or at least a tendency to disease of mucous structures. With such tendency as this, all that is necessary is that from cold or gastric irritation a simple inflammation of the throat be set up. This continues on until it produces the condition we are considering.

PATHOLOGY.—As just named, there is almost invariably a cachectic condition of the system, a tendency to disease of this character, which will manifest itself in different tissues, as they may suffer from some acute irritation or inflammation. Such children rarely live to adult years, and furnish subjects for the various forms of scrofula, hip-joint disease, white swellings, and finally phthisis. True this predisposition may be removed by a judicious and long continued course of training, but it is rarely adopted.

SYMPTOMS.—The attention will be drawn to the throat as the seat of disease, by continually recurring attacks of slight sore throat. Sometimes it will seem to involve the respiratory organs more; there will be a hacking cough, frequent endeavors to clear the throat, and slight difficulty in respiration. With such attacks the child is fretful, and has slight fever.

20

We notice it at all ages, from three months old to the close of life, and it presents somewhat similar symptoms. The evidences of general cachexia will vary in different cases, sometimes being marked, while at others the child seems quite healthy.

Upon examination of the throat we find the mucous membrane relaxed and thickened, the mucous follicles enlarged, and the circulation sluggish. There is also change in the color of the parts; they have become a dirty-red, dusky, livid, but more frequently a pallid red. The tonsils are frequently enlarged, and the mucous follicles about the base of the tongue are also enlarged, and keep the part constantly covered with their secretion.

In very bad cases the throat presents a peculiar appearance; the pillars of the fauces seem thickened, as is the velum and uvula, and with the enlarged tonsils, present a misshapen appearance. The voice in this case has changed, becoming *fluffy* or hollow, and lacks distinctness in articulation.

TREATMENT.—When the child presents evidence of a scrofulous cachexia, we proceed to treat that as named for scrofula. Usually it is not so marked, and our treatment will be a tonic and restorative general treatment.

The tonic for a child, as I have had occasion to name several times, is tincture of muriate of iron, with cod-liver oil if the child will take it. We make the usual prescription : ℞ Tincture of Muriate of Iron ʒj., Glycerine ʒiv., a teaspoonful three times a day. If the appetite is poor, or there is an infantile dyspepsia, a small portion of tincture of Nux Vomica may be added to it.

A very good means in this case is the use of Collinsonia. For a child two years old, I would prescribe : ℞ Fluid Extract of Collinsonia ʒss., Simple Syrup ʒiv., a half teaspoonful every four hours. If the child is old enough to gargle the throat, I direct an infusion of Hamamelis, or the fluid extract, in the proportion of one-half ounce to six ounces of water.

The vinegar pack to the throat, with the cold sponging in the morning, is an important part of the treatment. As is the case with all other chronic diseases of the throat and larynx, the patient seems to be continually taking cold, and this prevents that continuous amendment necessary to complete recovery. The use of the pack, even of cold water, with the cold sponging of the neck and shoulders, is the most efficient means to prevent this.

GASTRODYNIA.

I use the term gastrodynia to express the condition of pain in the stomach, whether the pain is the result of irritation of the gastric nerves, from cold, from irritative ingesta, or from spasmodic contraction of the muscular coat. In slight degree it is of very frequent occurrence, and is quite often seen in a severe form. With some children it is of daily recurrence, and is usually called colic by mothers and nurses. It not only causes much suffering to the child, but is source of very great annoyance to the parents; and when the mother is in feeble health, it becomes a serious matter, and if possible must be removed.

CAUSES.—The most common cause of pain in the stomach is indigestion. The child nurses well, and frequently thrives, but some portion of the food undergoes decomposition and proves irritant, or generates gas which unduly distends the stomach. In some of these cases there seems to be a peculiarly irritable condition of the gastric nerves, so that a very slight cause is sufficient to produce pain. In other cases it results from cold, and in these there is fever; and in others there is a real spasmodic action, or cramp of the stomach.

SYMPTOMS.—The child exhibits evidences of pain in its countenance, is uneasy, and will not remain in one position, and has violent paroxysms of crying. In some cases the body is drawn up, and the lower extremities are flexed upon the abdomen. Occasionally there are eructations and some of the food is thrown up, showing imperfect digestion; but at other times the milk will be curded and sweet, showing healthy digestion.

In the majority of cases the pain is paroxysmal, lasting for five or ten minutes, during which the child seems to suffer severely, and cries violently; it gradually abates until there is comparative ease, and as we are about to congratulate ourselves that the pain has wholly passed away another paroxysm comes on suddenly. Thus it may continue for an hour or two, or for a considerable portion of a day.

TREATMENT.—In many cases we find one or two drops of tincture Nux to a half glass of water, given in half teaspoonful doses, every fifteen minutes to one hour, will give speedy relief.

If there is irritation of the bowels, also with griping pain, Tinct. Colocynth gtt. j. to gtt. ij. to water ℥iv., a teaspoonful every half-hour to hour will give speedy relief. In some cases Aconite may be used with both of these, and aids in relieving the pain. Dioscorea in infusion, the tincture gtt. v. to gtt. x., to water ℥iv., is a good remedy. Apis answers a good purpose when the pain in the stomach is associated with an eruption like "heat," and Belladonna, if such an eruption has disappeared from the surface. The old remedies, after the following formula, may sometimes be given : ℞ Tincture of Lobelia ℨj., Compound Tincture of Lavender ℨiij., Simple Syrup ℥jss. ; the dose will be from one-fourth to one teaspoonful, and may be repeated as often as necessary. Another excellent preparation is : ℞ Chloroform gtt. xx., Glycerine ℥ij., in doses of one-fourth to one teaspoonful, as often as required.

When these attacks are of frequent occurrence in nursing children we may suspect a wrong in the mother's diet or digestion, and this being looked after, we will sometimes find that our little patient has more comfort. In other cases the pain is a symptom of infantile dyspepsia, and lime-water, phosphate of soda, or some of the simpler tonics, will give permanent relief.

The best local application, as a general rule, will be a hot dry flannel applied over the stomach and abdomen. Sometimes the use of the hot flannel to the spine will give speedier relief than when used over the stomach. In children of one year and older a mustard plaster may be used.

GASTRIC IRRITATION.

We are called to prescribe for cases in which there is marked irritation of the stomach. It occurs as a complication of many acute diseases, and sometimes alone, there being no other disease present. In the first case it increases the severity of the primary disease, prevents the absorption of remedies, and the taking and digesting of food It thus becomes, when severe, an unfavorable complication.

CAUSES.—We can not determine, in many cases, any cause for this irritation. In some it depends upon imperfect digestion of food, in others from cold and arrested secretion. In acute disease it is doubtless owing in part to decomposition of food, taken before the disease was fairly announced.

SYMPTOMS.—Prominent among the symptoms is the occasional nausea, retching, and effort to vomit. The child is thirsty, and desires drink frequently, but when any quantity is taken it is ejected by vomiting. If nursing, it wants the breast frequently, and after nursing it throws up its milk. When the child is weaned, it will generally reject all food.

If medicine is given, at least many kinds, it will be thrown up after a time. Thus, in the treatment of acute disease, we may be giving the sedatives, or any remedies, every hour; the child takes three or four doses, and then there is sickness of the stomach, and the whole is ejected. Thus, many times, the treatment will go on day after day without any appreciable influence upon the disease, none of the medicine having gained entrance into the circulation.

In this condition of irritation *osmose* is from the blood-vessels to the stomach, and of course there is no absorption either of food or of medicine.

Irritation of the stomach, without other disease may continue for several days, or in slight degree for two or three weeks. Necessarily there is sympathetic disturbance of the nervous system, and of other functions, and digestion being arrested the child is much prostrated.

There is an irritation of the stomach, arising from disease of the brain, that is of very serious import. This case is diagnosed by the contracted and pinched appearance of the countenance, the dullness of the eyes, and the increased frequency and hardness of the pulse (there are some exceptional cases in which the pulse is small and weak). The skin is dry, sometimes husky, the temperature of the trunk slightly increased, but the extremities are cold.

If the tongue is examined in irritation of the stomach, it will sometimes be found reddened, contracted, and pointed, in others pale and atonic; in the one there is irritation with determination of blood, in the other atony. If coated, it will be with a white fur, confined to the center.

TREATMENT.—When there is febrile reaction, or increase of temperature, I direct cold applications over the epigastrium. In the opposite cases, a sinapism or a spice plaster may be used. In all cases much benefit will follow the use of the hot foot-bath, which, if the extremities are inclined to be cold, may be rendered stimulant by the addition of mustard or capsicum.

Internally, if the tongue is elongated and red, we may administer small doses of Aconite and Ipecac, say : ℞ Tinct. Aconite gtt. j. to gtt. ij., Tinct. Ipecac gtt. v. water ℥iv., a half teaspoonful·every hour. If the pulse is sharp and the patient has frontal headache, the remedies will be Aconite and Rhus. In the second case,· when there is evidence of atony, the prescription will be : ℞ Tinct. Nux gtt. j: to gtt. ij. water ℥iv., a teaspoonful every half hour. Occasionally an infusion of our old fashioned compound powder of rhubarb given in small doses will give relief.

In place of this, and better adapted to severe cases, is an infusion of the bark of the peach-tree. We take the young and green limbs of the present year's growth, scrape the bark off, cover it with boiling water, and when cold it is ready for use. We give it at first in doses of one·fourth to one-half teaspoonful every few minutes until the irritation has partially subsided ; afterwards in teaspoonful doses for two or three hours.

Occasionally we will obtain benefit from the preparations of bismuth ; either the subnitrate, carbonate, or liquor bismuth may be used. The first I generally prescribe with mint water, in doses of from one to three grains.

When the irritation is very persistent, the bowels being constipated, an enema of a weak salt water to move the bowels, will aid in giving relief. If the temperature is high the injection may be cold, if the patient is exhausted the injection should be hot. In these severe cases we are very careful in administering remedies and giving food after the irritation seems to have been subdued.

GASTRITIS.

Acute inflammation of the stomach is rarely met with in the child, and only, as I believe, from the administration of irritant medicines. Subacute inflammation, however, is met with quite frequently, and is sometimes very persistent.

CAUSES.—The causes of inflammation of the stomach are twofold. It may be produced by the ordinary causes of inflammation in other organs, as from cold, arrested secretion, etc. It is also the result of irritant ingesta, or of such decomposition of food as gives rise to irritant products, or to irritant secretions.

SYMPTOMS.—Subacute gastritis may occur as an idiopathic affection, or as a complication of some other disease. In the first case it is announced by nausea, retching, and vomiting, following a short forming stage, in which for a few hours the child has been restless and uneasy, not caring for food, but wanting to drink freely. Occasionally there is a pretty well-marked chill lasting for an hour or two.

The retching and vomiting are not constant, but come on at longer or shorter intervals, are quite violent and attended with much straining. When the stomach is thus relieved for the time being, the child seems to lie easily, and sometimes falls into a short sleep. Presently the thirst increases; it wants drink continually, and having taken but a small amount, vomiting again commences.

There is always more or less fever. Usually it is paroxysmal, following the course of the vomiting, coming up after this has stopped and gradually increasing until the nausea and vomiting commence. After the nausea and vomiting have continued to relaxation, the fever passes off, to re-appear again, as before named. In some cases the nausea and retching simply increase the irritation of the nervous system and the circulation, and there is no abatement of the febrile action.

In some cases the irritation of the nervous system is very great, and manifests itself in uneasiness and restlessness, crying and other evidences of suffering. In a few cases the irritation is of such character as to produce convulsions, which are usually very severe and intractable.

Should the gastric irritation not be arrested, it will result in one of three ways. In nervous irritation, occasionally convulsions, and finally coma, gradually increasing until it proves fatal. In irritation of the intestinal canal, and diarrhœa, which, running the usual course, terminates in the same manner as muco-enteritis. In an irritative fever, which rapidly assumes typhoid symptoms, and proves one of the severest diseases we are called to treat.

TREATMENT.—The treatment of an inflammation of the stomach is quite simple, and usually very successful. The first object is to so modify the irritation as to stop nausea and vomiting, for these perpetuate if they do not increase the irritation; afterward we have time for the removal of the inflammation.

The patient has a general bath if febrile reaction is high, followed by a hot-mustard foot-bath, continued for twenty-five to thirty minutes. In some cases a general hot bath, using salt in the water, will prove beneficial; if this be used, the applications to the epigastrium should also be hot. In the majority of cases, I prefer cloths wrung out of cold water, over the stomach. Instead of this, a sinapism made of three parts of flour to one of mustard, may be used.

Internally I would prescribe an infusion of peach-tree bark, in doses of one-fourth to one-half teaspoonful, frequently repeated. The infusion is made of the bark of the limbs of the present season's growth. In place of this, bismuth with mint water may be employed, as named for irritation of the stomach. In some cases, those especially in which the inflammation is not high, we may use the compound powder of rhubarb in infusion.

To modify the febrile action small doses of Aconite are employed; five drops to four ounces of water; a teaspoonful every half hour or hour will be the proper quantity. If there is much irritation of the nervous system we may give the Gelseminum, in doses of from one to three drops, with the sedative. We find that these remedies not only answer their specific purpose, but they also have a beneficial influence in arresting gastric irritation.

In some cases we find the bowels inactive, and occasionally the gastric disease is increased by accumulations in the intestinal canal. As it is clearly out of place to give cathartics by mouth, we must depend upon enemas. An enema of salt watar will sometimes exert a marked influence in checking nausea and vomiting; not so much in gastritis, of course, as in simple irritation of the stomach, yet its influence in this respect is beneficial. Simple enemata are used when the object is to empty the lower bowels, but when the small intestine is to be reached, the enema should be of compound powder of Jalap, or of castor oil, in warm water.

INFANTILE DYSPEPSIA.

To many persons the name infantile dyspepsia will sound singular, as they have been accustomed to think of dyspepsia as associated with a long-abused and worn-out stomach. Yet dyspepsia is a very common complaint with the child, almost as frequent as in the adult, though not so persistent.

There are several causes of infantile dyspepsia, and we will obtain a better knowledge of the disease and of its therapeutics, if we study these. They have reference to the quality and quantity of the food, to the secretion of gastric juice, to the muscular action of the stomach, and to the condition of the upper intestinal canal.

We find this disease most commonly in children who have other food than the mother's milk. Here the trouble is in the difficulty with which the food is digested, and the imperfection of this process, in which some portions undergo decomposition. This may be corrected in some cases, to a limited extent, by changing the food, by giving it in a less concentrated form, and in such manner that the secretion of saliva may be increased. Thus in a case where the child was being fed on cow's milk, simply diluting the milk with one-fourth water has proved successful. Or, when it was fed with a spoon, or from a cup, the use of the bottle, so as to call for the act of sucking, would cause an increased flow of saliva, and the little patient would be greatly benefited.

A very common cause of infantile dyspepsia is over-feeding. This is the most common cause in nursing children, and depends not only upon the taking of too large a quantity at a time, but upon its too frequent nursing. In the one case the stomach is overdistended, and overworked, and gradually it loses its power of digestion, and evidences of dyspepsia are manifested. In the other case the stomach is allowed no repose, in which to regain its lost powers, or for its nutrition. The result is that its function becomes gradually impaired.

In another class of cases the dyspepsia is dependent upon a *bad* blood, from the many causes heretofore named for this deterioration. The child is cachectic, lacks power in every direction, and this, as well as other functions, are impaired.

Dyspepsia is not always a gastric lesion. The small intestines have as much to do with the digestion of food as the stomach, and a lesion here gives rise to dyspepsia. Generally this, so far as we can see, is a simple atony, at other times it is an irritation. The atony is not only of the small intestine, but also of the liver, the pancreas, and indeed of all associated parts. In such cases the circulation of blood and the innervation are deficient; there is a want of normal stimulus. So in cases of irritation, the associated organs are involved, to a greater or less extent, and there is too much blood and innervation to these parts.

There are cases in which a failure of power in the muscular

coat of the stomach is a cause of dyspepsia in infants, as well as adults. Others in which there is a defect in the secretion of gastric juice, either in quantity or quality. These in the child are principally of an atonic character, and hence there will not be so much difficulty in the treatment as in the disease of the adult.

SYMPTOMS.—The symptoms of dyspepsia in the child vary greatly ; in some cases being well marked, in others obscure. Usually the child is uneasy after taking its food, and is frequently troubled with colic. Vomiting of the food is of common occurrence, and it the matters ejected are examined, the food will be found acid, smelling sour and sickly, or undergoing incipient decomposition. These changes in the matter ejected are pathognomonic; for in the vomiting which follows mere repletion, in which case the overloaded stomach takes this easy mode of relieving itself, the food is sweet, the milk well curded, presenting evidences of a healthy condition of digestion.

With this digestive lesion we notice that the child becomes pale, the tissues soft and flabby, the circulation feeble and the extremities cold. There is also, in most cases, very marked irritation of the nervous system, the child being restless, irritable, and fretful, and at night tosses about, awakes frequently, and wants to drink.

In some cases there is the same desire for food, even a ravenous appetite at times. In other cases the appetite slowly fails, and the child cries neither for food nor for the breast. When this form continues for any considerable time, the child becomes anæmic and much debilitated.

The simplest division of these cases is into the *irritative* and *atonic*. In the first case the evidence of gastric irritation is pretty well marked. There is the red tongue, the irritability of the nervous system, the pinched face, and bright eyes, the harsh skin, and the greenish acrid discharges from the bowels.

In the *atonic* form of dyspepsia, the tongue pale, a whitish pasty fur accumulates at its base, the breath has a sickly odor, the child is feeble and languid, and likes to remain in one position, the face is pallid, the eyes dull, the face expressionless, the skin soft, relaxed, and cool, the pulse soft and feeble, and the discharges from the bowels light-colored.

DIAGNOSIS.—The diagnosis will be made in part by the symptoms above-named, which point to the stomach as the seat of a lesion, and in part by exclusion. The difference in the diagnosis between the adult and the child is, that the first points out the location of the unpleasant sensations, while in the second we have to determine this by careful observation. By making the examination thorough we are enabled to exclude one organ after another until the evidence of the gastric lesion becomes positive.

PROGNOSIS.—The prognosis is not always favorable. It will depend much upon the kind of food the child is taking; if nursing, it is favorable; if fed on cow's milk or other food, it is unfavorable. It is generally grave in proportion to the change in the matters vomited; if they are but little changed, or seem to be undergoing the process of digestion in a proper manner, we conclude that the lesion of the stomach is simple in its nature. But if they are undergoing decomposition, and have an unpleasant odor, the evidence of serious lesion is pretty strong.

TREATMENT.—The first thing that engages the attention of the physician is the food of the child, for as we have seen above, the lesion is frequently dependent upon a wrong in this. If the child is being raised on cow's milk, or other artificial food, we will see that this is the best of the kind, and properly prepared and given. When the dyspepsia is persistent we will have to take into consideration the propriety of changing it. Liebig's food will answer a good purpose in many of such cases, when the child is being fed upon milk. The directions, heretofore given under the head of food for the child, may be studied with advantage.

When the child is nursing we will inquire as to the mother's diet, for not unfrequently indiscretions upon her part will have caused, and may perpetuate, the disease. Usually it is the use of vegetables and fruits in too large quantity that has first occasioned it. In some cases the dyspeptic lesion of the child depends upon a similar gastric lesion upon the part of the mother, and it is her imperfect digestion which deranges its stomach.

In case of indiscretions in diet, we will represent the importance of care in this regard; and when this is not so manifest it will be well to change the mother's food until we have discovered what was giving rise to the trouble. When the mother suffers from dyspeptic symptoms, or has imperfect or feeble digestion, it will be proper to put her upon a course of treatment that will restore the functions of the stomach.

A very marked case of this kind came under my care in the earlier part of my practice. A lady had given birth to five children, of which three had died of infantile dyspepsia. One was saved by weaning, and the fifth, at the age of three months, was suffering severely. I treated the mother for chronic gastritis, by the use of the irritating plaster over the stomach, and an infusion of hydrastis taken internally, and as the irritation disappeared added iron and phosphorus. As the gastric symptoms of the mother disappeared and her health improved, the child became better, and in a few weeks was doing as well as other children.

In some cases we will have to take into consideration the propriety of weaning the child, as it becomes evident that the disease can not be controlled otherwise. In some, feeding the child with cow's milk is attended with benefit, though in others the child rapidly fails after it is put upon it. Liebig's food, when the child will take it, usually does well. When artificial food disagrees with the child it increases the pains in the bowels, and is either vomited or is carried off by a diarrhœa. In all such cases the chances of saving the child is much better if a suitable wet-nurse can be obtained, and this should be clearly stated to the parents at as early a period as possible.

In some cases the employment of restoratives and tonics with the mother, and the use of well-cooked animal food, will be the best treatment for the child. Tincture of muriate of iron and glycerine, with the addition of tincture of Nux Vomica, if there is atony of the stomach, answers very well. Hydrastis, with same preparation of iron, is a good bitter and stomachic. The hypophosphites have proven serviceable in my practice, either singly or in the form of compound syrup of the hypophosphites.

In some of the more simple forms of infantile dyspepsia we will find the administration of phosphate of soda in small doses, three or four times a day, either alone or with the food, will answer an excellent purpose (about ten grains daily for a child three to six months old). Common salt will also exert a marked beneficial effect with nursing children, in some of these cases.

Where the dyspepsia is of the irritative form, I prescribe an infusion of peach bark, as heretofore named, and some preparation of bismuth. The subnitrate or subcarbonate in powder, or the liquor bismuth, in small doses. With these, and proper hygienic means, the irritation will have so abated in one or two weeks, the child can be put upon the use of Collinsonia and tincture of muriate of iron. Where there is evidently accumulations in the small intestine, an infusion of the compound powder of rhubarb, given to produce its laxative effect, may precede the other remedies.

In some cases of irritative dyspepsia we will obtain much advantage from the use of small doses of Aconite, repeated every one or two hours for a day or two, and afterward at less frequent intervals. The wet pack to the abdomen may also be resorted to in more acute cases with advantage.

In the simpler forms of *atonic* dyspepsia, I put the child on the use of some simple bitter, as an infusion of Matricaria Chamomilla, Virginia snake root, or the Panax Quinquefolia. In others I prescribe Collinsonia alternated with tincture of muriate of iron in the following proportions: ℞ Tincture of Collinsonia ℥ss., Simple Syrup ℥iijss.; half a teaspoonful three times a day. ℞ Tincture of Muriate of Iron, ℥ij., Glycerine, Simple Syrup, aa. to ℥iv.; a teaspoonful three times a day. The tincture of Nux Vomica in small doses, or a solution of strychnia in doses of about the 1-200th to the 1-100th of a grain may be used with marked advantage.

In some of these cases Podophyllin, thoroughly triturated, will prove one of our best remedies. As named in the first part, it should be triturated in the proportion of one part to ten or twenty of sugar; have Ipecac triturated in the same way, and combine them in equal parts, making powders of one-half to one grain. This may be given three or four times a day, and associated with the tonics, as before named.

COLIC.

We may distinguish two forms of colic in children, the one being temporary and arising from slight indigestion, the other bearing a close relationship to the bilious colic of the adult, and due to spasmodic contraction of the muscular coat.

The first form is of very frequent occurrence, and in some cases seems to be natural to the child. It comes on from the slightest causes, and lasting an hour or two, passes away, and the child seems as well as ever. With many it will continue to recur frequently for the first three or four months, and will then cease.

The second form of colic is caused by cold, by indigestion, by some indiscretion in the diet of the mother, or occasionally by mental excitement, from anger, fear, grief, etc. The child not only suffers much pain, but there is functional disturbance, and some fever.

SYMPTOMS.— In either form of colic the child is uneasy, changes its position frequently, bends its body in various directions, and cries persistently. Simple crying, however prolonged, would not be evidence of colic, but with the contortions of the body, and frequent change of position, the diagnosis is pretty evident. It may continue but a few minutes, or for half an hour. When it lasts long it is remittent or paroxysmal.

The second form of the disease presents the same evidences of pain, but it is of longer duration, continuous, and severer. I determine its character principally from contraction of the features which expresses pain. The flexion of the trunk, and of the lower extremities upon the abdomen is more persistent.

TREATMENT.—In many cases the colic is speedily relieved by the administration of Nux Vomica, as—℞ Tinct. Nux gtt. ij.. water ℥iv.; a half teaspoonful every fifteen minutes or half hour. If the child is feverish, Aconite in the usual doses may be alternated with this. In other cases the indications will be stronger for Colocynth, and I usually administer it as follows: ℞ Tinct. Colocynth gtt. ij., Tinct. Aconite gtt. iij., water ℥iv.; a teaspoonful every half hour. An infusion of Dioscorea, or of Epilobium, will sometimes give prompt relief, and sometimes the tincture will serve a good purpose.

If the attack is very severe, and relief is not obtained in this way, an enema of hot water (sometimes cold is better), or water to which we have added grs. xx. of Compound Powder of Jalap and Senna may be used.

Sponging the abdomen with hot water, the general hot bath, hot packs to the abdomen, a cold pack, if the temperature of the child is increased, a sinapism, or the application of chloroform, will suggest themselves to the practitioner, and will be selected according to the indications.

In some cases the prescription of Lobelia with Comp. Tinct. Lavender may be used, more rarely chloroform may be given by mouth. If the pain is excessive chloroform may be used to produce partial anæsthesia and give relief. If we can not relieve the patient in any other way chloral may be given to relieve the pain until other means can be employed.

If the child suffers from gastric irritation or dyspepsia, we will treat the case as has been named for those affections. And if owing to errors or imprudence in the diet of the mother, we will endeavor to have them corrected.

DIARRHŒA.

Simple diarrhœa is of very frequent occurrence with children, and will arise from any cause deranging the processes of digestion, or from cold. We recognize the same forms of the disease as in the adult; first, from irritation of the intestinal canal; second, from increased secretion of bile; third, from atony of the intestines; fourth, from congestion of the portal veins and determination of blood; fifth, from increased secretion of mucus; sixth, from imperfect digestion. A simpler classification, however, and more easily recognized in practice, is the two forms, diarrhœa from *irritation* and diarrhœa from *atony*.

SYMPTOMS.—In diarrhœa from *irritation* the discharges from the bowels are dark-colored, usually a shade of green, or in some cases of a light or pea-green. They are sometimes acrid, so that when they have continued for some time, they excoriate and chafe the child. Usually, the child manifests some uneasiness before having a stool, and there is some straining with it. Occasionally, there is febrile action, the skin being harsh and the temperature elevated.

In diarrhœa from *atony*, the discharges are light-colored and watery, and passed without any pain or uneasiness. They are usually larger than in the other case, though the amount of solid matter is not increased. Here, if it persists for some time, the child seems relaxed, the skin soft and cool, extremities cold, face pallid, and circulation enfeebled.

In both cases the appetite is impaired if the diarrhœa continues, and what food is taken is not well digested, consequently the patient loses strength and flesh as the disease progresses.

TREATMENT.—The diarrhœa of irritation may be very frequently controlled by the use of Aconite alone, in the usual doses. Generally we continue Ipecac with it in the following proportion: ℞ Tinct. Aconite gtt. iij., Tinct. Ipecac gtt. v., water ℥iv.; a teaspoonful every hour. The Euphorbia Hypericifolia will sometimes prove a better remedy than Ipecac, and may be substituted for it in about double the proportion. If the patient complains of uneasy sensations in the abdomen, with tormina and some tenesmus, the prescription might be: ℞ Tinct. Aconite gtt. iij., Tinct. Colocynth gtt. ij., water ℥iv.; a teaspoonful every hour.

In atonic diarrhœa we frequently substitute Nux for the Aconite, as: ℞ Tinct. Nux Vomica gtt. iij., Tinct. Ipecac gtt. v., water ℥iv.; a teaspoonful every hour.

When there is evidence of irritant matters in the intestinal canal, an infusion of compound powder of rhubarb may be given to produce an action upon the bowels, and afterward in smaller doses until the diarrhœa ceases. Or, after obtaining this action, we may use the subnitrate of bismuth, in doses of one to three grains, every two or three hours.

In atonic diarrhœa, we may use the preparation of rhubarb, above named, to change the character of the secretions; or in its place: ℞ Neutralizing Cordial ℥ij., Chloroform ℥ss.; a teaspoonful every hour. Podophyllin thoroughly.triturated with prepared chalk, is an excellent remedy in these cases, when given in minute doses, say the 1-40 of a grain. If an astringent is thought desirable, tannic acid with glycerine will answer a good purpose, and will be readily taken by the child. It may be dissolved in the proportion of ℥j. to the ounce, and given in doses of one-fourth to one-half teaspoonful.

If the tongue is pallid, the patient should have bicarbonate of soda, or in some cases lime water, whatever else may be given. If the tongue is deep-red, an acid should be given, usually muriatic acid. If the tongue is red and dirty, the remedy will be sulphurous acid.

Leptandrin is an excellent remedy in those cases in which there is evident torpor of the liver, and would be preferable to the Podophyllin, were it not that the dose has to be large, and the remedy is very unpleasant. Stimulants, especially the aromatics, may be used in some of these cases, but as a general rule, they are not so beneficial as in the adult.

MUCO-ENTERITIS.

Subacute inflammation of the small intestine is most usually caused by cold, though it may arise from irritating ingesta, or from decomposition of the food and the production of irritant material from this. The inflammation is confined mostly to the mucous membrane, the secretions of which are increased, and an increased peristaltic action having been set up, there is more or less diarrhœa.

SYMPTOMS.—The child is restless and uneasy, and has considerable fever. The discharges from the bowels are not at first very frequent, but they are preceded by pain, the child flexing the body and the lower extremities to remove the pressure. The discharges are attended with some tenesmus, and in some cases the straining will be quite severe and prolonged. The stools vary in character; in nursing children they at first consist of curded milk, mixed with a greenish feculence, afterward they contain mucus, and are slimy and tenacious. In other cases the stools will be a green or brownish frothy material, mixed with mucus.

As the disease progresses the discharges from the bowels become more frequent, and are attended with greater uneasiness. They also contain more mucus, indeed sometimes seem to consist almost entirely of it. The febrile reaction also continues, and sometimes becomes quite severe. Generally it will run its course in from six to eight days, the fever ceases, the inflammation of the bowels passes away, and the diarrhœal evacuations stop.

DIAGNOSIS.—We diagnose a muco-enteritis by the febrile reaction which comes on with the diarrhœa, and by the tormina and tenesmus which attend the discharges. From dysentery it is distinguished by the feculence of the discharges. In some cases the diseases are associated, and we have a true dysenteric diarrhœa.

TREATMENT.—The treatment of this form of diarrhœa is very simple. We put the child upon the use of Tincture of Aconite gtt. ij. to gtt. v., Tinct. Ipecac gtt. v. to gtt. x., water ℥iv.; a teaspoonful every hour. In place of the Ipecac the Euphorbia may be used. If there is much griping pain, we will sometimes find Colocynth, the best remedy, as—℞ Tinct. Aconite, Tinct. Colocynth aa. gtt. ij., water ℥iv.; a teaspoonful every hour.

In some cases there will be nausea, not controlled with Aconite and Ipecac and in examining the patient we will note the expressionless mouth, and the pallid atonic tongue, and Nux in small doses will be suggested until these symptoms have passed away. When the tongue is broad, pallid and dirty, the patient

21

has sulphite of soda; if deep red, an acid is given. In some cases the marked periodicity in the disease calls for quinine, and it is given in the usual antiperiodic doses.

Among the older means that may be suggested are the white liquid physic, when there is much tormina and tenesmus; triturated Podophyllin, or the compound powder of rhubarb, when the discharges are light-colored, and the bowels tumid; bismuth, either in the form of sub-nitrate or liquor bismuth, or an infusion of Epilobium. In some cases where the fever runs high, I have given the Veratrum instead of Aconite, and where there was much irritation of the nervous system, adding Gelseminum.

CHOLERA INFANTUM.

Cholera infantum is a disease of the second year of childhood—a period which may be considered one of the climacterics of life, as there is now a change from the fluid food obtained from the mother, to the ordinary food of man. The change is a great one, but when there is strong vitality and good development, it is made without any appearance of disease. But if from any cause the child is enfeebled, or its development in other respects is tardy, it then suffers in the way we are about to describe.

It is not easy to say what the exact pathology of this disease is. In one case it seems to be almost wholly an irritation of the gastro-intestinal mucous membrane; in another it is an atonic condition, and a failure of power to digest food; in another still, the intestinal lesion seems to be secondary, and depends upon a want of assimilation and nutritive power; and in others the lesion is principally of the nervous system.

I conclude from all this, that the lesion is primarily one of the sympathetic system of nerves which govern all the processes of vegetative life. In all cases there is debility, though in one it may be manifested in irritative action, and in the other in atony.

Many have urged dentition as the cause, but as this is a physiological process, except when disturbed, we would expect to find the disease only in cases of dental irritation, whereas we find it in children who have no teeth, who are not cutting teeth at the time the disease commences, who have all their deciduous teeth except the four last molars, or when showing no swelling, tenderness or irritation of the gums.

Cholera infantum usually occurs in the summer, making its appearance in May, June, or July, and in the severer cases lasting until frost and cool nights in the fall. A continuous high temperature always increases the severity of the disease, while a moist cool atmosphere gives relief. As a general rule, it is more severe on low lands than on high ground, and is unfavorably influenced by the ordinary malarial causes.

SYMPTOMS.—We may divide the disease into two varieties, *acute* and *chronic*; the first bearing a very close resemblance to the cholera morbus of the adult, the second progressing slowly and occupying a period of weeks or even months.

Acute cholera infantum is usually preceded for a day or two by a slight diarrhœa, which gives the child but little uneasiness, and is scarcely noticed by the mother. But presently the child manifests an intense desire to drink and greedily takes whatever is offered. This causes nausea and vomiting, and the fluids are thrown up about as soon as taken. The diarrhœal discharges also become more frequent, and are large and exhausting.

In a very short time the child is much prostrated, there is frequent retching and vomiting, especially when any thing is taken upon the stomach. The thirst is a very marked feature, the child wanting water all the time, and crying for it whenever it sees a cup or glass; yet if even a tea-spoonful is taken, it is thrown up. The evacuations from the bowels have also increased in frequency.

In some cases the skin is harsh or dry, and the pulse hard and increased in frequency; in others the skin is soft and doughy, the extremities cold, and the pulse feeble and frequent. Occasionally the brain is affected, there being congestion or a low form of determination, or effusion. I recognize these cases by the extreme restlessness when there is determination, and a continued rolling of the head from side to side, and stupor in the other cases.

The disease may terminate fatally during the first twenty-four hours, or may continue for three or four days—the child recovering—or gradually pass into the chronic form.

The chronic form of cholera infantum may follow the acute attack, as above described. But in a majority of cases it is preceded for a week or two by a simple diarrhœa, which, as it does not make the child sick, is allowed to progress without attention.

The very gradual increase in the disease is probably the reason why an early treatment is not adopted.

The disease develops by an increase of diarrhœa, and the appearance of nausea and vomiting. The child has lost flesh and strength, its appetite is impaired, and it is thirsty, wanting to drink frequently and considerable quantities. As time passes, all the symptoms increase, but those that are especially marked are the intense thirst, irritability of the stomach, with frequent vomiting, especially when drink or food is taken, frequent evacuations from the bowels, and great loss of flesh and strength.

The discharges from the bowels vary much in character in different cases, and even in the same case at different times. Sometimes they are yellowish, with more or less stringy mucus mixed with them, showing disease of the mucous follicles; at others they are greenish, and have a sour smell; at others clayey; again almost white, and rarely a dark brown or black.*

In *febrile* cholera infantum the skin is harsh, dry and constricted, and in some cases seeming to be drawn upon the patient like parchment. There is great irritability of the nervous system, the patient being restless and uneasy, never satisfied, always changing its position, wanting everything, satisfied with nothing, and especially restless and wakeful at night. The child seems to be worse in the afterpart of the day and evening, and frequently every other day. When the disease becomes very severe, it is impossible to keep the child in bed at night, the heat seems to torture it, and it is only satisfied when laid where it can turn freely about, or when carried from place to place.

In the *non-febrile* form the skin is soft, relaxed, and flabby, the extremities cool, the bowels distended or pendulous, the tongue broad, flabby, and coated, and the pulse small, soft, and fluent. The child is not so restless as in the preceding case, seems stupid and dull when nursed in a comfortable position, but wants its

---

*Dr Golding Bird made an analysis of the *green evacuations of children.* and found, contrary to what was generally supposed, that they gave no indication of containing an excess of bile. He conceives that their color is due to the presence of altered blood, and that the state in which they are produced is analogous to that which exists in melæna, the portal system being generally in a congested state. In support of his views, he mentions that stools originally of a yellow or orange color, often turn green on exposure to the air, an occurrence very unlike what takes place with matters containing bile, though it is an ascertained fact that blood, under the influence of several oxidating agents, acquires a green color.

own way. In both cases the appetite is alike impaired, there is the same nausea, the same desire for drink, and the same prostration of strength.

We sometimes find the brain affected in those cases where there is a continued moving of the head from side to side, the child sleeping with its eyes partly open, and rolling the eyeballs upward. If the pupils are somewhat dilated and do not contract freely, upon exposure to light, I am satisfied there is congestion with effusion, and consider the patient's prospects very poor. Occasionally determination to the brain sets in, the head is hot, there is throbbing of the carotid arteries, contraction of the pupils, and great restlessness and uneasiness.

As the disease lasts for weeks, sometimes the whole summer, it does not present these active symptoms throughout. There are remissions and exacerbations; for a few days the vomiting will cease, the appetite improve, and the diarrhœal discharges are less frequent—consequently there is a marked improvement in the child. Then, from some slight indiscretion in diet, or over-exertion, but more frequently from alternations of temperature, there is another exacerbation.

DIAGNOSIS.—The acute attack is readily recognized by the sudden appearance of nausea, vomiting, and diarrhœa, the great prostration, and intense thirst. The chronic form, by its slow accession, and the same nausea and vomiting, the intense thirst, the persistent diarrhœa, and the impairment of digestion and nutrition, as marked by the rapidly wasting tissues.

PROGNOSIS.—The first element of prognosis with me is to determine the viability of the child, and I am guided in this by the depth of the base of the brain, as determined by the *life line* heretofore described. If the child has good viability, no matter how severe the disease, I am hopeful, and give a favorable prognosis. But if the life line falls below half an inch the prognosis is very doubtful.

The prognosis is less favorable when the brain becomes involved in the disease, presenting either irritation or congestion. Cases, also, which have been treated with irritant medicines, or by the persistent use of astringents, are difficult to manage.

TREATMENT.—The treatment of the *acute* form will be much like that for cholera morbus of the adult. The objects are to arrest the vomiting, to check the diarrhœa, and to restore tone to the system.

To fulfill the first indication we have a large-list of remedies, and, as is generally the case where there are so many remedies, there are none very certain. I think this will apply, however, to the medicines in ordinary use, and not to specifics. I will name the first in the order in which they are most commonly employed. ℞ Rhubarb, Bicarbonate of Soda, Spearmint, aa. This, as will be recollected, I recommend in place of the ordinary neutralizing powder of the Dispensatory. When the patient is first seen, add ten grains of the powder to ten teaspoonfuls of cold water, and allow it to settle; give a teaspoonful every ten minutes. Add a teaspoonful of the powder to one-third teacupful (two ounces) of boiling water; let it stand until cool, strain, sweeten, and if the child is somewhat exhausted, add a teaspoonful of brandy. As soon as prepared give half teaspoonful doses every half hour, and as it is retained, increase the dose to a teaspoonful every hour, until it acts slightly upon the bowels.

The first doses will be thrown up, but by repeating them it quiets the stomach and the nausea and vomiting cease; if not, after passing a reasonable length of time, select another remedy.

An infusion of the bark of the young twigs of the peach tree is a very reliable remedy in these cases, and may be given in small doses, frequently repeated, until the object is accomplished. The subnitrate of bismuth in mint water, will prove a good remedy in some cases; it is used in the ordinary doses.

In those cases where there is marked exhaustion, the extremities being cold, stimulants answer a better purpose than other remedies. The compound tincture of Cajeput is a good remedy in these cases. If not convenient, a tincture of the essential oils of cinnamon, cloves, peppermint, etc., singly or in combination, may be employed; or tincture of camphor may be used for the same purpose. Chlorodyne, the formula for which is given in the first part, may sometimes be given with advantage, in doses of about one to two drops. Chloroform may also be used, and will sometimes be tolerated when other remedies are not borne.

Whatever is given, or whether anything is given, the nursing is a most important part of the treatment. Place the child in the recumbent position, and wrap the feet in dry, hot flannels, keeping a plate, stove-lid, or any heat-retaining article in contact with them, so as to keep a continued heat. If there is feeble circulation, and marked coldness of the extremities, the powdered capsicum or mustard may be sprinkled upon these flannels. As an application over the stomach, I prefer the *cold* pack, either in the form of a wet towel folded, which may be renewed as often as may seem necessary, or an entire abdominal pack, which will remain for one or two hours, and will then be removed, the child rubbed thoroughly, and wrapped in dry flannel. The recumbent position should be insisted upon.

To quiet the thirst, I prefer giving small pieces of ice, or tying a piece in a cloth for the child to suck. If ice can not be had, then give cold water, a teaspoonful at a time, every few minutes. Sometimes a very little salt added to the water causes it to be retained better. When the vomiting is persistent, and the thirst excessive, I have employed an injection of salt water with good advantage. It is used in considerable quantity, and may be repeated when it passes away.

We have now to speak of *specific remedies* for this disease, and though they are not *cure-alls*, they deserve attention. When the skin is hot and dry, the pulse hard, and other evidences of a febrile condition, Aconite is the remedy. I use it in the ordinary proportion of two drops to water four ounces, but give one-fourth of a teaspoonful every fifteen minutes. The directions above as to nursing, must be implicitly followed.

In an ordinary case, I order Aconite and Ipecac alternately; they are prepared for use by adding two drops of Aconite and five drops of Ipecac to four ounces of water, and give a teaspoonful every hour. They may be a little slow but are very certain in their action.

When the face is pallid, and the mouth expressionless, the tongue pallid and atonic, Nux will serve our purpose best. It is given in quite small doses, gtt. j. to gtt. ij. to water ʒiv.; a teaspoonful every fifteen minutes until the nausea is quieted. The Ipecac is added in the usual proportion, and the two are continued until the diarrhœa stops. It is a little singular, but there is in the market much so-called "fluid extract" of Nux Vomica, that has no bitterness, and of course is absolutely worthless. Such stuff is not recommended.

In the acute form of the disease we have accomplished the greater part of the cure, when we have arrested the nausea and vomiting, and generally any simple remedy adapted to the case, will be sufficient to arrest the diarrhœa. In my practice I depend almost wholly upon the continued use of Aconite and Ipecac.

In true cholera infantum (chronic) it is well to make the classification of cases—irritative and atonic. Sometimes this is very readily done, as the symptoms are so marked, but in others we reach a half-way ground, a debatable land, and the classification can not be made.

In some cases the irritation of the stomach is extreme, so that the child can take neither food nor medicine with advantage. Aconite and Ipecac in very small doses, with restrictions in drink, (thirst is extreme,) will frequently relieve the stomach. Sometimes Euphorbia is substituted for the Ipecac, or if there is much tormina and straining at stool, Colocynth may be used. The dose of course must be very small. ℞ Tinct. Aconite, Tinct. Colocynth, aa. gtt. ij. water ℥iv.; a teaspoonful every hour. (The medicine should be kept cool, and prepared fresh every day.)

If the tongue is pallid, the face expressionless, and sallow or yellow about the mouth, Nux Vomica will be the remedy to relieve the nausea. One or two drops may be added to a half glass of water, and given in doses of one teaspoonful every half hour or hour. Ipecac may be added in some cases to assist in relieving the nausea, or after a day or two to control the diarrhœa.

The Compound Powder of Rhubarb, infusion of Peach bark, Bismuth, Hydrate of Chloral, and sometimes, but rarely, stimulants may be employed, as heretofore named.

In some cases we find marked irritation of the brain, and indications for Gelseminum. In others the patient is dull and stupid and Belladonna is wanted. There are cases in which the child is extremely restless, starts in his sleep, awakes frightened, cries out shrilly, and at the same time has a peculiar pinched expression about the eyes and base of the brain. The symptoms are characteristic, and the remedy is Rhus, and when thus indicated it will quiet irritation of the stomach, and check the diarrhœa.

There are cases in which sulphurous acid becomes an excellent remedy; cases that require sulphite of soda ; cases that want a simple alkali, bicarbonate of soda or lime water; and others that want muriatic acid. Here, as elsewhere, the indications for remedies should be followed.

Podophyllin thoroughly triturated with sugar of milk or loaf sugar, in minute doses, is sometimes an excellent remedy. Its use is in those cases in which there is atony of the stomach, and congestion of the liver and portal circle. Cases in which the tongue is broad, and coated, the abdomen swollen, the skin sallow, and the temperature lowered. Leptandrin is used in these cases, but is not so easily taken by children, and owing to its nauseous taste, is not so well tolerated by the stomach.

In cases where there is any evidence of periodicity, usually manifested by increase of nausea and vomiting and diarrahœa, quinia may be employed with advantage. I prescribe it as I do in intermittents, every two hours, until three doses are taken, or one dose of two grains, at the time when the patient is most free from nausea. In some cases its action is very decided—it checks the nausea and vomiting, lessens the thirst, allays the irritation of the nervous system, and indeed, seems to improve every function; and repeated for a few days, the patient is convalescent. In some cases minute doses of Arsenic with Veratrum will quiet the gastro-intestinal irritation, and strengthen digestion better than anything else.

When there is a soft compressible pulse, a relaxed and pale skin, and cold extremities, I have used the quinine inunction with benefit. I think in some very bad cases, it has given tone to the system, and so added to the patient's power to live, as to tide over the summer to frost, and convalescence.

Fatty inunction alone, answers a very good purpose. When the skin is dry and harsh, it is the best means to soften it and place it in condition for secretion. When it is soft and relaxed, by adding a stimulant to it, and using it with friction, it is a good means to stimulate capillary circulation. I do not like the free bathing sometimes recommended, as there is not the power of re-action, without which the use of water must do harm. If the child is bathed, let it be thoroughly and quickly washed with soap and water, or with salt and water, if the inunction is not practiced.

In some cases of cholera infantum a soft flannel bandage is applied to the abdomen, and gives as much relief as the medicine. A bag, a foot square, filled with good Fenugreek seed, sprinkled with oil of cloves, or some other of the essential oils, and quilted a half inch thick, and worn over the stomach, will give marked relief in some cases. It is after the order of the old fashioned spice plaster, but is better.

Much will depend upon the diet of the child in some cases. As a general rule, the mother's milk is preferable, and the case would be exceptional where I would recommend that it be weaned; still there are such cases. When the child has been weaned, however, we will have to take this into consideration. The directions heretofore given will enable us to determine whether the milk in use is good. If we see reason to change this, Liebig's food will be found the best, where it can be taken.

Raw or rare cooked beef will sometimes agree well with a child when it has been unable to take other food. Pickled pork will sometimes answer well, and be digested when the stomach will not tolerate lighter articles of food. The child's appetite may be gratified, and indeed it may be taken as a guide, with reference to all such things as constitute the healthy food of the country, only excepting some vegetables and fruits. In regard to the latter, we have difficulty sometimes in determining whether any, or what may be given with safety. As a general rule, *ripe* peaches and blackberries, *fresh* from the vines, will be the only ones we can give with safety.

The excessive thirst is an annoying feature of the disease, and we will be anxiously asked about gratifying it every time we see the patient. In some cases the child may have all the water it wants, even though it causes vomiting at first, for by giving drink freely in this way the irritation of the stomach is overcome, These cases, I am sorry to say, are few. The rule of practice is. that the child must be restricted to small quantities of water until the disease is arrested; it may be given frequently, however, and this is some alleviation of the thirst. In other cases, when the stomach refuses to tolerate water we order ice, and it may be given quite freely without danger, and sometimes with benefit. Recollect the injection of salt water in these cases, and the advantage that sometimes follows from the use of the wet-pack.

Convalescence must be managed with care, and indiscretions in diet avoided. I frequently put the little patient on the use of tincture of muriate of iron and glycerine, as heretofore named, and sometimes associate it with phosphate of soda, three grains, three times a day.

TABES MESENTERICA.

Tuberculous disease of the intestinal canal is most usually met with in childhood, though occasional cases will be seen even up to the age of twenty-five. It occurs only in those constitutions which we have before referred to as being tuberculous, and where, if the irritation had been of the lungs instead of the bowels, it would have been phthisis.

The pathology of the disease is well described by Habershon : " In disease of the mesenteric glands, a low organized product is effused into the glands themselves, probably because the chyliferous ducts become entirely obliterated, and the structure of the gland destroyed. Their extensive disease prevents the absorption of chyle into the system. The glands share the disease in various stages and gradations ; in some, but scanty abnormal product is found, in others the whole gland is destroyed and very much enlarged, constituting a whitish mass, the size of a pigeon's or hen's egg. The effused product consists of granular blastema, and imperfectly developed cells. The swollen and injected state of the glands less affected, appears to indicate that inflammation or hyperæmia is associated with the disease. The increase takes place by additions to the periphery of that already deposited, and degeneration occurs in the center from the scanty supply of nourishment afforded to the central part. The gland sometimes appears to be enveloped by a firm, fibrinous cyst, which consists of inflammatory product better organized, having assumed the character of fibrous tissue, while the center consists of calcareous deposit, the albuminous portion having been absorbed, and the organic only left. Degeneration of another character, however, takes place in the effused product, it is converted into a mass of granular molecules and highly refracting particles, constituting small, cheesy tubercles of a yellow color, or a softened and semi-diffluent mass. The lacteals between the glands become enlarged and distended with similar strumous product, or we can trace the distended ducts to the intestine, where they ramify on its surface, and at this part we generally find a cluster of tubercles and ulceration of the mucous membrane."

SYMPTOMS.—In children it is usually preceded by diarrhœa and gradually increasing prostration. The appetite is usually good, sometimes ravenous, but the patient receives no apparent benefit. The bowels are sometimes tumid, hot and tender, at others very much shrunken ; the evacuations consisting of a thin mucus, greenish, and frequently resembling the washings of meat. The countenance is contracted and pinched, the eyes set far back in the head, and the skin peculiarly dry, wrinkled, and sallow, giving the child a prematurely aged appearance. It is restless, irritable and fretful, and presents many of the symptoms of cholera infantum.

In the adult there may or may not be diarrhœa, frequently an alternation of diarrhœa and constipation, and sometimes severe pain. There is a marked marasmus, increasing day by day, though the appetite may be good, and the digestion seemingly well performed. The patient has an anxious expression of coun-tenance, a sallow, wrinkled skin, contracted abdomen, and is uneasy, restless and irritable. In the latter stage diarrhœa some-times sets in, and carries the patient off quickly, or disease of the brain or lungs comes on to assist the tabes. In both cases the enlarged mesenteric glands can frequently be felt through the abdominal wall.

DIAGNOSIS.—Tabes mesenterica is diagnosed with difficulty. The principal symptoms leading us to believe in strumous dis-ease of the mesentery are: the continuance of a good appetite, and seemingly good digestion, with continually increasing loss of strength and flesh, and the evidence of disordered bowels, and in the latter stages feeling the enlarged mesenteric glands through the abdominal walls. It will be seen that our diagnosis will have to be made principally by exclusion, a very common method, and possibly more correct than by direct symptoms.

PROGNOSIS.—The prognosis in well marked cases of this dis-ease is exceedingly unfavorable, as much so as any disease we are called to treat. In the earlier stages its progress may be arrested as it may also occasionally in the latter.

TREATMENT.—A tonic and restorative treatment would seem to offer the best results in these cases. Yet we find that it does not prove so serviceable as some other means. The general inunction with quinine once a day, and the use of the Uvedalia ointment to the abdomen is recommended. In some cases small doses of Arsenic exert a beneficial influence as : ℞ Fowler's Solution gtt. x., Tinct. Veratrum gtt. v., water ℥iv.; a teaspoon-ful every three or four hours. Hypophosphite of lime is a good remedy, as is the compound syrup of the hypophosphites ; cod-liver oil has given very good results when it could be taken well and digested. In some cases a good preparation of malt, taken after the foods, aids in digestion, and the child will improve on it.

Small doses of Ipecacuanha, alternated with Aconite, may be employed to relieve the irritation of the bowels, and check diarrhœa when it is present. Or, in place of these, we may use the extract of Hamamelis (Pond's), which I think very highly of. The dose will be about ten drops four times a day.

The use of Alnus, Rumex, Scrofularia, and others of our vegetable alteratives, has been recommended ; but I think they will not prove so serviceable as the means above named.

DYSENTERY.

Inflammation of the large intestine in the child arises from the same causes, and presents somewhat similar symptoms to the same disease of the adult. It occurs in both the *sporadic* and *epidemic* form, though in some epidemics children are remarkably exempt from it. Indeed, as a general rule, we will find that dysentery in the child is rarely the active inflammatory affection that we are called to treat in the adult.

Causes.—The common cause of sporadic dysentery is cold, though it may be produced by indigestible food, or such derangement of the digestive process as will permit the decomposition of food and formation of irritant products. Epidemic dysentery depends upon the influence of an animal poison ; what it is we do not know, nor why it acts upon this particular portion of the body.

Pathology.—Dysentery is a true inflammation of some portion of the large intestine—a *colitis.* The inflammation may involve but a small portion of the intestine, and be confined to the mucous membrane, or it may involve the greater part or all, and extend to the muscular and peritoneal coats. There are some cases in children in which the inflammatory character is not marked, but rather an irritation of the mucous coat, with increased secretion, and of the muscular coat, with increased peristaltic movement. This form of the disease is not attended by fever, and may continue for weeks or months without danger to life.

Symptoms.—In acute sporadic dysentery of the child, it is first noticed that the bowels are moving too frequently, and there is considerable straining at stool. The discharges, however, are feculant, though small, but if noticed closely they will look slimy, and after a day has passed they will be almost pure mucus, or mucus tinged with blood.

As the dysenteric character of the discharges becomes marked, febrile symptoms appear. The child is restless and uneasy, the skin is dry, the pulse is accelerated and hard, the appetite impaired, and indeed all the functions of the body deranged. These symptoms vary greatly in different cases—in some the fever runs high, in others it is hardly noticed.

But as the disease progresses the discharges become more frequent, every fifteen, ten, or even every five minutes; they are preceded by pain, and attended with tenesmus. This straining at stool is so marked a feature with the child, and so strongly expressed in its countenance, that it is an index to the character of the disease as soon as seen. Day by day the child loses strength, and the general and local disease seems to gain intensity, until the fifth or sixth day; then there is a gradual amendment, until convalescence is completely established. I speak of this as being the usual course of the disease when not influenced by medicine.

Epidemic dysentery presents the same symptoms that it does in the adult. Commencing with evidences of prostration, and impairment of the general health, there is usually a well marked chill, which is followed by febrile reaction. This general disease assumes so prominent a place at first, that it overshadows the local inflammation. Yet, at first, there is an irritation and increased activity of the bowels, and with the development of the fever, the small mucous or bloody dysenteric evacuations become marked.

It is not worth while in this place to enter into a lengthy description of the symptoms presented, as they are a continuation of those above described, with a remittent or continued fever. As the disease progresses, it develops typhoid symptoms, both general and local, and in the severer cases it becomes a very grave affection.

The dysenteric irritation referred to when speaking of the causes of this disease, presents but a part of the symptoms of dysentery. The discharges are attended with straining, are small, and in some are wholly mucus, in others mucus mixed with feculant matter. They are also frequent, and in some cases it seems that the child can not stoop or be placed in a constrained position without an action of the bowels. But the general health is less affected than we would suppose, not nearly so much in most cases as in ordinary diarrhœa. Still, as it continues for a long time, it is a cause of much annoyance to both parents and physician.

DIAGNOSIS.—The diagnosis of dysentery is made with ease. No one could see the peculiar expression of countenance during the tenesmus, without at once recognizing what was the matter, even if he had never seen a case of the disease.

PROGNOSIS.—The prognosis is favorable in the sporadic dysentery of children, but in those epidemics that reach children the mortality is greater than in the adult.

TREATMENT.—In the mild sporadic form of the disease, my practice resolves itself into the administration of Aconite with Ipecac in the usual doses. The general bath is used, and the hot foot-bath, and the child is kept still, and in some cases a cloth spread with lard or mutton suet is applied over the abdomen. The treatment is very simple and very satisfactory, and I should be glad if all treatment were as much so.

When there is persistent abdominal pain or tormina, I like the action of Colocynth, in the small doses heretofore named. In some cases Veratrum may be used, and if there is evidence of determination of blood Gelseminum is given with it.

In some cases the broad pallid and dirty tongue calls for sulphite of soda, and it is given in doses of from two to five grains every three hours until the tongue has cleaned. Sulphurous acid is an excellent remedy in typhoid dysentery, the tongue being red and full, and covered with a nasty glutinous coat.

In the olden time, the white liquid physic of our Dispensatory, held a prominent place in the treatment of dysentery. And it may yet be used with advantage in some seasons. I advise that the proportion of acids be diminished as follows: ℞ Sulphate of soda ℥viij., water ℥xxiv., Nitric acid, Muriatic acid, aa. ℥ss. Mix. The dose for a child two or three years old will be a teaspoonful every two hours, (with syrup), until it moves the bowels gently, then in smaller doses. In some cases olive oil may be used to move the upper intestine.

If in a malarial region, we look carefully for periodicity, which will be found in many cases. Here quinine is an indispensable remedy, the patient being prepared for it, and it is given in the usual doses.

In some cases Ipecacuanha may be prescribed alone, triturated with powdered gum arabic, which is a good form, or combined with bismuth, as : ℞ Subnitrate of Bismuth, ʒss.; Ipecac gr. iv.; mix and divide in eight powders, giving one every one or two hours.

The same treatment is employed for the local lesion in the epidemic form of the disease, and in addition such remedies as will antagonize the septic tendency, and give tone and support to the system. These have been fully described under the head of continued fever, to which the reader is referred. In some of these cases the Ipecacuanha seems to fulfill nearly all the indications, and it is given with the best results. In others the use of white liquid physic to the extent of gentle purgation, afterward in smaller doses, answers best; and in still others we obtain the best results from bismuth.

In dysenteric irritation we find a very stubborn case, and one that sometimes refuses to yield to remedies, literally wearing out. I place most dependence upon the continued use of minute doses of Aconite, for several days ; next upon the use of the subnitrate of bismuth. Occasionally we find a case in which there is evident relaxation of the rectum, with congestion, and in such the fluid extract of Hamamelis will prove very good.

The local means will consist of cold packs to the abdomen, the disease being sthenic and the patient carrying a high temperature ; or hot packs, or sponging the abdomen, if the disease is of a lower grade. In some cases a hot sitz bath gives relief, and the patient has rest and sleep for a few hours.

In some cases we are obliged to use enemata to get rest for the inflamed bowels, and exhausted patient. Sometimes water as warm as it can be borne serves a good purpose, but usually we think of tinct. Opium gtt. v. to x., starch water ʒij. to ʒss.—this is repeated after each evacuation.

CONSTIPATION.

Constipation is an attendant on many acute diseases, and is generally to be regarded as a favorable indication, though we may have to use means to overcome it. It is also met with as an idiopathic affection, and it is this form that we wish to consider.

Infantile constipation seems, in a majority of cases, to be a congenital defect in the irritability of the intestinal canal, rather than an acquired lesion. Thus we will find it manifested early, the child's bowels not moving for two or three days after birth, and continued on until by change of food, during the second year, a normal activity is induced. In some cases it seems to be dependent upon constipation of the mother, the acquired vice in her constitution manifesting itself in the condition of the child. In some others it is dependent upon the character of her food; but in the majority, we can give no reason for it, and we have simply the fact that constipation exists.

Sometimes the constipation is attended with uneasiness in the abdomen and colic; but in the majority of cases the child does not suffer at all, but nurses well, and thrives like other children, and were our attention not called to it by the mother, we would not suspect anything wrong. In some cases the child will make forcible efforts to stool, and may pass a small portion of hardened feces, at such times, or nothing may pass. Such attacks of tenesmus may come on three or four times a day, and are very unpleasant. Indeed, this is the only case that demands special attention.

TREATMENT.—If the mother is of a constipated habit, we will be more successful in removing the torpor of the child, by such means as will restore normal activity to her, rather than by giving it medicine. The means I commonly employ are—that on first rising in the morning, a glass of cold water be taken, adding to it one or two drops of tincture Nux Vomica; that the abdomen be thoroughly rubbed with the hand dipped in salt water; and that a regular time be set apart (immediately after breakfast is best) to evacuate the bowels. Should this not be sufficient, ten or twenty grains of phosphate of soda may be taken in a glass of water, on going to bed.

As remarked above, the medication of children to overcome constipation is not very successful, sometimes we will accomplish the object, at others we will be partially successful, and in others we will fail entirely. Phosphate of soda, in doses of two or three grains, three times a day, will occasionally answer the purpose, and the influence is usually permanent. Small doses of Belladonna may also be given, with the expectation of good results, in some cases. In those cases where the abdomen is re-

22

laxed and pendulous, with a torpid circulation, small doses of Nux Vomica will answer the purpose; and where the liver is inactive, Leptandrin, thoroughly triturated, may be given in doses of one-fourth to one-half grain.

When the child suffers from tenesmus, it is well to move the bowels at first with olive oil or castor oil, and then follow with the means named above.

Dr. Tanner remarks, that "since reading Corvisart's views on the action of pepsin, he has extensively used the remedy; which may be given to the youngest infant, in doses of three to six grains, dissolved in milk, twice daily. It seems to relieve constipation in the same way that it checks some forms of diarrhœa, by enabling the stomach to do its work thoroughly." If we use pepsin for this purpose, it will be quite as well to use ordinary *rennet*, giving the water in which it is soaked in quantities of one or two teaspoonfuls, each time after nursing.

I think the use of enemas and suppositories, in these cases, should be abandoned, as they are likely to do more injury to the rectum, than they do good in relieving the constipation. I have omitted to mention one thing which I regard as important in some of these cases, to see that the child's clothing is worn loose. Not unfrequently the *band* is put on so tightly as to paralyze the abdominal muscles, and I have no doubt that this is the principal cause in some cases. Thoroughly rubbing the abdomen with the hand, and a little lard, will prove beneficial in these cases.

PROLAPSUS ANI.

Prolapsus ani is met with most frequently from the age of two to four years, though it is sometimes seen during the first months of life. It comes on slowly, but as it is not noticed in its early stages, it seems to the mother as if it had wholly developed within a few hours.

In almost all cases there is a softness and relaxation of the perineal tissues, which will be noticed as soon as the child is examined. In some the sphincter ani seems mostly at fault, having so lost its power of contraction that it is not able to support the bowel. We may recognize two varieties of the disease: In the one there is this failure of the sphincter, and the entire bowel is permitted to descend, yet retaining its natural relation. In the other, the connective tissue between the mucous and muscular coats is relaxed, or has given way, and the mucous coat is extruded through the sphincter, which grasps it tightly. In this case the bowel seems as if turned inside out.

SYMPTOMS.—In the early stages the child complains of uneasiness about the rectum, after going to stool, and if walking will be observed to keep the legs separated; but in a few minutes the bowel naturally retracts, and the uneasiness passes away.

In the severer cases there is tenesmus during and after going to stool, and occasionally the child suffers severely. The bowel requires to be returned artificially in order to give relief. If this is neglected it becomes congested and swollen, sometimes so much so, that days will be required to so reduce its size as to permit a return through the sphincter. It is very easy to determine what is the matter in these cases, as an inspection of the part shows the bluish or dusky discolored tumor at the site of the rectum.

When the disease is of long duration, the skin and mucous membrane seem too large for their purpose, and the first hangs in folds outside, while the second is gathered in folds inside. In some cases there is also an enlargement of the veins of the mucous coat, so that in consequence it is considerably thickened.

TREATMENT.—The first object, when called to a case of prolapsus ani, if the bowel is still prolapsed, is to replace it. This is not always a pleasant job, yet it can be accomplished, even when considerably swollen, if care is used and the physician has patience. Have a soft cotton or linen cloth spread with lard, and place upon the protruding gut, the little patient being upon his hands and knees, or more frequently lying across its mother's lap, face down, and feet depending. Now placing both hands so that one or two fingers may rest on the gut, the other fingers and thumb press upon the perineum in such direction as to relax it. Gentle and continued pressure being made upward, in a short time reduction is effected.

For the radical cure, if the case is mild, I direct that the part shall be washed with an infusion of Hamamelis, and that the distilled extract of the same (Pond's) be given in doses of five or ten drops three times a day; or in place of this, we may give the Collinsonia in doses of one drop three times a day.

In some cases small doses of Nux, in others of Colocynth, and in still others of Aloes, will relieve the tenesmus that causes the prolapse, and will strengthen the lower bowel and femoral muscles.

In severe cases, when of long duration, I use the persulphate of iron as a local application. In most cases one part of the solution to two or three parts of glycerine, will be as strong as it can be borne, but in others it may be used of full strength.

The hypodermic injection of ergotin near the sphincter has given excellent results in cases otherwise incurable.

There are a few cases which will require operative interference. Two methods are practiced. In one the folds of skin alone are removed, in the other the entire projecting part—mucous membrane and skin. In the first case the operation is simple—the ordinary silk ligature having been passed through the base of the fold in two or three places, this is cut away with a pair of curved scissors or a bistoury, and the ligatures tied, bringing the cut edges together. Thus four, five, or six folds are cut away, and when the part has cicatrized, it is much smaller and tense, and gives the needed support. In the other case the ligatures are passed through the projecting rectum, at distances of about one-half inch apart, leaving the ends sufficiently long to tie. The gut is then cut away outside of the ligatures, and when these are, tied the cut surfaces are again in apposition, and cicatrization occurring, a very firm support is given, and the disease is readily cured.

INTESTINAL WORMS.

The presence of worms in the intestinal canal is not always attended by symptoms of disease; indeed children may enjoy very good health while infested with these parasites. It has been claimed that an entirely healthy person will not have worms, as that an entirely healthy and well-taken-care-of child will not have *pediculi*. I think this is true, to the extent that the naturally healthy child will not suffer from them, and not furnishing a comfortable home they are not likely to remain.

A depraved condition of the intestinal canal, with increased mucous secretion, seems to furnish the conditions necessary to the propagation of worms, and as digestion is necessarily interfered with by this condition, there will be more or less impairment of the general health.

CAUSES.—In olden times it was thought that there was *spontaneous generation* of worms, and there are some who yet hold to this opinion. The fact is, there is no *new creation* in this world; every organized body, from the simplest cell or monad to the highest animal life, being the product of parents who possessed the same form and functions. Intestinal worms, therefore, have a parentage, and are propagated only from the same species.

The subject has been carefully studied for many years by Kuchenmeister, a very persevering German physician and naturalist, and the facts are very clearly established. The two varieties of *ascaris* and the *trichocephalus* are produced directly from ova taken into the intestinal canal, which, finding the conditions for incubation, are developed into the fully formed worm. Previous to this, it was believed that these vareties were viviparous—giving birth directly to the young. This was the opinion of Dr. Good, of Watson, and others.

The tænia are oviparous, producing eggs from which the worms are developed. The development, however, requires two animals of different species, in one of which the worm attains its *pupa* state, and in the second attains its perfect development. The process is very much like the development of a butterfly—from the egg to a grub, from that to a fully developed insect.

Kuchenmeister traced the tænia solium from the *cysticercus* of the pig, through all its gradations up to the fully formed worm. These cysticerci are very tenacious of life, and may get into the intestinal canal by eating raw or partially cooked fresh pork, or even bacon.

When once introduced into the human body, a new development commences, and from these small grub-like bodies are produced the perfect tape-worm. Each joint of one of them contains a multitude of eggs, which being discharged with the intestinal contents, regains its original habitat, the hog, is developed into a cysticercus, which in turn by transplantation becomes a tænia. Thus from the fully developed worm in the human intestine is thrown off the eggs, which, being taken by the hog with its food or drink, gains access to its tissues and is developed into the cysticercus or grub, and this being eaten by men produces the perfect worm.

This being the origin and mode of development, we find that certain sections of country are *verminous*, while others are comparatively free. This has been remarked in case of the tape-worm, and I think it just as well established with the other forms. When the conditions are favorable for the reception of the discharged ova in water or food, we will find that intestinal worms are common, and where these do not exist they are rarely seen.

The principal varieties of intestinal worms are: the *ascaris lumbricoides, the ascaris* or *oxyurus vermicularis, the trichocephalus dispar,* and the *tœnia solium* and *vulgaris.*

The *ascaris lumbricoides,* or long round worm, is described by Dr. Good as having a slightly incurvated head, with a transverse contraction beneath it; mouth triangular; body transparent; color, light yellow, with a faint line down the side; gregarious, viviparous; from six to fifteen inches long; inhabits principally the ilium, but sometimes ascends into the stomach, and creeps out of the mouth and nostrils; occasionally travels to the rectum, and passes away at the anus.

The *ascaris vermicularis,* or small thread worm, has its habitat in the rectum, though it sometimes gets into the intestines, and occasionally in the female, into the vagina. "The head is subulate, nodose, and divided into three vesicles, in the middle of which it receives nourishment; skin at the sides of the body finely crenate or wrinkled; tail finely tapering and terminating in a point; gregarious, viviparous, and about half an inch long."

The *trichocephalus dispar,* or long thread worm, is found in the intestines, both large and small, and in the stomach, and especially in sickly children, and those who are poorly nourished.

" The body is obese, slightly crenate, beneath smooth, finely striated on the fore part; the head obtuse, and furnished with a slender retractile proboscis; tail or thinner part twice as long as the thicker, terminating in a fine hair-like point; about two inches long, and its color light yellow."

The *tœnia solium,* or long tape worm, is described by the same author "as having long and narrow articulations, with marginous pores, by which it attaches itself to the intestines; one on each joint, generally alternate; ovaries arborescent; head with a terminal mouth, surrounded with two rows of radiate hooks or holders; and a little below, on the flattened surface, four tuberculate orifices, or suckers, two on each side; it is from thirty to forty feet long, and has been found sixty. Inhabits the intestines of mankind, generally at the upper part, where it feeds on the chyle and juices already animalized. Is sometimes solitary, but commonly in considerable numbers; and adheres so firmly to the intestines that it is removed with great difficulty. It is said to have the power of reproducing that which has been broken off; but this assertion wants proof. The animal is oviparous, and discharges its numerous eggs from the apertures in the joints.' The articulations are from four to six lines in length, and nearly as much in width, and resemble gourd or melon seeds.

" The articulations of the broad tape worm are short and broad, with a pore in the center of each joint, and stellate ovaries around them; body broader in the middle, and tapering toward both ends; head resembling the last; inhabits the upper part of the intestines, and feeds on chyle; from three to fifteen feet long; usually in familes of three or four."

SYMPTOMS.—With many if not all forms of worms, it is necessary that the bowels be in a condition to furnish a comfortable habitation. This condition is essentially one of want of tone, with, in many, increased secretion of intestinal mucus. We observe in many cases that the child or person is poorly nourished, the muscles are soft and flabby, there is a loaded tongue, bad breath, and derangement of the secretions. We are not inclined to believe that this is the result of worms, but simply coincident with them, and in some cases the patient has what is termed *worm fever*, usually of an intermittent character, the paroxysms occuring in the afternoon and evening, at which time we find the skin hot and dry, the pulse frequent, the head hot, and marked irritability and restlessness, and occasionally convulsions. Or the fever may be more obscure, the child is fretful and nervous, sleeps poorly, its breath is fetid, tongue coated, bowels irregular, abdomen tumid, is frequently picking its nose, the upper lip swells, a white line appears around its mouth and it seems to be out of order generally. These are the symptoms of the first named varieties, though not nearly so well marked in the case of the ascaris vermicularis. Though seeming to be very plain, yet all these symptoms may be present and no worms; or worms present, and but few of these symptoms. The only certain evidence of worms is their presence in the feces, and even then we can not be certain but that all have passed. The ascaris vermicularis makes itself known by an intolerable itching and crawling sensation about the anus. At first it generally comes on after the little patient gets warm in bed, the irritation being so great that sleep is impossible; at last, they are more or less troublesome all the time. The irritation is occasionally so great as to impair the health, and occasionally gives rise to convulsions.

As regards the symptoms of tape-worm, they are very decep-tive. In one hundred cases recorded by Seeger, in sixty-eight instances nervous affections, or general or partial convulsions, occurred—epilepsy, hysteria, abdominal spasms, convulsive cough, dyspnœa, melancholy, and hypochondriasis; in forty-two, vari-ous pains in the abdomen; in thirty-three, disordered digestion and irregular states of the evacuations; in thirty-one, irregular appetite and voracity; in nineteen, habitual or periodical hemi-cranias; in seventeen, sudden colic; in sixteen, sensations of undulatory movements in the abdomen up to the chest; in fifteen, vertigo, delusions of the senses, and defects of speech; and in eleven, shifting pains in various parts. The only definite evi-dence of the presence of tape-worm is the passage of portions of it with the feces, and as this usually occurs with this worm, the non-appearance of the joints in the evacuations during a consid-erable time, may be considered as good evidence that the worms do not exist in the intestinal canal. .

TREATMENT.—The treatment of the ascaris lumbricoides and trichocephalus will be very similar, the object being to remove the worms, and break up the predisposition to them by remov-ing the condition on which they depend. Very many vermifuge remedies have been recommended and used with success, so that the trouble will be, not that we have no remedies, but that we have too many. The old fashioned remedy, " pink and senna " in infusion, seemed to be about as certain as any other agent, and I am satisfied that if it were as disgusting to the worm as it is to the child, it would readily leave its nest in the bowels, rather than take the dose. Still it is not more nauseous than the oil of wormseed, which is an ingredient of all the principal vermifuges, as—℞ Oleum Chenopodii ℥x., Oleum Terebinthinæ ℥ij., Oleum Ricinii ℥iij., Aqua Calcis ℥x., Syrupus Limonis ℥vj.; Mix; the dose being two teaspoonfuls three or four times a day. Kuchen meister recommends the santonine, and the santonate of soda, for the ascaris lumbricoides; he considers it to be best adminis-tered in oil, in order to bring it into solution as readily as possi-ble, and thus combines it with castor oil, or sprinkles it on bread and butter, and follow it with Jalap and Senna. Troublesome effects sometimes follow the administration of this remedy, as severe irritation of the nervous system, convulsions, tenesmus, bloody stools, and the minor disturbances, green or bluish vision, and discoloration of the urine. The santonate of soda he gave in doses of from two to six grains on a Friday night, and the same dose on Saturday and Sunday mornings, fasting; half an hour after this last powder, confection of Senna and Jalap is taken in sufficient doses to produce several fluid evacuations, the worms passing alive, and sometimes wandering forth without any operation, the intestines having become unpleasant for them.

I now employ a trituration of Santonine with Podophyllin in the following proportion—℞ Podophyllin gr. j. to ij., Santonine gr. x. to gr. xx., White Sugar ʒij.; triturate and make twenty powders. One of these may be given morning and night, until the object is accomplished. When there are no worms, but only the atonic condition of the mucous membrane with increased secretion of mucus, the patient will be benefited, and sometimes cured by the remedy.

A judicious tonic course of medicine, the bowels being kept regular, and the other secretions free, with an avoidance of all grease or indigestible food, the daily use of the bath, and exercise in the open air, are the only means by which we can break up the tendency to the formation of worms.

Many remedies have been recommended for the ascaris vermicularis, but in my opinion all vermifuge medicines should be discarded. If the patient's bowels are irregular, proper means should be taken to overcome the difficulty, and if necessary a tonic and bracing treatment adopted. For the worms I have always directed an injection of salt and cold water, in the proportion of a teaspoonful to half a teacupful, and so far with invariable success.

Should this fail, we will be able to relieve the patient by the use of an aloetic purgative, as in the following formula: ℞ Tincture of Aloes, Compound Tincture of Lavender, aa. ʒss., Simple Syrup ʒj.; a teaspoonful four times a day.

Tape-worm is of very rare occurrence in childhood, and is possibly never met with during the first two or three years. The mildest treatment will be best here, and I would recommend the emulsion of *pumkin seed*, or the oil of male fern. The emulsion is made by depriving two ounces of the seed of their capsules, and beating them into a pulp, with sugar and water; this is the dose for an adult, and is taken upon an empty stomach, in the morning; and is followed in from two to four hours by a full dose of castor oil. The oil of male fern is also taken, fasting, the dose being from gtt. xx. to ʒj. in mucilage of milk.

The most effective means for the removal of tape-worm will be found in the Pomegranate bark, according to the directions of Prof. Locke, a full description of which may be found in my Practice of Medicine.

DISEASES OF THE LIVER.

In the olden time, diseases of the liver were the most prominent affections of both adults and children, and would occupy a conspicuous place in treatises upon the subject. But as we have learned more and theorized less, the liver has played a less important part in pathology, and has received less attention from the therapeutist.

We have reason to believe that the intra-uterine function of the liver is of more importance than its function after birth. For, of all the organs of the body, it alone had its complete development before birth, and in size was much larger than any other organ in proportion to the body. What its functions were there, we are as much unable to say as what its functions are in the adult; but of one thing we are convinced, that it is so admirably adapted to its office that it rarely suffers from serious disease.

CONGESTION OF THE LIVER.

We notice a singular disease, occurring during the first three or four months of infancy, in which there is enlargement of the right hypochondrium and the epigastrium. It is known by nurses and old ladies as *liver-grown*, and they propose its removal by rubbing with some fatty material—generally goose-grease. That there is such enlargement there can be no doubt, and that it is relieved by these frictions is also certain, but what the exact condition is, I have never been able to determine. The symptoms would indicate congestion of the liver, hence I have given it that name.

SYMPTOMS.—For some days the child has been fretful and uneasy, seeming to suffer from indigestion and colic. The discharges from the bowels are more frequent, and have an unhealthy appearance, generally light-colored. At last the child is undoubtedly sick, has some fever, does not sleep well, is startled in its sleep and cries out as if in pain. It does not nurse well, and usually feels uneasy after taking the breast, and relieves itself by throwing up its milk. This has an unpleasant sour smell, and is not curded as in healthy digestion.

The symptoms named may continue for some weeks, the child losing flesh and becoming quite feeble; and I doubt not, though I have never observed it, that this condition might continue to a fatal termination.

DIAGNOSIS.—With the symptoms above given, which resemble those of infantile dyspepsia, we will find a distinct enlargement of the right hypochondrium. A careful examination, with palpation, will show the liver to be really enlarged, as its outlines can be distinctly traced.

TREATMENT.—As named above, the treatment of the nurse is usually quite successful, and this is simply the thorough rubbing of the part with some oily material. In the hands of an old lady, the treatment is very vigorous, with rubbing, kneading, and palpation, the child being turned from side to side, now on its abdomen, then on its back, and this continued for a quarter or half hour, and repeated every day, and sometimes twice a day.

On first witnessing it I thought it pretty rough usage for a delicate and sick infant, and protested against it. But the old nurse insisted that doctors were *fools*, and would have her own way, and the result justified her claim to be able to treat *liver-grown*, for the child made a quick recovery.

I learned one lesson from this, that I have put in practice many times since with success—that torpid organs could be stimulated to action by passive movements, and their circulation and nutrition decidedly improved. This is the principle of the Swedish "movement cure," which is claimed to be as well adapted to a dyspeptic stomach, as to an enfeebled limb.

The remedies I would advise in this case are—℞ Tinct. Belladonna, Tinc. Aconite, *aa.* gtt. iij., water ℥iv.; one-fourth to one-half teaspoonful every hour. If it was somewhat chronic, having lasted for some weeks, small doses of Nux might be given, or if there was still irritation, the tincture of Chionanthus in drop doses every three hours.

If the child is feeble, and especially if suffering from a malarial influence, quinine inunction would be advisable. In all cases I would advise the thorough rubbing of the part, as first named.

STRUMOUS ENLARGEMENT OF THE LIVER.

This is not of frequent occurrence, but may be occasionally met with in the children of the poor in large cities, as the result of poor and scanty food, bad air, want of sunlight, want of cleanliness, proper clothing, warmth, and indeed all that we are accustomed to regard as the comforts of life. Rarely a case occurs in a family in good circumstances, from scrofula or bad blood, or from the depression of other sickness.

Dr. Tanner observes "that a peculiar enlargement of this gland not unfrequently occurs in feeble and delicate children. The abdomen gradually enlarges, so that the little patient is said to be 'pot-bellied,' and on examination one or more well defined tumors are discovered. These tumors are formed by the enlarged liver, with perhaps an enlarged spleen; the increase in size in both cases being due to the interstitial deposit of albuminous matter. As this foreign matter is soft, and has no tendency to contract like the lymph poured out in ordinary inflammation, it does not much impede the passage of the blood through the liver, or the escape of bile through the ducts; and hence it very seldom gives rise to serious disturbance. Should the same material, however, become deposited in the structure of the kidneys, then the function of these glands becomes so completely interfered with that the cases cease to be amenable to treatment, and albuminuria, ascites, and anasarca ensue, and death ultimately results."

SYMPTOMS.—The child has lost flesh in a very marked degree, though it is usually not so feeble as we should suppose by its appearance. Its appetite is variable; sometimes craving, especially for fruits, sweets, pastry, and things it should not have; at others it wants to eat but little. Digestion is accomplished tolerably well, so far as we can observe from any derangement of stomach or bowels, following the taking of food. But it is imperfect in that the material is deteriorated during the digestive act, and can not make good blood or good tissue.

Generally, there is more or less derangement of the bowels, the discharges being too frequent, attended with uneasiness, and are light-colored. The urinary secretion is somewhat scanty, highly colored, and sometimes presenting distinct evidence of the coloring matter of bile. The skin is generally sallow, relaxed, and hangs in folds, the circulation being sluggish, and secretion imperfect.

In the majority of cases there is slight febrile action in the after part of the day, and the child is restless and fretful at night, wanting to drink often. In some cases an unpleasant irritation of the nervous system, causing the child to be restless and fretful and sleeping badly, is the most unpleasant symptom.

DIAGNOSIS.—The distended abdomen, so large as compared with the general loss of flesh, and the distinct enlargement in the region of the liver, its lobes being felt as distinct tumors, is sufficient. In tabes mesenterica there is a somewhat similar enlargement of the abdomen, but it is most prominent in the region of the umbilicus, and the nodulated glands may also be felt in the mesentery.

PROGNOSIS.—When of long duration, and attended with great marasmus, the prognosis is unfavorable. But if the child has considerable strength, and digestion is not much impaired, we may expect favorable results from the use of remedies.

TREATMENT.—I have much faith in a judicious tonic and restorative treatment, in all these cases, paying particular attention to the clothing, food, and that it has an abundant supply of sunshine and fresh air.

The internal administration of Collinsonia and Hamamelis alternated, in the doses heretofore named, will be found to quiet irritation of the bowels, check the frequency of the discharges and change their character. Associated with this, I would give the tincture of muriate of iron with glycerine morning and night.

Let the child be thoroughly rubbed twice a day with the quinia inunction, freeing the surface from any surplus material by rubbing with clean flannel, and occasionally by washing with soap and water. This I consider one of the most important parts of the treatment.

The child should be clothed in flannel throughout, both winter and summer, and in chilly or cold weather, especial care should ·

be used to keep it warm, and in an atmosphere of uniform temperature. The room where it sleeps should be thorougly ventilated, and have a good supply of sunshine, and in addition the
child should be carried in the open air.

The food may be the same as recommended in cholera infantum,
and should be carefully prepared. Sometimes more will depend
upon this than upon the treatment. For a poor and bad blood
must increase the disease, while from good food, well digested,
good blood is formed, capable of supplying the tissues, and giving no waste for struma.

JAUNDICE.

We have two varieties of jaundice in childhood which require
separate description. The first, or *icterus neonatorum*—the jaundice of infancy—is usually a very simple affection, and though of
frequent occurrence, it passes away with simple treatment. The
second variety resembles the jaundice of the adult, both in its
pathology and symptoms, and while of rare occurrence it is
sometimes a severe affection.

CAUSES.—The causes of the first form are very obscure, and
while occasionally it seems to arise from some irritation of the
primæ viæ, in other cases there is no apparent cause for it.

The second form of jaundice may be caused by cold, producing congestion of the internal organs, and especially of the portal circle. Or it may be produced by a malarial poison, which,
impairing the circulation of blood, causes congestion of the abdominal viscera.

PATHOLOGY.—Jaundice is but a symptom of some lesion of
the liver or of the blood. It consists in a change of the coloring
material of the blood, and its being set free and deposited in the
tissues; or more frequently in the retention and absorption of
the coloring matters of the bile, and their deposit in various
structures, principally the skin and conjunctiva. Occasionally
it is deposited in the deeper structures, as of the eye, giving
rise to yellow vision ; in the nails, and in the internal organs.

It is singular that the mildest and the severest form of jaundice are alike, in that they are dependent upon change in the
coloring material of the blood. In *icterus neonatorum* I regard
the coloring matter deposited in the skin, as arising from the
natural retrograde metamorphosis of the red globules, but fail-

ing to be removed, from some temporary inactivity of the excretory organs. All the symptoms point to this as the true condition, for in the majority of cases, there is no lesion of the digestive canal or its associate organs, all their functions being carried on regularly and well; while in those cases in which there
is such irritation, it seems rather from arrest of excretion than
from any other cause.

In those extreme cases of jaundice, running a regular and
very certain course to a fatal termination, the coloring material
is changed hematine and not bile. In this respect it resembles
the bronzed discoloration in Addison's disease (disease of the su-
prarenal capsules).

The other form of jaundice arises from—a, hypersecretion of
bile; b, from congestion of the liver and portal system; c, from
chronic alterations of the structure of the liver, preventing se-
secretion or the free discharge of bile; d, from spasm or temporary obstruction of the biliary ducts; e, from obliteration or
compression of the biliary ducts or gall-bladder; f, and lastly,
from disease of the duodenum, partially or entirely occluding the
ductus communis.

SYMPTOMS.—Icterus neonatorum is frequently attended at first
by slight febrile action. Occasionally this is quite marked, and
the child is restless and uneasy, and sleeps badly, continuing for
from one to four or five days, in a remittent form; it gradually
passes away, and the secretions being restored, the child seems as
well as ever, but for the yellow discoloration. It will be recollected that in the majority of cases the fever is not high, and
passes off entirely by the end of the second day.

The yellow discoloration very gradually fades, seeming to be
removed more by the natural removal of tissue than by special
absorption. Rarely continuing less than two weeks, it may not
be entirely removed for as much as three months.

The other form of the disease rarely occurs during the first or
even the second year. Usually, there is distention of the bowels, colicky pains, constipation, the feces being clayey, pale, and
scanty. The mouth is dry, has a bad taste, tongue coated, and
sometimes nausea and pain in the head. The yellow tinge usually makes its appearance in the eyes, and gradually extends to
all parts of the body, the color being deepest in the folds of the
skin.

For practical purposes, we may divide this variety into two forms, the symptoms being distinctive and the treatment different. In the one case the patient is irritable and restless, the skin is dry and harsh, the urine high-colored, the pulse hard and increased in frequency, and the temperature elevated. In some cases the patient complains of pain in the right hypochondria and in the shoulders, and sometimes in the small of the back.

In the second case the tongue is broad, palid, and covered with a uniform yellowish-white coat. The patient is dull and sluggish, and has an inclination to sleep. The bowels are swollen, but not tender, the extremities cold, the skin inelastic, and the discoloration not bright, as in the preceding case, the pulse soft and oppressed. In the first form there was febrile action, in this the opposite state.

DIAGNOSIS.—The diagnosis of jaundice is easily made. The yellow discoloration is so unlike anything else met with, if we except Addison's disease, which never occurs in childhood, that no mistake can be made. Icterus neonatorum occurs during the first few weeks of life, and may be determined by this. The pathological conditions, giving rise to jaundice proper, will be determined by the symptoms named above.

PROGNOSIS.—The prognosis of the jaundice of children is very favorable. While it is sometimes very persistent, especially the first form, there is but very little, if any danger, except from injudicious medication. The second form yields readily to appropriate treatment.

TREATMENT.—For *icterus neonatorum* I should prescribe tincture of Aconite in the usual doses, aiding it with the general bath, the general hot bath, or the hot foot-bath. When the fever declines I should direct that an infusion of Asclepias be given in addition. It will be recollected that saffron has been the remedy of old women and nurses for centuries, and there is no doubt but that benefit has been obtained from it. Its action was diaphoretic. The treatment I have named restores all the secretions, but acts especially upon the skin.

In the first variety of jaundice proper, with fever, partial arrest of the secretions, and with an irritation and determination to the liver, I would advise the following course: Put the patient upon the use of tincture of Veratrum or Aconite, the usual doses, adding Gelseminum, if there is irritation of the urinary apparatus, and uneasiness in passing urine. Associated with this, the tincture of Chionanthus may be given in doses of from one to five drops every three hours, according to the age of the patient. The Chionanthus is as nearly a specific for jaundice as a remedy can be for a named disease. The patient should have the general bath and hot-foot bath, to aid the action of the sedatives, and the wet pack to the right hypochondrium, or, better, an entire abdominal pack to relieve the irritation of the liver.

This plan should be followed up, without addition, until the pulse has lost its frequency and hardness, and the skin is becoming soft and active, and urinary secretion more free. If need be we now add to this a solution of acetate of potash in doses proportioned to the age of the patient. We watch its action at first, that it does not irritate the kidneys, and as it acts kindly increase its quantity.

We observe as the febrile action passes off, and the urinary secretion is re-established, that the discoloration commences to fade. The reason is obvious, as the urine is highly charged with coloring matter, so much so as to stain the linen. Indeed, in all cases of jaundice, the coloring material is principally removed by the kidneys, and not, as generally thought, by the liver.

In the second form there is an atonic condition, with congestion of the liver, and a failure on its part to remove the biliary material from the blood; or, having secreted it, it is prevented from passing to its usual destination by the engorgement of the viscus, and is re-absorbed into the circulation, and from thence deposited in the tissues. In this case we may give mild cathartics, especially such remedies as act as special stimulants to the liver. Small doses of podophyllin and leptandrin, thoroughly triturated with sugar, or bitartrate of potash, may be given with advantage, until they have produced the desired effect. In place of this, small doses of Nux may be given, and the region of the liver thoroughly rubbed with quinine and lard, or if there is marked fullness, or the patient has suffered with malarial disease the inunction will be with ointment of Uvedalia.

This is followed by the bitter tonics, hydrastine, quinine, or remedies of like character, and some pleasant preparation of iron. Stimulant baths with friction, are employed to obtain an equal circulation, and stimulant frictions to the region of the liver, to excite it to action.

23

Having obtained these influences and restored the functions of the liver and intestinal canal, we have still to promote the removal of the coloring matter from the skin. This is accomplished, as in the preceding case, by the use of a saline diuretic, acetate of potash or citrate of potash being preferable.

PERITONITIS.

Peritonitis is of rare occurrence as an idiopathic disease, both in the adult and child. As a secondary disease it may arise during the progress of fevers, or from diseases of the intestinal canal, and is met with more frequently.

Usually but a comparatively small portion of the peritoneum is involved in the disease; in some, however, a large portion is involved, and the disease is consequently very severe.

CAUSES.—The cause of idiopathic peritonitis is most commonly cold and arrest of secretion. In febrile diseases it is difficult to determine why a particular part is selected as the seat of the local disease, and we must be satisfied with the simple knowledge of the fact. In diseases of the intestinal canal, or abdominal viscera, the peritonitis is an extension of the inflammation of the part first diseased. We have still the exceptional cases where, from perforation of the intestinal canal or stomach by ulcers, there is an escape of their contents into the peritoneal cavity, and inflammation as the result.

PATHOLOGY.—In inflammation of serous membranes the exudation is upon the free surface, and the change of structure noticed, is from organization and change in this. The inflammation is usually active in its character, and the exudation plastic. We have as yet no reasonable theory why this extremely thin and delicate membrane, into which nerves can scarcely be traced, and which ordinarily has no sensation, becomes so exquisitely sensitive and painful.

Post-mortem examination shows the peritoneum thickened and opaque, having lost its smooth and even appearance. The lymph thrown out is in every state of organization, from a thin film upon the surface to strong bands, which tie parts together, or agglutinate adjacent parts. In some cases there is an abundant serous fluid in the peritoneal cavity, at others more or less pus.

SYMPTOMS.—In idiopathic peritonitis the inflammation is ushered in with a chill, followed by high febrile action. The surface is hot and dry, the pulse frequent, small and hard, with much irritation of the nervous system. The countenance has an anxious and restless expression, and is frequently contracted as if from severe pain. The thighs are flexed upon the abdomen, and the body flexed to take off the extension of the abdominal muscles.

In all cases the abdomen is tense and hot, and more or less tympanitic. There is marked tenderness on pressure, even the weight of the covering producing pain in the severe cases, and slight movements elicit cries from suffering.

As the disease approaches a fatal termination, the pulse becomes very frequent, small, and thready; the countenance is contracted and anxious, and the abdomen much distended, though there is less sensibility to pressure.

DIAGNOSIS.—The diagnosis of peritonitis is made from the symptoms of acute inflammation within the abdominal cavity, with tenderness on pressure, and tympanitis; at the same time the fever being active, and the pulse frequent, and unnaturally small and hard.

PROGNOSIS.—The prognosis is not unfavorable, except in those cases in which it is associated with some other severe lesion.

TREATMENT.—The patient is put upon the use of Aconite or Veratrum, as the pulse is large or small, with Rhus, Bryonia, Asclepias, Viburnum, Colocynth, etc., according to special indications. If the bowels are loose, Ipecac or Euphorbia are combined with the Aconite. If there is the expressionless mouth with nausea and pain simulating colic, Nux may be alternated with Aconite.

The hot foot-bath may be used at first, but in severe cases it is better to use hot bricks, wrapped in cloths wrung out of vinegar and water. When we can depend upon good nursing we order hot hop fomentations to the bowels, but if we can not depend upon their being continuously and carefully applied, it will be better to order a mush poultice.

If the pain is very severe, and is not controlled by the means named, enemas containing opium may be employed. Five to twenty drops of tincture of opium may be used with two tablespoonfuls of starch water or mucilage, and will give great relief. I think it much better than to give opium by mouth, and we can not well resort to the hypodermic use of morphia, which I prefer in this disease of the adult.

HERNIA.

Hernia, or *rupture*, in the child is of frequent occurrence, and may be through any of the natural openings in the abdominal wall, or may be *direct* from the feebleness of the structures. Hernia is of more frequent occurrence in the male, the intestine passing down through the inguinal canal. It is of rare occurrence in the female through the femoral ring.

Inguinal Hernia.— Inguinal hernia is of most frequent occurrence. In the majority of cases the bowel passing through the internal abdominal ring, traverses the inguinal canal, passing through the external abdominal ring, and finally makes its way into the scrotum.

Attention is drawn to the child by its cries and evidence of severe suffering, and to the abdomen by its contortions, and by the lower extremities being forcibly flexed upon it. A careful examination detects the enlargement in the inguinal region, and the hand placed upon it detects the succussion as the child cries.

Direct inguinal hernia is that form in which the bowel is forced through the abdominal wall immediately above the external ring, through which it may pass, and descend into the scrotum, as in the preceding case.

The distinction between the two is made by tracing the course of the protrusion. In the first there is the oblique distension in the course of the inguinal canal, while in the second the enlargement ceases immediately above Poupart's ligament.

Femoral Hernia.—In femoral hernia, the intestine forces its way under Poupart's ligament through the femoral ring, and passing up through the saphenous opening, the tumor is formed in the groin. If the hernia continues to increase, the bowel passes upward over Poupart's ligament, assuming very nearly the position of an inguinal hernia. The symptoms are the same as in the preceding case, and it will be diagnosed by a careful examination.

UMBILICAL HERNIA.—Umbilical hernia is of more frequent occurrence in the child than in the adult. This is owing to imperfect closure of the umbilicus, after the detachment of the cord, the intestine being forced through this natural opening. It varies in extent, being sometimes but slight, not larger than a good-sized cherry, but in other cases it may attain the size of an egg.

The diagnosis is easily effected. The child suffers pain and cries severely; draws its feet upwards, and contorts its body as if from colic. An examination of the abdomen detects the seat and character of the injury.

TREATMENT.—A hernia being diagnosed, the first object of treatment is to return it. This is done by the *taxis*—pressure being made on the bowel in the direction that it has passed down. This pressure should be gentle, yet continued, the whole protrusion being well supported with the hand, while one or two fingers are engaged in dislodging and carrying upwards small portions of the gut. As a general rule, but little difficulty will be experienced in replacement, if the pressure is well directed.

If the taxis fails from the straining or resistance of the child, it should be brought under the influence of chloroform, when pressure will readily reduce it.

After reduction is effected, we will have to select an appropriate apparatus to prevent the reproducement of the hernia. For infants and young children, I prefer a well fitted bandage, with a soft pad large enough and thick enough to give efficient support. I think in the majority of cases a radical cure will follow its use.

In children over two years of age, the ordinary spring truss may be used. The hard pad is decidedly the best, and should be adapted to the injury, so as to excite the part and promote adhesion.

CHAPTER VIII.

DISEASES OF THE URINARY APPARATUS.

The urinary organs of the child do not suffer so frequently from disease as do those in the adult; yet when such diseases do occur, they are quite severe and treated with difficulty. The urinary secretion removes materials from the blood which are highly noxious when retained, and diseases impairing this function are always followed by marked constitutional disturbance.

ACUTE NEPHRITIS.

Acute inflammation of the kidneys may be produced by cold, or may arise from other causes of inflammation, but in the majority of cases it arises during the progress of eruptive fevers, especially scarlatina; and is doubtless dependent on some action of the fever poison, with which we are unacquainted. It is possible that the arrest of secretion from the skin in these cases throws an additional amount of work upon the kidneys, from the irritation of which the inflammation arises.

SYMPTOMS.—Inflammation of the kidneys is announced by increased excitement of the nervous system, and an increase in all the febrile symptoms, if it occurs during the progress of another disease. The child passes urine frequently, but in very small quantity, and seems to suffer much at these times. In some cases the urine will be tinged with blood, sufficient to discolor the diaper. The pain and uneasiness are attended with contortions of the body, flexure of the limbs, and straining, so that it may be easily mistaken for colic.

These symptoms, as well as the duration of the period of excitement, will vary as one or both kidneys are involved in the inflammation. In the first case they will not be so severe, and may last two or three days, or even more; but in the second they will not last longer than twenty-four hours. Occasionally convulsions ensue early in the disease.

The second stage of an inflammation of the kidneys is that in which, from retention of urea, the nervous system is depressed, and coma results. A very short time is necessary for this change of symptoms (two to four hours); the nervous irritation passing off, the patient becoming dull, drowsy and comatose; this gradually deepening until the patient dies.

In the first stage the pulse is frequent and hard, the skin dry and harsh, the mouth and tongue dry, and the temperature quite high. In the second stage the pulse becomes small, its frequency still more increased, but it loses its hardness, the temperature of the surface is diminished, and the extremities become cold, yet the thermometer shows a higher temperature than before.

In the severe cases the duration of the disease will not be more than from two to four days, unless the patient is relieved. In mild cases it may continue a week or more.

DIAGNOSIS.—The increased nervous excitement will attract our attention, and cause us to make a very close examination. The pain that the child suffers is undoubtedly in the abdomen, and would be mistaken for colic, were it not that it comes on in passing urine, and is very closely associated with this. The scanty urine, tinged with blood, passed a few drops at a time, and with straining and pain as named, is the most striking diagnostic feature.

PROGNOSIS.—The prognosis is not favorable when both kidneys are involved, unless the lesion is noticed early and prompt treatment is adopted. When but one is involved we expect a favorable termination. The prognosis is of course more favorable in the stage of excitement, than in that of depression and coma.

TREATMENT.—In the first stage of the disease the child is put upon the use of Veratrum and Gelseminum in full doses, the object being to arrest the fever and inflammation at once. I think there are no remedies that will take the place of these, at least there are none that I would like to trust. I would give them in the following proportion: ℞ Tincture of Veratrum gtt. v., Tincture of Gelseminum gtt. x., water, ℥iv. The nurse is directed to give a teaspoonful every hour, until she observes an abatement of the fever, then reduce the dose one-half and continue.

In place of Gelseminum we sometimes employ Macrotys, as there is muscular pain or soreness, or Cannabis Indica, there being irritation of the urinary passages, even to the meatus urinarius. Bryonia is sometimes a good remedy if there is marked contraction of the abdominal muscles; Belladonna if there is fullness of the abdomen with stupor and disposition to sleep, or if there has been retrocession of an eruption. Eryngium is a good remedy when there is a constant desire to pass urine; Apocynum when the eyelids are swollen and the feet œdematous, or the slightest evidence of dropsy.

Cathartics are absolutely contra-indicated, as are diuretics, and diaphoretics are of no value until we have first obtained the influence of the remedies named.

As a local application I like the influence of the hot hop fomentation better than anything else. I order the application across the loins, the child lying upon the abdomen, but when tired in this position it is placed upon its back, and the fomentation applied over the lower part of the abdomen.

The child should always be kept in bed, between blankets, and the action of the skin may be solicited by the use of hot bricks near the feet, wrapped in wet cloths.

This treatment is persisted in until the febrile action is partially checked, and secretion of urine is increased, the patient meanwhile being allowed as much fluid as desired, unless it causes nausea. As soon as we have the patient fairly under the influence of the sedatives, we commence the administration of the milder vegetable diuretics, as haircap moss or mentha viridis, and continue them until the secretion is wholly restored.

In the second stage of the disease the patient is suffering from uræmic poisoning, and the treatment will be wholly different. We give at once a brisk cathartic, as of jalap and bitartrate of potash, and if the case is one of emergency, we may aid its action by a cathartic enema.

Dry cups are applied to the neck and over the loins, and a hot stimulant application over the kidneys follows their use. Stimulants and heat are also applied to the extremities.

The internal remedy that I rely upon with most confidence is Belladonna, which may be given in the following proportions: R Tincture of Belladonna gtt. v., Tincture of Aconite gtt. v., water ℥iv.; a teaspoonful every hour. If the reader can not make up his mind to the use of such small doses in such a grave affection, I would advise in its place a prompt and very thorough emetic, followed by the cathartic as first named.

In the use of diuretics we will have to be very careful as the patient convalesces, for the kidneys are in a very irritable condition, and a slight irritant is sufficient to arrest the flow of urine. It is fortunate that the tincture of muriate of iron, in quite small doses, is a good tonic to the kidneys, as it is the restorative we most frequently employ.

CHRONIC NEPHRITIS.

Chronic nephritis is never met with during the first year, is very rare during the second, but is occasionally met with from this up to the sixth year. After this time it is hardly ever seen until the person has reached middle life. It is a very insidious disease, coming on slowly, and presents so few prominent symptoms that in many cases it is not detected.

CAUSES.—The causes of chronic nephritis are very obscure. It may arise from cold, from irritant diuretics, but probably is most frequently the result of some morbid condition of the blood, or change in the process of retrograde metamorphosis. We know the irritant effect that the presence of the *urates, triple phosphates* and *oxalates* have on the bladder, and we can well imagine that their excretion by the kidneys may set up a slow process of inflammation in these organs.

PATHOLOGY.—The inflammation may be confined to one, or may affect both organs, or it may involve but the cortical structure of the kidney, the tubular remaining comparatively free.

Post-mortem examination shows the kidneys enlarged, with evidences of enfeebled circulation. The structure is softened and friable, and the capsule readily detached. In some cases there is a free deposit of a white amorphous material. In others the inflammation has progressed to suppuration, and more or less of the structure has been destroyed by it.

SYMPTOMS.—The child has lost flesh and strength, in a marked degree; its appetite is impaired, digestion feeble, the bowels irregular and the skin dry and harsh. Irritability and restlessness is a marked feature, and is sometimes developed to that extent that the child becomes a burden to itself and all around it.

If old enough to locate its sufferings it will complain of pain in the back and across the lower part of the abdomen. Occasionally it experiences uneasiness before urinating, and sometimes after its discharge. Frequently, however, no complaint of this kind is made, and though the child undoubtedly suffers uneasiness and pain, it is of such a fugitive character that he can not locate it.

These symptoms are constant, and gradually increase, but in addition there are exacerbations in the disease, coming on at intervals of two or three weeks. At these times the patient suffers from fever, the skin is dry and hot, the urine very scanty, the bowels constipated, while the face has a pinched and anxious expression. It is evident at these times that the child suffers from pain in the back; and frequently the lameness is quite marked.

Thus the disease progresses, until finally the periodic exacerbation of the inflammation is much worse. The fever is active, with great excitation of the nervous system, occasionally with convulsions. The urinary secretion being arrested, the symptoms of *uræmia* are rapidly developed, and within twenty-four hours the patient is comatose, and by another day it terminates fatally.

DIAGNOSIS.—The diagnosis is principally made by exclusion. The examination is made with reference to disease of special parts, as of the brain, lungs, digestive apparatus, etc. One after another being excluded, we finally trace the diseased action to the kidneys, and the symptoms given above will be found sufficient to designate the character of the disease.

PROGNOSIS.—The prognosis is favorable in the early stage of the disease; but when it has continued for some months, having lost flesh to a great degree, with frequent pulse, harsh dry skin, scanty high-colored urine, great irritability of the nervous system, with a pinched and anxious expression of countenance, the prognosis is doubtful.

TREATMENT.—The treatment of this case is one of much difficulty, and will have to be varied from that we would use in the adult. As a general rule, I should give but few medicines, indeed experience will teach that the stomach will not tolerate much, and that all irritants are injurious.

Small doses of Aconite with Gelseminum, Rhus, Bryonia, Apocynum, Eryngium, Cannabis Indica, or Agrimonia, as indicated, will serve the purpose well. Apocynum should be especially noted as the remedy if there is the slightest evidence of œdema. Agrimonia is indicated by the pain which sometimes extends up into the hypochondria. If the child is nervous and cries at every little thing with sobbing, small doses of Pulsatilla may be alternated with the other remedies.

The child may have a hot salt-water bath every day, being rubbed with dry flannels afterward and clothed warm. Or, in place of this, we may use fatty inunction, rubbing the surface thoroughly. Inunction with quinine answers an excellent purpose in some cases, especially those in which there is a malarial influence.

The best lcoal application for the child is the vinegar pack, applied upon going to bed. A flannel cloth, six inches wide, is wrung out of tepid vinegar, and applied around the body at the loins, and this is again covered by a dry strip; it is removed on getting up, and the part thoroughly rubbed with flannel.

As the severer symptoms pass away, the patient may have some of the tonic and astringent diuretics, if thought best, or it may be continued upon the remedies already named, at less frequent intervals. The compound syrup of the hypophosphites, some of the preparations of malt, occasionally iron, Collinsonia, Hydrangea, or Eryngium, or the quinine inunction, may be continued to complete recovery.

ALBUMINURIA.

While albuminuria in the child is most frequently the sequel of the eruptive fevers, it is occasionally seen from other causes. It is not dependent, however, on *granular degeneration*, or Bright's disease, in either of these cases, and is, therefore, a much less grave affection.

PATHOLOGY.—It has been supposed that the albuminuria following the eruptive fevers was dependent upon irritation, or a low grade of inflammation of the kidneys. While this may be the cause, to a certain extent, we have reason to believe that the changes effected in the blood by the febrile poison is an important element of the disease.

Post-mortem examination shows the kidneys swollen, the *glomeruli* congested, and the *tubuli uriniferi* more or less loaded with albuminoid matter. Casts from the tubules may be detected in the urine during the progress of the disease.

SYMPTOMS.—Occurring almost always after the eruptive fevers, on exposure to cold, by which the surface is suddenly chilled, it commences generally with a well-defined chill; symptomatic fever follows, the pulse being hard and frequent, the skin hot, dry, and constricted, the tongue coated white, the mouth dry, frequently nausea and vomiting, bowels constipated, pain in the back, and marked restlessness and nervous irritation. With these symptoms the patient complains of a sense of weight and constriction in the region of the kidneys, never, as is said, extending to or causing retraction of the testicles. The pain may be confined to one side, but one kidney being affected, or it may be equally in both sides.

With the occurrence of these symptoms the urine becomes scant, almost suppressed, and highly albuminous, of a reddish color, and occasionally bloody. Its specific gravity is almost always above that of healthy urine, and it gives an acid reaction. When allowed to rest, it deposits a filamentous substance, and when examined with the microscope it will present blood-globules, mucus, epithelium, and in some cases complete casts of the urinary tubules. A dirty-white sediment is frequently deposited from the urine, not unlike mucus, and easily diffused by agitation. The urine is frequently passed with difficulty, and sometimes with pain, the calls to urinate being frequent and distressing.

In the course of the second or third day dropsical symptoms make their appearance, most frequently as anasarca of the eyelids, face, and at last of the whole body. The skin at this time is hot, and does not pit except upon firm pressure. If properly treated, in a majority of cases, we find that the symptoms are much mitigated in the first three or four days, and the disease terminates in recovery by the twelfth to the fifteenth day. In other cases coma comes on by the second, third, or fourth day, and the disease terminates fatally within the first week. Occasionally convulsions appear, and continue until the patient is exhausted. In other cases the disease seems to give way slowly, until it reaches the chronic stage, in which it continues.

TREATMENT.—If the temperature is above the normal standard and the pulse is too frequent, we give Aconite with Apocynum, (gtt. iij., gtt. v. to gtt. x., water ℥iv.) In some cases, following the eruptive fevers, there has been something omitted from the treatment of the cases, and we endeavor to find it. It is possibly Belladonna, Rhus, Apis, Phytolacca, or Macrotys, but whatever it is, it should now be given if the indications still persist.

In some cases the hydragogue cathartic, like jalap with bitartrate of potash, is employed, to increase secretion by way of the bowels, and remove dropsical accumulations, but I prefer the more direct remedies, Apocynum and Aralia Hispida.

If the dropsy is not reached by these remedies, we will sometimes find the small dose of sulphate of manganese a good remedy. It may be given in the following proportion : ℞ Sulphate of manganese gr. v., white sugar ℥j.; make twenty powders, and give one every three hours until it acts upon the bowels.

The hot foot-bath may be used, and dry cups, followed by hot fomentations, applied to the loins. If the case seems to be urgent, we may use the general hot-bath with advantage, continued for one or two hours, the surface being rubbed dry, and the child clothed or wrapped in flannel.

In some cases quinine inunction answers an excellent purpose, but sometimes the fatty matter seems to close the skin, and we are obliged to have the child thoroughly washed with soap and water to remove it. In other cases the inactivity of the skin is relieved by a sponge bath of hydrochlorate of ammonia and water.

DIABETES.

I have never seen but one case of diabetes in children giving the evidence of sugar on analysis, and this was in quite small quantity, and for a short time ; neither do I recollect seeing any such cases recorded. Diabetes *insipidus*, on the contrary, is met with quite as frequently in childhood as in the adult, and is some times very persistent.

CAUSES.—The causes giving rise to this increased flow of urine are very obscure. In some cases the exciting cause seems to be cold, but more frequently it will be found to be dependent upon, or at least associated with some disease of the nervous centers.

PATHOLOGY.—I think, in all of these cases, there is an impairment of circulation, with tendency to congestion. The nervous system suffers first, and the enfeebled innervation seems to be one cause of the lesion of the kidneys. The condition of the kidneys seems to be one of atony rather than excitation—one which permits the water of the blood to filter off, rather than increased activity of secretion. Indeed, in some cases, the patient suffers first the nervous irritation from retained urea, and afterward from uræmic coma. Occasionally there is the same profluvia from the bowels, but in less degree.

SYMPTOMS.—It will have been noticed that the child has lost flesh rapidly for two or three weeks, or may be, for as many months, and that it is becoming quite feeble; yet it has a pretty good appetite, and digestion is well performed, and the child does not complain of any particular suffering. If associated with diarrhœa, the loss of flesh and strength will be attributed to this. When we examine the patient, we find the pulse feeble, small, and soft, and somewhat increased in frequency; the surface is pale, cool, and wanting in tone, and the child suffers from cold feet and feeble circulation. The tongue and mucous membranes are pallid, but the tongue is not coated. Usually there are two, three, or four fluid movements from the bowels daily, but they are not large or unhealthy.

Upon close inquiry, we will learn that the child passes water frequently at night, and the bedding is very wet, and when the mother's attention is called to it, she will notice the frequent passage of large quantities of clear urine. If we examine a portion of the urine we will find it limpid, very light-colored, and of low specific gravity, frequently not more than 1008. I have seen cases, in which the child would urinate on an average every fifteen minutes during the day, and every half hour at night.

In the severer cases the marasmus is as great as in the diabetes mellitus of the adult. In such cases the irritation of the nervous system is extreme, and the child suffers intensely from nervous excitement. In some cases it seems almost impossible for it to obtain sleep. Continuing on in this way for a short time, this is replaced by coma, and a few hours will terminate its life.

DIAGNOSIS.—The diagnosis of diabetes of the child will be easily made, if our attention is directed to the kidneys as the seat of disease. This is the advantage of forming a habit to correct the diagnosis of every case by *exclusion;* as each organ is thus passed under review, it is almost impossible to fall into serious error. The frequent and abundant discharge of urine may have escaped notice until attention is drawn to it, or the mother may have considered it an unimportant circumstance.

PROGNOSIS.—The prognosis is favorable in the majority of cases, as the disease yields readily to remedies. When it has continued for a considerable time, the child being wasted almost to a skeleton, the nervous system being very irritable, the prognosis is unfavorable.

TREATMENT.—In the milder cases, the treatment is very simple and direct. The child is put upon the use of—℞ Tincture of Belladonna gtt. v., water ℥iv.; a teaspoonful every two hours. To restore the action of the skin, and at the same time, get its action as a tonic, I employ the quinine inunction once daily. When there is atony of the digestive apparatus, and torpor of the liver, minute doses of podophyllin may be given. The following combination will answer well: ℞ Podophyllin gr. j., Hydrastine gr. v., white sugar gr. xx.; tritutate thoroughly and divide in twenty parts; one may be given in the morning and one at night.

In some cases we may alternate the Belladonna with Hamamelis (Pond's extract), in doses of ten drops, giving each every four hours; and when diarrhœa accompanies it, this may be controlled with Ipecac.

As the disease is brought under the control of remedies, we may alternate them with the Collinsonia, tincture of phosphorus, Rhus Aromatica, and the tincture of muriate of iron, as a restorative. But if the patient gets along well, we will, in the majority of cases, find no cause for change.

ISCHURIA.

Ischuria, or *arrested* discharge of urine, is divided into two varieties—ischuria renalis, suppression of urine, and ischuria vesicalis, retention of urine. Both of these should be regarded as but symptoms, as they may arise from various and widely different causes.

SUPPRESSION OF URINE.—Suppression of urine may be the result of an active inflammation of the kidneys, as we have already seen. But the cases that will come under our notice most frequently, depend upon exhaustion or enfeebled innervation of the kidney, or from congestion. It occurs most frequently during the progress of acute disease, especially in such as are attended with serious blood-poisoning.

SYMPTOMS.—The symptoms of suppression of urine vary in different cases. Where the suppression is partial, as is generally the case, we find the first evidence in great excitation of the nervous system, and an increase of the febrile symptoms. Not unfrequently, we will find the desire to pass urine increased, and the frequency of the effort, but small quantities passing, will first attract the attention.

As it continues, the symptoms of the uræmic coma begin to be developed, the child is dull, has a tendency to sleep with its eyes partly open, and starts frequently from its sleep as if alarmed. Occasionally convulsions precede the coma, which seem to be dependent upon it?

Nausea and vomiting occasionally occur in the early part of the disease, and the irritability of the stomach is one of its most troublesome features. From the commencement of coma, we find it becoming deeper as time passes, until at last it is impossible to arouse the patient. The pulse is feeble and irregular, the extremities cold, the features contracted and pinched, with occasionally more or less convulsive movement, until death ensues.

RETENTION OF URINE.—Retention of urine is most frequently due to paralysis of the bladder, though it may be produced by irritation of and contraction of the neck of the bladder or urethra, or in some rare cases, it is dependent upon the presence of a calculus.

Retention from exhaustion of the muscular coat is of common occurrence in low forms of disease, and is sometimes a source of much trouble. Retention from irritation of the urethra is attended with much pain, and a frequent desire to pass water.

SYMPTOMS.—In retention of urine occurring during the progress of low forms of fever and imflammation, it will be noticed that the prostration is much greater, and the symptoms more

grave, than would have been anticipated in the ordinary progress of the disease. Frequently the attention is drawn to the
difficulty by the marked uneasiness of the child, and the movement of the hands to the region of the bladder.

In retention from contraction of the neck of the bladder or
urethra, there is the evident desire to pass urine, attended with
straining and pain, but the want of power to overcome the obstruction.

DIAGNOSIS.—Having learned that the urine has not been
passed, we have to determine whether it is a suppression or retention. This may be determined in many cases, by the symptoms above described, or by an examination of the bladder with
the hand. If this is not satisfactory, a catheter may be introduced into the bladder. This, however, is attended with difficulty, and may generally be dispensed with.

TREATMENT.—In the treatment of suppression of urine in the
child, I rely more upon the use of hot fomentations across the
loins, than on any other means. They should be assiduously
employed, and continued until the desired result is produced.

If the patient has fever, Tinct. Veratrum gtt. iij., Tinct. Gelseminum gtt. x., water ℥iv., a teaspoonful every hour, will be
good treatment. If the face is pinched and the patient is restless, starting in sleep, Rhus may be given with Aconite. If the
patient is dull and wants to sleep most of the time, Belladonna
will be suggested. Apocynum answers an excellent purpose, if
there is a puffy condition of the eye-lids, or swollen feet.

To relieve the system from the symptoms of uræmic poisoning,
a cathartic may sometimes be given, if there are no circumstances
to contra-indicate it; jalap and bitartrate of potash will answer
this purpose very well.

As soon as urine commences to pass in small quantities, diuretics may be administered, the remedy being selected to meet the
condition of the patient, being demulcent, stimulant, or tonic, as
the case may require.

In some cases, dry cups may precede the use of the hot
fomentations, and will be found of much service.

In the treatment of retention of urine in children, I prefer the
use of santonine to any other remedy I have employed. I have
used it in doses of from one-fourth to one grain, triturated with
sugar, repeated every two hours. This may be assisted by hot

24

fomentations applied over the bladder and genitals, or in some cases by a hot sitz bath.

The use of diuretics is contra-indicated in this case, as they rather diminish than increase the power of the bladder to evacuate urine, while they certainly increase the distension.

When the retention is dependent upon stricture of the urethra, or contraction from irritation, we will obtain the best results from an enema of equal parts of Lobelia and Gelseminum ʒss to water ʒj.

ENURESIS.

Incontinence of urine is most commonly a disease of childhood, though we occasionally meet with it in the adult, but usually as the result of injury, or from disease. In the child it varies from the slight form of nocturnal incontinence, to the severer cases in which the patient is wholly unable to retain the urine, and it flows away all the time.

CAUSES.—The causes of incontinence are somewhat obscure. In some of the milder cases it seems to be a habit more than a disease. The severer cases follow the eruptive fevers, diseases of the nervous system, and occasionally diseases of the urinary apparatus.

PATHOLOGY.—In some cases the incontinence is dependent upon an irritable condition of the mucous membrane of the bladder, slight distension of the viscus giving rise to involuntary muscular contractions and expulsion of urine. In other cases the muscular fiber is in an irritable condition, and contracts from slight excitement. In other cases, and by far the most numerous ones, the lesion is of the nervous system. In one it is of the spinal system, an atonic or feeble condition, and as the result of this, there is imperfect contraction of the circular fibers at the neck of the bladder, and the first portion of the urethra, which serves the purpose of a sphincter. In the other it depends upon irritation of the lumbar spine, with involuntary contraction of the bladder from the disordered innervation.

SYMPTOMS.—The simplest form of the disease is that in which the child *wets* the bed. Frequently, in these cases, the child drinks freely in the evening, and the bladder being distended at night, the urine is passed unconsciously, or under the influence of a dream. This may continue to the age of puberty, or even longer, becoming a source of very great annoyance and mortification.

In other cases there is not only the nocturnal incontinence, but there is a frequent desire to evacuate the bladder through the day, and if not promptly attended to, involuntary passage of urine occurs.

Then there is the severe form of the disease, in which the person has no command over the passage of urine, but it dribbles away continually, keeping the clothes soiled, and making it quite impossible to keep the child free from a disgusting urinous odor, irritating the surface about the parts, and requiring much care to prevent its suffering from cutaneous irritation.

TREATMENT.—In the simpler forms of nocturnal incontinence, it is frequently sufficient to prevent the child's drinking in the evening, see that it passes urine before going to bed, and have it waked once in the night for the same purpose. What we want is to break up the habit of involuntary passage, and this may be done in the majority of cases in this way.

In this, and other forms of the disease, when there is no special lesion, I rely upon the Belladonna as a *specific*. I usually prescribe it in the following proportions. ℞ Tincture of Belladonna gtt. x., water ℥iv.; a teaspoonful three times a day. In severe cases, those dependent upon atony of the spinal cord, or consequent want of contraction of the sphincter fibres, the Belladonna is our best remedy, and will give relief when the person has been supposed incurable. Ergot has a very similar influence, and may sometimes be alternated with it. I would use it in the proportion of—℞ Tincture of Ergot gtt. xx., water ℥iv.; a teaspoonful four times a day.

The Rhus Aromatica has been highly recomended in these cases, and may be tried, though thus far I have not seen the marked results that have been described. In feeble children the syrup of iodide of iron has been given with advantage in doses of five or ten drops three times a day.

Irritability of the bladdder should be treated by the use of Agrimonia, Hydrangea, or Collinsonia, combined with a tonic and restorative treatment, as it is usually associated with impaired health. A Belladonna plaster applied over the bladder will give present relief, but its continued use will sometimes produce unpleasant symptoms. Frictions over the bladder with the liniment of Stillingia will answer a better purpose.

In those rare cases in which the incontinence is epileptic in its character, I would advise the bromide of ammonium alternated with Belladonna.

URINARY CALCULI.

Stone in the bladder is met with even during the first year of life, and from this up to adult age. It is not of as frequent occurrence in the child as in the adult, yet it is met with more frequently than many suppose. There is no doubt but that many children suffer for months, and finally lose their lives, without the detection of the lesion.

Previous to the last quarter of a century we find but few cases of stone in the bladder of children recorded, and the cases of cutting for stone in children under six could be counted upon the fingers. In the last few years a large number of cases have been reported, with successful operations for their relief. One surgeon alone reports some thirty cases in ten years. We must, therefore, conclude that the disease was not recognized, as it is not probably of more frequent occurrence now than in the centuries past. Knowing these facts, we will not allow ourselves to make the same error in diagnosis.

If stone in the bladder in the child was incurable, this error in diagnosis would make less difference; but the experience of some of the most skillful surgeons shows that the operation for its removal is much more successful than in the adult or aged.

PATHOLOGY.—Urinary calculi may result from lesions of digestion, but mostly from lesions in the waste of tissue and retrograde metamorphosis. All are composed of the normal elements of tissue, except in those rare cases in which the material is furnished by maldigestion. The urinary organs do not seem at all at fault, indeed they perform the additional function of removing a material from the blood which would prove injurious there.

The calculous formations in childhood are principally of three kinds—*uric acid and urates, oxalates,* and *phosphates.* These may be deposited in the form of minute crystals, being washed from the bladder by the flow of urine, giving rise to no lesion but irritation of the bladder and urethra; or they may accumulate and coalesce to form those masses of variable size, which we know as calculi.

The *uric* or *lithic* calculus is of a brownish mahogany color, oval or flattened in form, and finely tuberculated or smooth, though not polished. It is perfectly dissolved in caustic potash, and disappears with effervescence in hot nitric acid, the solution affording, when evaporated to dryness, a bright carmine residue. It becomes black and is gradually consumed before the blow-pipe, leaving a minute quantity of white alkaline ashes.

Oxalic calculi are generally of a dark brown color, rough and tuberculated, hard, compact and imperfectly laminated. It is insoluble in the alkalies, dissolves slowly in nitric and hydrochloric acids, if previously well broken up, and under the blow-pipe, expands and effloresces into a white powder.

Phosphatic calculi are of a lighter color, less clear and friable. Phosphate of lime more frequently forms layers with other matter; when it occurs alone, they are usually small, of a pale brown color, of a loosely laminated structure, not fusible with the blow-pipe, but readily soluble in hydrochloric acid with effervescence. The amoniaco-magnesian calculus is of a white color and friable, looking a good deal like a mass of chalk. It exhales an ammoniacal odor before the blow-pipe, is not affected with caustic potash, but is easily dissolved in dilute acids. There is a form of phosphatic calculi that is fusible. It is white, extremely brittle, easily separated into layers, and leaves a white dust on the fingers. It is not affected by caustic potash, is soluble in hydrochloric acid, and is melted into a transparent pearly glass under the blow-pipe.

SYMPTOMS.—The general symptoms attending the presence of urinary calculi in the child are pretty well marked. The appetite is irregular, digestion is impaired, and frequently there are well marked dyspeptic symptoms. The secretions are deranged, the skin being very susceptible to external impressions, at times dry and harsh, at others soft, relaxed and flabby, with cold extremities. The bowels are torpid and sluggish in most cases, but in some the irritation of the bladder extends to the bowels, and there are frequent discharges. The patient is irritable and fretful, and extremely restless, and dissatisfied with everything done for it.

The local symptoms are sometimes very marked, in other cases the patient will have to be observed for some time before we have such evidence as will lead us to make the necessary examination.

There is some difficulty in passing water, but it is not constant, as at one time the urine may be evacuated without any trouble ; at another time the urine will be stopped in its passage, with more or less pain and tenesmus. Occasionally at such times the urine will be stained with blood.

As the case advances the difficulty in passing urine becomes more frequent, and is attended with greater suffering. The patient suffers from sense of weight and pain in the bladder, loins, and small of the back, sometimes so severely that the countenance presents that peculiar contracted anxious expression, which we observe in long-continued painful diseases.

Progressing in this way the child becomes exhausted from the long continuance of suffering, and the functional derangements that follow it, and will succumb to this, or to some local disease of the pelvic viscera induced by the calculus.

DIAGNOSIS.—The uneasy sensations in the bladder, with difficulty in passing urine, point to the bladder as the seat of disease, while the long duration of it and its intractability would lead us to suspect the presence of a calculus. To determine this a sound is introduced into the bladder, and careful manipulation will detect its presence.

PROGNOSIS.—The prognosis may be regarded as favorable, whether we conclude to practice lithotrity or lithotomy—both these operations being successful in the young.

TREATMENT.—Having determined upon an operation, we prepare the patient for it as best we may, in the short time we have. I think it of advantage to keep the patient very still for a few days, giving small doses of Aconite and Gelseminum, with some demulcent diuretic to relieve the irritation of the bladder. In some cases much relief will be obtained if the child lies upon its abdomen a portion of the time, as it removes the pressure from the base of the bladder.

In the meanwhile we see that the stomach is placed in condition to take and digest some food; and we may use the tincture

of muriate of iron and glycerine as a restorative. Quinine inunctions will prove beneficial in some cases, giving strength to
the nervous system.

Lithotrity. is preferred by many surgeons at the present time,
as giving good results, notwithstanding the smallness of the
urethra. The calculi of children are generally more friable than
in the adult, and are much easier crushed, and the urethra may
be dilated so as to admit a good sized instrument.

Lithotomy is performed in the usual way, the *lateral* operation
being preferred. Though it is not prudent to make the large
incision permitted in the adult, but with a little care in its dilation with the finger, and in seizing the stone so as to present its
smallest diameter, there will be no trouble in its extraction.

The principal points in the after treatment are to keep the
patient quiet, keep down the febrile action, and the stomach in
condition to receive food.

When the child has recovered from the operation, it should be
put upon such a course of treatment as was named for dyscrasias.

In addition to this tonic and alterative plan, it may be possible to employ special means to antidote the tendency to calculous formations. The information necessary for this will be
found in my work on Practice, pp. 357 to 366

IMPERFORATE URETHRA.

Rarely a case of imperforate urethra will be met with in the
new-born child. The malformation will be brought to the physician's notice by the child not passing urine, and suffering pain
in consequence of it. An examination will determine the character of the lesion.

In other cases the urethra terminates by false openings in the
upper or under surface of the penis—*hypospadias* and *epispadias.*
These are rarely more than one inch from the meatus, frequently
not more than half an inch.

TREATMENT.—In imperforate urethra the difficulty may be
remedied by passing a large exploring needle, or small trocar,
through the meatus and carrying it upward until it reaches the
free portion of the canal. A silver catheter being now introduced it is worn until the part heals.

In hypospadias, and epispadias, though there is no danger to life, I think that the malformation should be remedied in infancy, as it can be best done at this time and with the least danger. Open up the canal of the urethra in the manner above named, and passing a catheter, retain it for the urine to pass through. Then freshen the edges of the artificial opening, and draw them together with a silver suture. If, as in some cases, the canal really terminates with the false opening, I would operate by detaching the skin at that part, if not too far up, bring it down to the glans and attach it with silver ligatures, retaining a catheter until the part had healed.

PHYMOSIS.

The natural condition of the penis in the child is, with the elongated and contracted foreskin entirely covering the glans. It is only in exceptional cases that this is a source of trouble.

In one of these cases an inflammation of the prepuce and glans is set up, and there is a free discharge of pus. Sometimes the inflammation runs high, and the prepuce is so swollen as to render the passage of urine very difficult and painful. This is really a case of infantile *balanitis*, and being difficult of cure, or recurring frequently, it may require removal of the prepuce.

In the other case, the prepuce being quite long and much contracted, is irritated and inflamed by the urine, so as to be a source of very great annoyance. I have seen two cases in which this was a source of serious disease, and there was but the one way for its removal, and that by removal of the prepuce.

TREATMENT.—Circumcision is very readily performed in the child, and is not attended with any danger. The method I would recommend is to take the foreskin between the thumb and finger for half an inch, then drawing it down well to clear the glans penis, excise it with a bistoury, cutting toward the finger. But if afraid that you may wound the glans, draw down the prepuce with the fingers, and have an assistant grasp it next the glans with a pair of dressing forceps, or the arms of a pair of scissors, then cut between the fingers and it. Retract the foreskin and apply a water dressing until the part heals.

Some may make objection to circumcision, as was once the case in my practice. The mother declared she would not have her child made a Jew of, even though the penis had to be cut off. In this case, simple incision of the prepuce on a grooved director, its entire length, retracting it over the glans, and retaining it with adhesive straps, was all that was required.

PARAPHYMOSIS.

Paraphymosis is the retraction of a tight prepuce above the glans penis   It is occasionally met with in boys of from six to ten years, who have, out of curiosity, or prompted by those older, pushed the foreskin back and have not been able to return it. The glans becomes swollen and painful, and if the constriction continues long it may go on to inflammation and finally to sloughing.

In many cases the application of cold water or ice for a short time will reduce the swelling and permit the foreskin to be drawn down. It is also claimed that if a couple of hairpins are used, pressing the bent end up under the prepuce on each side of the glans, the prepuce can be drawn down over them. If this does not succeed, or if so long a time has already elapsed that we do not think it prudent to try these means, we will incise the prepuce and thus free it. This is easily accomplished by passing a grooved director under the foreskin on the dorsum of the penis, and cutting outward with a sharp-pointed bistoury.

URETHRITIS.

Urethritis in the child is of very unfrequent occurrence, yet occasionally a case will be met with. It is pretty difficult to determine the cause of the inflammation. Sometimes it is undoubtedly venereal, and has been contracted from the clothing of some persons about the house affected with the disease, or sometimes by being nursed by such persons. The only case I ever met with was as marked a case of gonorrhœa as ever I saw in the adult. The boy was about four years old, not yet in pants, and was nursed frequently by an uncle who was suffering from the disease.

SYMPTOMS.—The child will complain of pain in passing urine, and if the disease is severe, will suffer intensely from this cause. Upon examination, the prepuce will be found much swollen, the penis tender, and a more or less abundant discharge will pass from the urethra.

TREATMENT.—Of course, it is hardly possible to determine the cause of the difficulty, and we will treat it as a simple inflammation. The parts should be kept scrupulously clean, and if the inflammation runs high, a warm hip-bath may be used once or twice during the day, and some soft poultice at night.

Internally I would prescribe, ℞ Tincture of Canabis Indica, Tincture Veratrum, *aa.* gtt. v., water ℥iv.; a teaspoonful every two hours. The bowels might be opened with a saline cathartic every three or four days, but active catharsis is to be avoided.

HYDROCELE.

Hydrocele is occasionally met with in young children, and gives rise to considerable alarm upon the part of the parents. In some cases it is congenital, the inguinal canal not having been closed after the descent of the testes. In these it may sometimes be associated with congenital hernia.

Our attention having been called to the enlarged scrotum, we will endeavor to determine its cause by an examination. If the enlargement is caused by the distension of the tunica vaginalis with water, palpation will detect the movement of the fluid, and when the child cries we will find no movement communicated to the swollen scrotum. On the contrary, if a scrotal hernia exist, the movements of the diaphragm in the act of crying, will be communicated to the hand supporting the scrotum. In this case also, there will be no movement from side to side, as when distended with fluid. The diagnosis between hydrocele and hematocele is readily effected in the child by the marked change of color in the last.

TREATMENT.—In some cases we will be successful in removing the fluid and curing the disease by a stimulating application and compression. The use of a strong infusion of Hamamelis, with a well adjusted bandage, in the usual form of a *suspensory* bandage, was attended with success in one case that came under my notice. Mayer's ointment would prove a good application, as would also a soft piece of leather, cut to fit the part and spread with our ordinary strengthening plaster. Tincture of iodine has been recommended as an application to the part, to stimulate absorption, but I think it would be too irritating.

If these means should not succeed, the scrotum may be punctured with an exploring needle or small trocar, and the fluid drawn off. If now the part is supported by a well adjusted bandage, we may expect a radical cure.

INFANTILE LEUCORRHŒA.

A slight discharge from the vulva is not of unfrequent occurrence in children from one to twelve years of age, but it is only in exceptional cases that it is severe, and produced by an active inflammation at first, but which, as it declines, becomes chronic and very stubborn. These last are very troublesome cases, and are a source of great annoyance both to the physician and the family.

CAUSES.—The slight discharge is the effect of cold, after overexertion in the use of the lower limbs, or occasionally from sitting on the damp ground, or a stone. This may also be a cause of the active inflammation, or it may be produced by acridity of the urine, irritating and excoriating the vulva, the inflammation arising in and extending from this part. In some cases it may arise from want of cleanliness, especially in those who are badly nourished.

Then we have exceptional cases, in which the inflammation is specific in its character, and is produced by contact with gonorrhœal virus. This may occur in a house where some person has gonorrhœa, from sitting on the privy seat, or by the use of cloths impregnated with the virus.

PATHOLOGY.—In the slight cases there is simply an increased circulation in the parts, and activity of the mucous follicles. Usually this is confined to the glands about the vulva, and the vulvo-vaginal glands.

In the severer form, there is a well marked inflammation, involving the parts about the commencement of the vagina, the vulva, and sometimes extending up the entire length of the vaginal canal. The inflammation loses its active character after the first ten days, and the parts present a swollen, deep-red appearance, bathed with muco-pus. There is no means of determining from the appearance of the parts, whether it has arisen from gonorrhœal virus, or not.

TREATMENT.—The treatment is unsatisfactory in many cases, from the difficulty we experience in getting a proper application of local remedies.   Indeed, with the best assistance the mother can give, we make but slow progress.

I put the child upon the use of Aconite with Cannabis Indica in the usual doses; or in place of the last we might give the Gelseminum.   The bowels are gently opened by citrate of magnesia or phosphate of soda every two or three days.   If the child is old enough to take it, she may use an infusion of equal parts of Althæa and Uva Ursi as a drink.

I now use salicylic acid and borax or chlorate of potash as the local application, as—℞ Salicylic Acid, Borax (or Chlorate of Potash), aa. gr. xx., water Oj.   This may be applied to the vulva only, or if the discharge is vaginal it may sometimes be used with a syringe.   Bathing the parts with hot water alone (as hot as can be borne comfortably) will sometimes effect a cure.   The hot bath may extend to the abdomen and spine.

Perseverance in these means will relieve the acute inflammation, and many times will effect a radical cure.   But if a chronic discharge continues, we will obtain the best results from the local use of a solution of Sulphate of Hydrastia grs. iv., to water ℥iv.

There are many other local remedies that have been used in this case, but they are the same as recommended for the adult, and for these the reader is referred to treatises on disease of women.

The general treatment in the chronic form of the disease, will be of a tonic and restorative character.   Further than this, it is doubtful if any internal remedies can be used with benefit to the local disease.

ONANISM.

The practice of onanism is of very common occurrence after the twelfth year, or about the commencement of puberty.   But at an earlier age than this, we will find occasional instances of the vice, which will give us much trouble.   I have seen it as early as the second year, and in one case, the boy being in his fourth year, it had proven a cause of severe constitutional disturbance.

It is difficult to determine the cause of the habit in children of this age, as the organs are yet undeveloped, and we are taught to believe that the venereal appetite is not present until the age of puberty.   It is my impression that the habit of handling the parts, is sometimes attained at a very early age, and attended with erections, is found pleasurable, and is repeated for this reason. I have noticed this occasionally, and even in this slight degree, if continued but for two or three months, it will give rise to marked irritability of the nervous system and impaired health.

In the severe cases we find the child has lost flesh and is much debilitated. There is marked depression of the nervous system, and the child has a furtive and uneasy look. It is easy to see that the mind is affected, that the child does not receive natural impressions, and that its reasoning power is impaired. Sometimes it seems to suffer from spinal disease, at others from disease of the lower extremities, and again from disease of the brain. Some cases seem to be dependent upon inactivity of the stomach and bowels—dyspepsia—and treatment is directed to this with the expectation of restoring the health and strength, by the use of bitter tonics and restoratives.

I recognize the difficulty by the furtive uneasy look of the patient, and by the disposition to rub the thighs together, and move the lower part of the body. In the milder cases, the child does not seem to appreciate but what it is all right, and will carry the hand to the penis while you are conversing with him. In the protracted cases, the child seems as much impressed with the offense as the adult, especially if he has been reproved for it; and practises it secretly whenever he has an opportunity. If the penis is examined in these cases, the prepuce will be found red and irritated, and occasionally the organ will be more developed than is usual at the age.

If continued, the practice will lead to epilepsy, idiocy, or to such impairment of the general health that the child will die early of some cachectic disease.

TREATMENT.—In the milder cases, those which have continued but a month or two, we may break up the habit by close attention and mild reproof. Gentle means are much better than harsh, persuasion than force, and if the matter is properly presented to the parents, and they will take advice, we will get along with the patient well.

When the habit is well-established, I put it beyond the child's power to continue it, by vesicating the surface so that handling will be painful, and an erection impossible. Tincture of Cantharides, or Cantharidal Collodion will answer this purpose well. It should always be applied without the patient's knowing of the purpose for which it is used, which can very readily be done, and it should be so repeated that for one or two months the organ must remain at rest.

In the meanwhile, by the use of tonics and restoratives, nutritious diet, baths, and exercise in the open air, we restore the general health, and at the end of a couple of months we find the habit broken and the health restored.

If the difficulty should still continue circumcision is a most effectual remedy. If parents will not consent to the complete removal of the prepuce, it may be slit up on a grooved director, and the glans exposed.

CHAPTER IX.

DISEASES OF THE NERVOUS SYSTEM.

Lesions of the nervous system exert a very important influence in disease, and we rarely find a case in which they do not form a part, and must be estimated in the diagnosis. They also exert a more or less marked influence on all pathological changes, and it is through the nervous system, to a considerable extent, that we are enabled to modify and change the various processes of life.

The nervous system of the child is peculiarly susceptible to the causes of disease, and as susceptible to the influence of remedies. Hence, minute doses of *direct* medicines are found preferable to large doses of such as act in an indirect manner.

It will be recollected that the nervous system is divisible into three parts—the brain, the spinal cord, and the sympathetic nervous system.

The brain is the organ of the mind, and gives us conscious existence, and from it the will calls into action the voluntary muscles to serve its purposes. As a whole it is an element of weakness, and draws upon the vegetative functions for its life, rather than giving strength or aid to them. The popular opinion "that a child may have too much brain," like many such, is founded in fact. The large brain, as compared with the body, appropriates an excess of blood, nutritive material, and vital or formative power, and the remainder of the body suffers in consequence.

It is essential to success, in some cases, to see that this morbid growth of the brain be arrested. This is accomplished by adopting such a course of training as will develop the body and keep the mind at rest.

The basilar portion of the brain is an expansion of the spinal cord, and influences vital processes. The tenacity of life and power of living depend, to a very considerable extent, upon its development and perfect condition. We have already seen that this power may be very closely estimated by a measurement determining the depth of the basilar portion of the brain.

The spinal cord is the center of reflex action, or automatic movements. It carries on certain functions when the will is in abeyance, and others that are but partially under the influence of the will. The functions of respiration, deglutition, defecation, etc., and all the involuntary movements, are carried on under its control. Nerves from this source pass to all parts of the body, and we know that they may take complete control of all the voluntary functions, usurping the place of the brain.

The sympathetic nervous system presides over the functions of vegetative life, controlling digestion, assimilation, blood-making, the circulation of the blood, nutrition, waste of tissue, secretion and excretion. Though the entire amount of gray nerve substance in the sympathetic ganglia of the body would not be the size of a pigeon's egg, it is probable there is no structure in the body, however minute, but what receives nervous supply from this source.

DETERMINATION OF BLOOD TO THE BRAIN.

Determination of blood to the brain is of frequent occurrence in acute disease. We understand that determination is the result of irritation, but we can only account for this by the supposition that it follows from the accelerated circulation and increase of temperature. The immature brain is very susceptible to irritation, and but slight change in its vital condition is required to induce it.

SYMPTOMS.—Determination to the brain is announced by the increased restlessness and irritability of the patient. The face is flushed, the eyes bright, the pupils contracted, and the head hotter than usual. These symptoms gradually increase until they become very marked, the child suffering severely from it, and occasionally having convulsions.

With determination to the brain all the other symptoms are increased. The pulse is more frequent, the temperature higher, greater arrest of the secretions, less disposition to sleep, and further impairment of the digestive organs.

It may continue to increase until an inflammation is established, or after some days terminate in the opposite condition of congestion, from exhaustion of the irritability of the brain.

TREATMENT.—The treatment of determination of blood to the brain in childhood is direct and certain. The use of Gelseminum, associated with the special sedatives, will meet the indications in almost every case. But the specific action of the remedy is only obtained from the one preparation—a tincture made from the *green* root ; the most that is sold is wholly worthless for any purpose in medicine. In this form I prescribe it in the following proportions : ℞ Tincture Gelseminum gtt. v. to gtt. x., Tinct. Aconite gtt. v., water ℨiv.; a teaspoonful every one or two hours.

If the features are contracted, expressive of pain, and the child starts in its sleep, and cries out shrilly, Rhus will take the place of Gelseminum, in the proportion of gtt. v. to water ℨiv.

With this treatment we employ the general bath and the hot foot-bath ; the last frequently repeated, and if the head is hot, it is sponged with warm water, and evaporation is promoted by fanning. Catharsis, counter-irritation, and ice to the scalp, should be avoided.

CONGESTION OF THE BRAIN.

Congestion is that condition in which there is excess of blood in a part, with its movement impaired. It may arise from irritation of the brain at first, but usually depends upon some cause which depresses the vitality of the organ. The circulation is feeble and sluggish, while the vessels are unduly distended, and necessarily the function of the organ is impaired.

SYMPTOMS.—The child is dull and inactive, wants to sleep much, and usually sleeps with its eyes partially open. The dullness and hebetude are so marked that they cannot escape notice. As the disease progresses we find this condition increased; the child sleeps a considerable part of the time, but it is not so much sleep as stupor, and gradually this passes into coma, which at last becomes so profound that the child can not be aroused from it.

With this condition of the brain the whole system sympathizes, and there is atony with tendency to sluggish circulation, and consequent arrest of function in every part.

TREATMENT.—This condition of the child, like the preceding, I prefer to treat with a *specific*, rather than to depend upon the old means—counter-irritation and catharsis. Belladonna is the remedy, and I would prescribe it in the following form: ℞ Tincture of Belladonna gtt. v., Tincture Aconite gtt. ij. to gtt. v., water ℥iv., a teaspoonful every hour. I think the remedy will very rarely disappoint expectation, as I have successfully employed it for this purpose for many years, and it has been used very extensively by others with like results.

It may be aided by the use of the hot mustard foot-bath, continued for thirty minutes at a time, and repeated two or three times a day, and in very severe cases by dry cups to the neck.

When the condition is so marked, that it will evidently prove fatal if not speedily removed, I have been accustomed to employ a prompt and thorough emetic, and follow with the Belladonna. This treatment may be recommended in those very severe cases sometimes met with in the eruptive fevers, in which the emetic will throw off the profound coma, and give freedom to the circulation and the respiration. I trust to the Belladonna to accomplish this result in children under one year, and it has so far proved satisfactory. Still there are very many who will not have faith in such small doses of a single remedy, for the relief of so grave a condition.

PHRENITIS.

Inflammation of the brain is of more frequent occurrence in childhood than in the adult. It is rarely an idiopathic disease, but when seen it is a complication of some other affection, usu-

25

ally a fever or inflammation. As it is preceded by determination of blood, the symptoms of which are well marked, and the remedies employed almost specific, an imflammation of the brain should be of very rare occurrence.

CAUSES.—An inflammation of the brain may arise from any of the causes that would produce inflammation of any other organ or part; but, as we have seen above, such origin is very rare. Occurring, in a majority of cases, during the progress of an acute fever or inflammation; it is dependent upon an irritation produced by the accelerated circulation and increase of heat. It occurs in such persons as have a natural irritability of the brain and nervous system. These changes prove sufficient to establish an irritation, which is followed by determination and inflammation.

PATHOLOGY.—The inflammation may be confined to the dura mater and arachnoid—*cerebral meningitis;* or it may affect the substance of the brain itself—*cerebritis.* But it is impossible, by the symptoms presented, to determine the difference between the two during life. The inflammation presents its usual features; a first stage, in which there is an active circulation, the structures being filled with blood, and presenting similar appearances to those seen in superficial imflammation; and a second stage, in which there is stasis of blood and exudation into the connective tissue.

Post-mortem examination shows the seat of the lesion. If of the membranes, the dura mater and the arachnoid will be found injected in patches of greater or less extent. There is also an increased quantity of fluid, sometimes but little changed, at others more or less viscid, or containing flocculi of coagulable lymph. When the acute stage has continued for three or four days, we sometimes find adhesions between the free surfaces. When the substance of the brain has been involved, the vesicles of the pia mater are distended, and on making an incision into the convolutions, the cut surface will present a more uniform red color than natural, and the puncta vasculosa are more numerous and larger.

SYMPTOMS.—The invasion of the disease is indicated by a sense of fullness and pain in the head, the integuments being suffused, and sometimes a marked sense of heat. Frequently the patient complains of dullness, with confusion of ideas and forgetfulness, and unquiet sleep. Extreme irritability and fretfulness, with indisposition to sleep, and frequent startings during rest, the cry being sharp and quick, as if terrified, are the precursory symptoms in children. The disease is usually ushered in with a marked rigor or chill, continuing for an hour or two, or sometimes for nearly a whole day. Following this, there is in most cases high febrile reaction, the skin is hot and flushed, the pulse frequent and hard, tongue coated white, bowels constipated, and urine scanty and high-colored. The head is turgid and hot, the eyes more prominent and suffused, the pupils contracted and fixed, and a deep-seated, heavy pulsating and tensive pain in the head.

As the disease progresses, the patient becomes more irritable and restless, the pain in the head increases, there is intolerance of light, ringing in the ears, and intolerance of sound, sleeplessness and delirium. Up to the third or fourth day the fever is usually continuous, though sometimes there is a slight remission in the forenoon, and the head symptoms increase or continue without abatement. A marked change is now observed, the acute sensibility gives way to torpor, and the delirium becomes low and muttering, or is replaced by coma. The pulse becomes fuller, softer, or slow, or in some cases very hard and frequent. The head and trunk are still hot, the face turgid and of a deeper color, or in some cases blanched and contracted, the pupils dilated, the extremities cool, respiration difficult and sometimes stertorous, and more or less involuntary movement and starting of the tendons. The coma gradually becomes deeper, and the insensibility more marked; all the functions are feebly performed, the patient lies on his back, slips down to the foot of the bed, grasps at imaginary objects, and thus slowly sinks. According to Copland: "In some cases, particularly those in which the cerebral substance is early and generally inflamed and turgid, instead of phrenitic delirium, an apoplectic sopor, often preceded by convulsions, quickly supervenes; with a slow pulse, stertorous, slow, or labored breathing, turgid or bloated countenance, startings of the tendons, involuntary evacuations, torpor of the senses, and flaccidity of the limbs." Here the first stage is very short, or not noticed, and the disease passes rapidly to a fatal termination.

In children we frequently find inflammation of the brain

making its appearance during the progress of other diseases.
The head becomes hot, the face turgid, the pupils contracted,
with great restlessness and constant movement of the head.
Though not very marked on account of age, the child is evident·
ly delirious, and the frequent movement of the head, and putting
the hands up to it, shows that it suffers pain. In other cases the
acute stage has passed without notice, the face is blanched and
contracted or white and puffy, the pulse is small and very frequent, the extremities cool, bowels loose, the operations being
unnatural and offensive, there is continued movement of the
head and restlessness, or a deep stupor or coma. Sometimes the
symptoms will continue for three or four days, but at other
times the disease will terminate fatally within forty-eight hours.

DIAGNOSIS.—It is not difficult in the most of cases to determine the presence of phrenitis. The heat and turgidity of the
face and scalp, the deep-seated and tensive pain, contracted pupils, and the great irritability and restlessness, with the high
grade of fever, are sufficient for the diagnosis. In those other
cases in which coma, difficult respiration, full but oppressed
pulse, coldness of the extremities, dilated pupils, etc., are the attendant symptoms, the diagnosis will be very difficult, and if we
can not have the previous history of the case, almost impossible.

PROGNOSIS.—In the first stage of the disease, the prognosis is
usually favorable, if prompt measures are adopted for the arrest
of the inflammation. In the second stage the lesions are so great
that we will have to be guarded in our prognosis, though a considerable number will recover.

TREATMENT.—If called to a case in the first stage of the disease,
we have the patient thoroughly bathed with the alkaline wash,
drying with brisk friction; this is followed by a hot mustard
foot-bath, continued for half an hour, and both are repeated once
or twice daily. Internally, I prescribe—℞ Tinct. of Aconite
gtt. v., Tinct. Gelseminum gtt. x., water ℥iv.; a teaspoonful every
hour. If the fever runs high, I would add to this, Tinct. of Veratrum gtt. x., for a few hours.

To lessen the heat of the head, we direct that it be sponged
with warm water and fanned, to produce evaporation. We will
find quite warm water is very agreeable to the little patient, re-

lieving the excitation of the nervous system, and lessening the temperature. We never apply cloths wrung out of cold water, as is the common practice, or bladders of pounded ice, as recommended in most works. This is not nature's method for removing surplus heat, as is the case with the plan first named. On the contrary, it is directly depressant in its first influence, and if not persistently applied there will be a corresponding reaction; if, therefore, it is continued, the depression may cause death; if suspended, the reaction is higher than before, and the patient in greater danger.

I do not approve of the use of a cathartic in all cases, for while in some it will prove beneficial, in others it exerts an unfavorable influence. Never give a cathartic when the tongue is contracted, reddened around its border and white in the center, or when it is elongated and pointed. In this case it is quite certain to irritate the stomach and upper intestine, increasing the febrile action and the disease of the brain. It may, however, be used with good effect when the tongue is moist, somewhat broad, and tolerably uniformly coated with a yellowish or grayish fur. I would prefer a small portion of jalap with a saline, as—℞ Jalap grs. iij. to grs. v., Bitartrate of Potash gr. v. to grs. x.

If the tongue is thus coated, sulphite of soda, in doses of two to five grains, may be given every two or three hours, until the tongue cleans.

So soon as the influence of the special sedative is observed, we may prescribe a solution of acetate of potash, to be taken largely diluted with water. I usually give it as a drink, and as the child is thirsty it will be taken pretty freely.

In some cases Rhus is a very important remedy, and replaces the Gelseminum. The indications are usually very clear: the pulse is sharp, the tissues about the eyes and base of brain contracted, with evident frontal headache, and the tongue shows the red papillæ at its tip.

If the child is wildly delirious, and clutches at its mouth and throat, Stramonium may be given with Veratrum, aa. gtt. v., water ℥iv.; a teaspoonful every hour.

There is one point in the treatment that should not be neglected—*keeping the child quiet*—rest to the nervous system being essential to complete recovery. I prefer that the child shall remain in the recumbent position in its crib or bed, with the room darkened, and as little noise as is possible.

In some cases there is extreme irritation of the stomach, and it becomes important that this should be relieved. The Aconite and Gelseminum will have to be in very small dose, and we may need to add portions of Aconite and Gelseminum to the water that we sponge the child's head with. If this is not sufficient to relieve the irritation, peach bark may be used, or a small portion of a Seidlitz powder may be given with the drink. If the temperature is high, cold packs may be used over the stomach and bowels; if the child is feeble, hot packs or sponging will be better. With a temperature of 105° I should think of an enema of of four to eight ounces of *cold* salt water; with cool extremities, the water should be as hot as the patient can bear it.

In the *second stage*, the condition is wholly changed. Instead of the active circulation and excitement of the brain, we have a stasis of blood and coma.

Locally, I order dry cups to the neck, and sometimes to the spine, with the hot mustard foot-bath; the first being repeated if the case is severe, and the second used two or three times a day. Evaporating lotions are applied to the scalp, as sulphuric ether, alcohol, or cologne; sometimes a camphor or arnica lotion will answer the purpose well.

Internally, I prescribe—℞ Tinct. Aconite gtt. v., Tinct. Belladonna gtt. x., water ℥iv.; a teaspoonful every hour. Or if there is fullness of the fontanelles or swollen eyelids, we might add Tinct. Apocynum gtt. x. to water ℥iv., and give in teaspoonful doses alternately with the Belladonna. If there is retention of urine, or sluggish passage, Santonine may be given in doses of one-fourth to one grain, as heretofore named.

If there is the condition of the tongue named above as permitting the use of a cathartic, I should give—℞ Jalap grs. iij., Capsicum gr. ½, Bitartrate of Potash, gr. x., and repeat it every four hours until it acted freely.

To act upon the kidneys, I would prefer sweet spirits of nitre in doses of ten or fifteen drops, every two hours, as it is also an excellent stimulant.

In some cases we will find quinine inunction to answer a very good purpose. If there is much prostration and feeble circulation in the skin, some rubefacient may be added to it, as one of the essential oils, or the oil of mustard in *very* small quantity. If the quinine is not used in this way, it may be given internally to the extent of two grains a day, during convalescence.

ACUTE HYDROCEPHALUS.

We use this term, for the want of a better one, to distinguish a certain class of cases in which the disease of the brain is a principal lesion, though the symptoms would indicate something else. The disease is confined almost entirely to children, occuring most frequently from the ages of one to three years, and being rare after twelve. It is very difficult to determine the cause, though we are of the opinion that it is frequently dependent upon irritation of the digestive apparatus, or upon any cause that will enfeeble the system. At the age of ten or twelve it is usually brought on by over mental exertion.

PATHOLOGY.—The lesions observed in this disease are by no means constant. In some cases there is considerable effusion into the ventricles and the cavity of the arachnoid, but in others there is very little or none. Sometimes there is evidence of determination of blood, and occasionally small patches of lymph or flocculi in the effused fluid; rarely the brain exhibits evidence of slow inflammation.

"The nature of acute hydrocephalus," says Dr. Bennet, "has been keenly disputed, and whether it be inflammatory or non-inflammatory, and should be treated with antiphlogistics or nutrients, will be found to be discussed at great length in systematic works and numerous monographs. The fact is, that the group of symptoms indicating the occurrence of water on the brain is altogether insufficient to prove the existence of this morbid product in acute cases. What we observe are symptoms of excitement, gradually passing into those of depression, occasionally passing into paroxysms of pain, restlessness or screaming, alternately with drowsiness and coma. These symptoms are common to various lesions of the brain, and may be the result of congestion, or of this state terminating in effusion and frequently in exudation. Hence, why sometimes after death we find no lesion whatever, at others more or less distension of the ventricles with serum, and very commonly, in addition, exudation at the base of the cranium. In every case, the symptoms are referable not so much to one or the other of these lesions as to something which they all have in common, and this undoubtedly is more or less pressure on various portions of the brain, causing,

first, irritation and then perversion of function, or so operating
as to excite some parts and to depress others.   In the great majority of cases, the fluid distending the ventricles is more allied
to the dropsies than exudations.   Nay, even when lymph is
thrown out at the base of the brain, the amount of serum in the
ventricles is altogether disproportioned to the quantity of coagu-
lated fibrin deposited.   Hence, I am disposed to think that even
when evidence of so-called inflammation exists, still the fluid
that distends the ventricles is owing to a mechanical obstruction
of the vessels, causing dropsical effusion."

SYMPTOMS.—At an early age, we find the disease commencing
as an obscure remittent fever, the languor, or more properly
stupor, being the most prominent symptom.   The fever usually
has an exacerbation in the afternoon, the child being restless and
fretful at this time.   Nausea and vomiting are very frequently
present, especially if there is irritation of the bowels, forming
one variety of cholera infantum.   In a longer or shorter time,
usually not more than from two to ten days, the patient becomes
almost entirely unconscious, though from the occasional glance
of intelligence it is not believed by the parents.   Still it is restless and uneasy, turning its head from side to side, putting its
hands to its head, and uttering those sharp, piercing cries indicative of pain; if the tongue can be seen, it will be found dry, its
tip and edges red, and center covered with a white coat.   The
countenance is now pallid and pinched; the eyes want expression,
and are sunk in the head, the pupils generally dilated; the head
is not above normal temperature, frequently dry, although the
forehead is covered with a clammy perspiration.   If there was
not diarrhœa at the commencement, there is now, the stools
being of a dirty-yellowish or greenish color, mixed with slimy
matter and having an offensive odor.

A very common grouping of symptoms is thus reported in a
clinical case by Dr. Bennet: "Unconsciousness of surrounding
objects, not recognizing even her mother; pupils not contractile
to light; slight strabismus of right eye; frequently puts her
hands to the head, which is rolled about uneasily; continual
grinding of the teeth, low moaning, and occasional muttering.
Tip of tongue, which is all that can be seen, very dry and of a
scarlet color; loss of appetite; constant thirst; vomiting; involuntary discharge of feces and urine; on pressing the abdomen

uneasiness evidently experienced, and moaning increased; skin hot and dry; no eruption; a small abscess at the back of the neck, with a sanious discharge; action of the heart feeble and fluttering; pulse one hundred and forty, small, and occasionally intermittent. Breathing short and hurried; no rales." These symptoms were developed in a child aged six, commencing fourteen days previously with diarrhœa.

In older children the first symptoms will be a more or less severe headache, with intellectual stupor, the child being restless and uneasy, and passing bad nights; an obscure fever may be recognized in the after part of the day, the skin being dry and husky, and the pulse frequent and hard. For days, and even for two or three weeks, the symptoms continue in this way, the child being occasionally better for a few hours or sometimes for a day or two. Suddenly the pain in the head becomes intense; the face is pinched and expressive of great suffering; the tongue is red at its tip and edges, dry, and its center covered with a white coat; the bowels constipated, or there is diarrhœa; urine scanty; the pupils dilated, and immobile on exposure to light. The child does not like to be disturbed, is constantly dozing, though its nights are restless. The pulse may be either frequent and sharp, or in some cases slow and feeble. Very frequently there is nausea and vomiting, sometimes very persistent and intractable. These symptoms becoming very severe, deep coma results, from which the child never recovers, but two or three days elapsing from its accession until the fatal termination.

DIAGNOSIS.—I diagnose this disease by the pinched expression of the countenance, every part seeming to be contracted, the dilatation of the pupil, the stupor and at the same time restlessness of the patient. These symptoms may be confounded with the second stage of inflammation of the brain, but the prior symptoms are usually sufficient for the diagnosis.

PROGNOSIS.—It is very difficult to so describe this disease, that the reader may determine which cases will recover, and which will unavoidably prove fatal. Usually, if the child is still conscious, and there is not such marked contraction of the countenance, as to render it hippocratic, we may hope for a favorable result; if the contrary is the case, it will in all probability prove fatal.

TREATMENT.—The treatment will have to be much modified
to suit each individual case; yet in every one it must be decidedly
stimulaut and sustaining. If there is continuous nausea, with
evidence of morbid accumulations, the stomach should be relieved by the administration of a prompt emetic, very marked
benefit following its action. In other cases the nausea may be
arrested by a sinapism over the epigastrium, and the administration of au infusion of peach-tree bark, or compound powder of
rhubarb and potassa. Injections of salt and water, an even teaspoonful to four ounces of water, will sometimes prove very
efficient in irritation of the stomach ; chloroform and glycerine
may be used for the same purpose, and may afterwards be continued in doses of five drops every hour or two, for its stimulant
and at the same time soothing influence. If there is diarrhœa,
it is the generally received opinion that it should be checked by
the use of astringents ; but this is bad practice. as almost invariably on the arrest of the discharges, the coma becomes complete,
and the child dies.

Remedies should be very carefully selected in these cases.
Aconite and Ipecac, in small doses, will sometimes quiet irritation of the stomach and relieve this irritation. Gelseminum
may be indicated as in inflammation of the brain. Belladonna
is called for when dullness and stupor become marked symptoms.
Rhus will sometimes prove a most valuable remedy, there being
a sharp frequent pulse, contraction about the eyes and base of the
brain, sudden startings in sleep with shrill cry.

With the relief of irritation of the stomach, and to a certain
extent of the brain, Apocynum becomes a prominent remedy,
though it must be used in small doses. ℞ Tinct. Aconite gtt. iij.,
Tinct. Apocynum gtt. v., water ℥iv.; a teaspoonful every one or
two hours. If indicated, Rhus or Belladonna should be substituted for the Aconite.

If the tongue is pallid and dirty, sulphate of soda should be
given in doses of two to five grains every three hours. If red
and dirty, sulphurous acid will be the remedy. If dry and harsh,
the child's drink may be acidulated with muriatic acid, or it can
have small portions of good sharp cider, or in some cases, of
whey. The hot mustard foot-bath will be sufficient in some
cases, if thoroughly used, but if the case is severe, I prefer a tub
of water as hot as the child can bear it, and rendered stimulant by the addition of mustard or capsicum, into which I put the

child, letting it remain for half an hour, being well covered with a blanket. In place of this, sponging the surface with hot water, will answer a good purpose. Dry cupping to the neck, and even sometimes to the entire spine, is among our most important measures; sometimes the cups to the neck may be scarified, especially if there is much heat of the head. A sinapism to the neck and spine will sometimes answer the purpose, but is not as good as the cups, and I think not as useful as friction with strong salt and water hot. If the kidneys fail to act freely, I would administer an infusion of hair-cap moss or of marsh-mallows, with a suitable portion of acetate of potash or sweet spirits of nitre.

CHRONIC HYDROCEPHALUS.

Dropsy of the brain is almost exclusively a disease of childhood, and occurs most frequently before the third year. It is difficult to determine the causes that give rise to the effusion of water from the arachnoid, but as it occurs almost invariably in children of feeble vitality, and in families whose children die during infancy of acute hydrocephalus, cholera infantum, or this, we are led to believe that it depends upon hereditary debility. The exciting cause may be the exanthemata, whooping cough, disease of the bowels, or inflammatory disease, or it may arise from depression, produced by cold and other causes.

SYMPTOMS.—The symptoms vary very greatly, the disease running a tolerably rapid course in some cases, and a very slow one in others. The child usually complains of its head, if it can talk, or moves it from side to side, putting its hands to it frequently. The face is pallid and contracted, or in some cases puffy and without expression; the circulation is feeble, the extremities being cold, and the surface easily chilled; the appetite is irregular, sometimes good, at others very poor, and digestion seems to be feeble; the bowels are torpid and constipated, though sometimes irregular. As the disease progresses, we notice that the child is very stupid, and that at times it has difficulty in controlling the voluntary muscles; there may be temporary or permanent strabismus, and an involuntary rolling about of the eyes, with a dilated and fixed pupil.

Occasionally we observe a marked irritability of the stomach, that is with difficulty controlled, and in some cases an extreme irritability and restlessness, though the intellectual functions are greatly impaired. As the disease progresses the torpor becomes deeper, and the child does not exhibit the symptoms of pain above named. The pulse is now seen to be getting perceptibly weaker, and occasionally irregular; the hands are tremulous and unsteady, and frequently raised to the back of the head. When the child sleeps, its eyes are half open, and the eyeballs are constantly moving and usually drawn upward. When the torpor is not so great, the child is in some cases constantly picking its nose or lips, and is extremely irritable, having paroxysms of rage from the slightest supposed offense. The disease may continue this way for months, or in some rare cases for years, finally terminating fatally by the development of some ataxic disease, or of acute hydrocephalus, or with a gradually developed marasmus.

DIAGNOSIS.—In very young children, and sometimes up to the age of three years, there will be found a perceptible enlargement and distension of the fontanelles, and separation of the sutures, and the child's head is appreciably larger. After this, we are guided entirely by the general symptoms above named.

PROGNOSIS.—The prognosis is usually unfavorable, though some cases may be cured, and in others life may be prolonged for a considerable period. Cases are recorded in which the persons lived to adult age, and in four cases to twenty-seven, thirty-two, forty-five and fifty-four years.

POST-MORTEM-EXAMINATION.—If the disease has been of long duration, the bones of the cranium will be found thin and transparent, and occasionally separated from each other by very considerable intervals. The effused fluid is found in the sac of the arachnoid, and in the ventricles; if in the ventricles to a great extent, the convolutions are unfolded, and the medullary and cineritious substances can with difficulty be distinguished. The brain is often denser than usual, and is not diminished in weight.

TREATMENT.—I am satisfied that the greatest success in the treatment of dropsy will follow the use of specific remedies, and especially in the case under consideration. The remedy I prefer in this case is the Apocynum cannibinum—its substitute, Aralia hispida. I prescribe the first in the following form: ℞. Tinct.

Apocynum gtt. x., water 3iv.; give a teaspoonful every two or three hours. The Aralia I have used in infusion, associated with the tonics named below. A number of cases of hydrocephalus cured by Apocynum are on record, and I have had three cases in my practice which have recovered under its use.

After the use of these means for some days, their influence in removing the accumulation being observed, we put the patient upon the use of tonics, alternating them.

As a tonic, some of the preparations of hydrastia answer the best purpose, and when there is constipation of the bowels, may be given with podophyllin in small doses; the combination of these remedies heretofore given may be employed. The Collinsonia canadensis is a favorite remedy in my practice, especially in cases where there is irritation of the nervous system, and we may associate with it the Ptelea, Cornus, Euonymus, or other remedies of this class. Rye whiskey and cod-liver oil are excellent when the stomach bears them kindly, and should it reject the oil, its place may be supplied by sweet cream or beef-suet.

The child should have a daily salt-water bath; sometimes the entire bath will be best, and may be used either warm or cold, the first being generally preferable, or the sponge-bath may be used; in either case it should be followed with brisk friction. The child should be taken out in the open air every day, being warmly clad in flannel. If possible it should be removed to a high locality in the country, where it can have pure air and sunshine, exercise and pure milk.

SPINAL MENINGITIS.

Inflammation of the meninges of the spinal column is not an uncommon disease, though sometimes, from the obscurity of its symptoms, it may be mistaken for other affections. It occurs in two forms, as a distinct sporadic inflammation, and as an epidemic or endemic fever, which involves the spinal cord. It is in the last named cases that mistakes in diagnosis are most usually made. The causes of this affection are those which give rise to other inflammations, as cold, sudden changes of temperature, injuries, and especially a sudden chilling of the surface after active exertion. It occurs most frequently in the young and vigorous, and is very rare after middle life.

SYMPTOMS.—Spinal meningitis usually commences with a well-marked chill, lasting for several hours, though sometimes with a severe rigor of considerable duration. I have seen cases in which the chill was of twenty-four hour's duration, the later part of it being alternated with flushes of heat. Following this there is marked febrile reaction, with hot, dry skin, hard and frequent pulse, tongue coated white, the edges and tip being red, constipation of the bowels, and scanty and high-colored urine. The patient complains greatly of pain in the back, which is so increased on movement, that he dislikes to change his position for any purpose; though in some cases, when not so severe, they are constantly shifting their position to give them ease. By the second or third day the fever usually becomes high, the pulse running some thirty or forty beats higher than in health, the skin being very dry and constricted, and the irritability and restlessness marked. These symptoms may be so prominent as to completely overshadow the symptoms of spinal inflammation, the patient not even complaining of pain, unless his attention is directly called to it. It will be noticed, however, that the slightest movement or changing the position of the body gives rise to pain, and when the attention is thus drawn to it the soreness of the spine will be continually noticed. Deep pressure usually elicits tenderness, and sometimes the sensibility is so exquisite that the patient cannot bear to be touched.

As the disease progresses, the fever assumes an irritative or typhoid type. The tongue soon becomes brown, and sordes appear on the teeth. Typhomania occurs about the sixth or seventh day, and is frequently attended with looseness of the bowels. Sometimes there is marked irritation of the brain and delirium, at others a stupor which soon passes into deep coma. As the local disease progresses, it is found that the lower extremities are subject to involuntary movement, and that the patient has but partial command over them; and that the bladder and rectum are evacuated without the knowledge of the patient, or there is retention of urine without the power of discharging it. At last, in severe cases, paralysis of the parts below the seat of inflammation is complete. The fever is usually continued, though sometimes remittent, and is invariably ataxic, presenting well marked typhoid symptoms, with the exception of diarrhœa, by the tenth to the twelfth day. It is usually protracted, lasting from two to eight or ten weeks.

DIAGNOSIS.—We diagnose inflammation of the spinal cord by the marked tenderness of the spine and inability to move, the constant pain in the back, with the severe attendant fever. It is almost impossible to overlook these local symptoms, and yet in many cases they have been disregarded, to the great detriment of the patient.

PROGNOSIS.—The prognosis is usually favorable, if treatment is commenced in time, but is unfavorable after it has made progress for several days, in many cases terminating fatally, or in paralysis.

POST-MORTEM EXAMINATION.—In some cases there is marked evidence of determination of blood to the membranes, and enlargement of the vessels. Sometimes the membranes are thickened, and fragments of organized lymph on the free surface; there may also be flocculi in the fluid of the spinal cord, which is increased in quantity. In other cases the disease seems to be confined to the pia-mater and the substance of the cord, the former being slightly reddened, and sometimes thickened, and the latter softened, sometimes so much as to have lost all traces of organization.

TREATMENT.—The treatment of this disease will vary according to the indications. In some cases Aconite or Veratrum, with Gelseminum, will be the remedy, and as the pulse and temperature come down, we find the symptoms of spinal irritation reduced. In other cases the indications for Rhus will be marked, and it will be associated with Aconite. Bryonia is indicated by the steady pain in the spine, with possibly a flushing of the right cheek. Macrotys will be indicated by muscular pain, as in other cases; Sticta by the pain in shoulders and neck to the occiput; Phytolacca by soreness of the mouth and throat, and fullness of cervical glands.

In some cases, the attack being sudden and severe, and the patient suffering from nausea or disgust for food, drink, or medicine, we find a full, dirty tongue, and conclude that the treatment had best be commenced with an emetic. The Acetous Emetic Tincture acts well, but in many cases we would prefer Ipecac in doses of two or three grains every fifteen minutes to free emesis. With a broad, pallid and dirty tongue, sulphite of soda will be indicated in the usual doses. If the tongue is red, and covered with a glutinous yellow coat (like fecal material)

sulphurous acid is the remedy. When the face is full and purplish, like one who has been exposed to severe cold, Baptisia is
given.

When the temperature is high we may sometimes obtain excellent results from the cold wet-sheet pack. It requires a little
courage to use it, for both child and parents are afraid; still, if
we have a severe case, and the stomach is irritable, so that we
do not see our way clearly with other remedies, we will use it.
If the case is severe, and there is marked impairment of the skin
with exhaustion, I think well of the hot-blanket pack. In place
of this, rapid sponging of the surface with hot water answers
well, (it should be done before a fire, and the surface protected
against chill by a blanket.) Any part that the patient complains
of may be thus sponged.

The local applications to the spine (if we use any) will vary
with the case. Sometimes dry cups give relief. In other cases
hot spongings, or hot packs, hot bran, or a bag of hot salt, to the
part, is of benefit. I do not have much faith in liniments, and
am inclined to believe that rest is better than the rubbings.

As the inflammation passes away the child may have some of
the simpler bitters, quinine inunctions, the hypophosphites, cod-
liver oil, or some of the preparations of malt.

CURVATURE OF THE SPINE.

Curvature of the spine occurs most generally in the young,
and is rare after the age of twenty-five. In all cases it is the
result of enfeebled vitality, either congenital or induced by des-
titution, over mental exertion, or sexual excitement. In some
cases this manifests itself in the form of scrofula or tuberculosis,
and in such case we may expect disease of the bones. Two varieties of curvature are met with, lateral and posterior, both occurring most frequently in the dorsal region, though at last always
compensated by curvature of the lumbar and cervical portions.

Lateral curvature may be dependent upon affections of the
muscles, as hypertrophy, atrophy, spasmodic contraction or
inflammation; upon general debility, the body not being sufficiently strong to support itself in the erect position; upon obli-
quity of the pelvis, the result of injury or disease of the lower
extremities; upon altered capacity of one side of the chest; upon
rachitis or softening of the bones, or defective development of

the vertebra. Posterior curvature is most generally dependent upon disease of the bodies of the vertebra, though in some cases it undoubtedly results from debility, and the habit of throwing the head and shoulders forward in sitting and walking; in the last case being very mild. Practically we have to study the case, first, with reference as to whether it depends upon disease of the muscles or bones; second, whether its continuance depends upon determination of blood or upon feeble circulation; and third, as regards the general health, whether there is simple debility from imperfect digestion and assimilation, or a scrofulous or tubercular cachexia. The success of the treatment will depend upon accurate diagnosis as regards these points, as in many respects it must differ in different cases.

SYMPTOMS.—The symptoms of curvature of the spine vary greatly in different cases, in some being very marked, and in others obscure. Usually the child's health is noticed to be feeble, its appetite variable, and digestion and assimilation imperfect. It may or may not complain of pain in the back, but it will be noticed that the back is weak, and that it makes unusual efforts to rest it. In lateral curvature, the disease is most usually dependent upon local debility of the erector muscles of the spine, and there is frequently no complaint, except from weakness of the back, and the symptoms of general debility above named. If partially owing to spasmodic action, pain would be a constant attendant, though usually there would be no tenderness on pressure. If the result of disease of the bones, as in most cases of posterior curvature, in addition to more or less pain, there will be tenderness on deep pressure. In these cases the disease of the bone causes irritation of the spinal cord, and we have the symptoms heretofore named.

DIAGNOSIS.—An examination of the spine will determine the existence of curvature, and it is usually not difficult to determine which is the primary and which is the curvature of compensation. In almost all cases of lateral curvature we will find the fault to exist principally in the muscles at first, though as the disease progresses, irritation is frequently developed, resulting in spasmodic action, and finally in atrophy or softening of the bones; hence spinal tenderness will usually result in the latter part of the disease, and not at is commencement. In posterior curvature, we sometimes have the most marked evidence of scrofulous cachexia, and in most cases we have marked general debility. It will be recollected that the disease of the bodies of the vertebra may be a true inflammation, or result from deposit of tubercles and scrofulous material, or may be simple softening from rachitis. In the first case the pain will be marked and decided, in the second there is simple irritation and aching of the part, with tenderness on pressure ; and in the last we will have the previous curvature and deformity of the legs and pelvis, in addition to the absence of pain and tenderness, to aid us in the diagnosis. Mr. Solly believed that softening of the bones might be entirely local, and might be dependent upon nervous exhaustion ; in such case the symptoms would be obscure.

PROGNOSIS.—In lateral curvature a favorable prognosis may be given in many cases, the deformity being nearly entirely removed, or it may be simply arrested, the body so accommodating itself to it as to give rise to but little subsequent trouble. In posterior curvature the best results usually obtainable is to stop the disease and prevent further curvature. It is true that in some cases we may partially correct the deformity, but in a large majority the attempt is attended with injury rather than benefit. If there has been destruction of the bodies of the vertebra the best result is anchylosis of the bones and of course permanence of the curvature ; and if this is prevented by instruments for extension, the life of the patient will almost surely be sacrificed.

TREATMENT.—In all forms of spinal curvature attention to the general health is one of the most important points in the treatment. Those bitter tonics that improve the tone of the stomach, and give the patient a good appetite and power of digestion are applicable. For the selection of these remedies, the reader is referred to the first part of this work, "restoratives," which will enable him to select the best remedy. If there is disease of the bones assuming the form of softening, phosphoric acid has been recommended ; and from the little experience I have had with it, I am inclined to believe that it will generally be found advantageous ; we would commence its administration in doses of two drops of the dilute acid, four or five times a day, and increase it if deemed best. Phosphate of soda may be given with the food in such quantities as will keep the bowels soluble.

Simple lime water is sometimes useful, and children will improve when it is added to their milk. Even common salt becomes an important remedy when children have not had it in sufficient quantity (bottle babies), and added to their milk there is a decided improvement. Hypophosphite of lime, the compound syrup of the hypophosphites, and some of the phosphates, may also be thought of. Occasionally veratrum and arsenic in small doses will improve the appetite, digestion and bloodmaking.

When the symptoms would lead us to believe there was scrofulous disease of the bones, the vegetable alteratives may be brought into requisition. A combination of yellow dock and tag alder, with small portions of acetate of potash has answered my purpose well. If there is great irritability of the nervous system I would substitute the bromide of ammonium for the preparations of potassa. These remedies should not take the place of tonics and restoratives, but should be associated with them in such manner as that normal digestion and assimilation shall be the first object in view. A nutritious and easily digested diet should be prescribed, and frequently a small amount of malt liquor is advisable. The sponge bath should be used daily, sometimes of simple water, salt and water, or stimulants, as capsicum or mustard, or the mineral acid baths, or of a decoction of the bitter tonics and astringents.

If there is simple loss of muscular power, as in many cases of lataral curvature we would recommend open air exercise, and friction of the spine with cold salt water, and sometimes the use of electricity. These are the only cases in which exercise is permissible, and then it should be so regulated as not to prove exhaustive. Sir B. Brodie recommends that the muscles of the back be strengthened by climbing and other exercises, for which, in delicate girls, friction or shampooing for an hour or two daily might be substituted; and the patient should lie down for a part or a whole of the time she was not engaged in exercise. Mechanical support may be used in these cases, but it should always permit free movement. If in any case there is irritation and pain, with tenderness on pressure, the child should maintain the recumbent position, and especially is this the case in posterior curvature. Rest is all-important in these cases, until the disease is entirely arrested, and though it will sometimes seem as if the child could not bear the continued confinement, we will find that it absolutely improves in every respect, while maintaining the most perfect quiet. Counter-irritation is of much importance in these cases, but we must be careful not to carry it so far as to unduly irritate the nervous system, or induce debility by the excessive discharge. The irritating plaster is a favorite application, and will usually be found the best of any. It may, in severe cases, be replaced by the issue, and in others by two, three or four small setons, as common surgeons' silk, passed through a fold of the skin on each side of the spine.

In cases of the diseases of the bones, Dr. Pirrie remarks, "That any attempt to remove the curvature would be injudicious. Anchylosis is the only favorable termination to be hoped for, and therefore the object to be aimed at in treatment should be to place the patient under circumstances most likely to conduce to that result. With that view, it is indispensable, first to keep the patient in a recumbent position, so as to remove from the diseased parts the pressure of the superimposed weight, and to preserve the parts in a state of perfect quietude in that position; and secondly, to use all means, judicious and available in the circumstances of the case, for maintaining the general health. One particular advantage which results from preserving the parts at rest in the horizontal position, is that the removal of the irritation caused by the superincumbent weight from the diseased parts diminishes the danger of the formation of abscess, which is a most unpromising occurrence, and must induce the gloomiest apprehensions as to the ultimate result."

A most excellent means of attaining perfect rest is afforded by a common camp cot, with the head elevated about a foot, and covered with a soft hair mattress; two crutches softly padded, should pass from the foot up to the arm-pits, and an india-rubber webbing attached to the arms of these to support the trunk. In this apparatus there is constant gentle extension; the body is supported by the webbing, the patient lying on the back, or face downward, as seems best suited to the case. For full description the reader would do well to consult Bigg on Deformities, the second volume containing most explicit descriptions of apparatus and well-drawn wood-cuts.

EPILEPSY.

Epilepsy is one of the most serious of the diseases of the nervous system, not because of its fatality, for it runs a very chronic course, but because there is no tendency to spontaneous arrest, and medicine has heretofore had very little influence upon it. One of the most distressing features of the disease is, that it gradually impairs the mind, until the person, once bright and of sound mind, becomes a driveling idiot or a raving maniac. The disease usually commences in childhood, most frequently between the ages of six and twelve.

The causes of epilepsy are various, and not very well understood. They may be divided into *intrinsic* and *extrinsic*, in the first case existing in the cerebro-spinal nervous centers, or their immediate surroundings, and in the second existing at a distance, and affecting the spinal cord through the nerves. Of the first, we may instance inflammation and determination of blood to the cerebro-spinal centers, disease of the meninges and of the bowels, and injuries of the bones, giving rise to compression; or continued irritation, as by the presence of a spicula pressing the nerve-substance. Derangements of the blood may sometimes give rise to epilepsy, as in the retention of the solids of the urine and other changes that we are not cognizant of. By an *extrinsic* cause, we understand one in which the irritation being set up at a distance is propagated along the nerve trunks to the spinal cord, where, setting up an irritation, it manifests itself through the excito-motory system of nerves. The most simple instance of this action is witnessed in the case of cramps of the muscles of the extremities from irritation of the intestinal canal, as in cholera morbus, and in the case of infantile convulsions from teething or from gastro-intestinal irritation. Epilepsy may in this way arise from irritation of the stomach from crude indigestible food, from worms, from irritation of the bowels, the kidneys, or bladder, or genital organs. The cause being sufficient to set the disease going, may disappear entirely in a few days or weeks, and yet the epileptic attacks continue. It would seem that when this abnormal action is once set up, the tendency to its continuance is the same as in healthy functions; but why this is we know not, and neither can we give any probable theory.

As regards the *pathology* of epilepsy, we are much in the dark.

In some cases it would seem to be dependent on a too free circulation of blood in the nervous centers—determination of blood; in other cases upon a sluggish circulation—congestion; and in still others, upon some defect in nutrition. There are cases in which it is very manifest that the condition of the blood is the exciting cause of the epileptiform seizure, though we must still imagine an unnatural irritability of the nerve centers to be so impressed. Thus, I have seen cases in which every convulsion was preceded by deficient secretion of urine ; and so long as this secretion could be maintained in the normal condition, so long would the patient be free from its seizure. Cases in which the disease is dependent upon the amount and character of the menstrual discharge, have come under the notice of almost every one. Experience, however, has proven to me, that epilepsy is eminently a disease of debility of the nervous system, even in cases in which there seems to be the most evident symptoms of irritation and determination of blood.

Dr. Radcliffe has written a most interesting paper on the pathology of convulsions, and draws the following conclusions: "1st. The epileptic and epileptiform paroxysm is not unfrequently preceded by signs of defective respiration. 2d. It is usually accompanied by a state of unmistakable suffocation. 3d. The condition of respiration during convulsion is one which supports the notion that the convulsion is connected with depressed and not with exalted vital action. 4th. In the chronic form of convulsive disorders, the inter-paroxysmal condition is usually marked by evident signs of feeble circulation. 5th. The epileptic and epileptiform paroxysm is usually if not invariably preceded by signs of failure in the circulation. 6th. In the fully developed paroxysm, the pulse is sometimes aroused to a considerable degree of activity, not because the arteries are receiving a largely increased supply of *red* blood, but because they are then laboring under a load of *black* blood, as they are found to labor during suffocation. 7th. Convulsion is never co-incident with a state of active febrile excitement of the circulation. 8th. Epileptiform convulsion is a direct consequence of sudden and copious loss of blood. 9th. The condition of the circulation during convulsion is one which supports the notion that the convulsion is connected with depressed and not with exalted vital action."

It is of but little use to try to study the original cause in many cases of epilepsy, for as has been remarked, it has possibly passed away months before our examination. There is always, however, an exciting cause, which it is necessary to determine, if possible, as upon its removal, the success of our treatment will in great measure depend. I have known it to be a failure of excretion, an imperfection in digestion, derangement of the menstrual function, excessive mental emotion, and not unfrequently excessive sexual excitement.

SYMPTOMS.—In some cases there are brief premonitory symptoms of the approaching seizure, and rarely, the patient has notice of it for hours. The sensations differ in different cases: sometimes a sense of weight and oppression in the head, with giddiness and loss of voluntary power; in others, a coldness passing from the feet upwards, and terminating in the epileptic seizure when it reaches the head. In the more protracted cases, there is usually a marked dullness and hebetude, noticed by the friends, and the patient feels a loss of consciousness that is very unpleasant.

In an attack of epilepsy the patient becomes suddenly unconscious and falls to the floor, or wherever he may be situated. Involuntary movement from spasmodic contraction and relaxation, is characteristic of the disease, and may be very intense or mild. If severe, the limbs are thrown in various positions, the trunk contorted, and the features remarkably changed. First one group of muscles contract and then another, so that parts are kept in constant movement. The lower jaw and tongue being also affected, we find that usually the latter organ is severely bitten, if means are not taken to avoid it. The patient usually froths at the mouth; respiration is normal in frequency, and the pulse but little changed, except that it is smaller and feebler. The countenance is not only distorted by the convulsion, but in some cases is turgid and purplish, or almost black. Frequently the urine, and sometimes the feces, are passed involuntarily during its continuance.

The duration of the epileptic seizure is very variable, sometimes lasting but a few seconds, and at others for fifteen or twenty minutes. The patient may have but one attack at a time, or they may succeed one another at short intervals, until quite a large number have passed. When the attack ceases, the patient

becomes completely relaxed, and usually falls into a deep, comatose sleep, from which it is almost impossible to arouse him for
an hour or two. The frequency of their recurrence varies in
different cases; in some they do not appear oftener than once a
month; in others every week, or almost every day. Sometimes
they are so distinctly periodic that the return can be closely calculated, but at others they are very erratic in their course. In
many cases there are slight seizures during the intervals between
the principal attacks; in these the patient seems to lose consciousness for but a moment, and stares vacantly at persons present;
passing off, he has no recollection of it, nor of the epileptic
attack.

DIAGNOSIS.—We diagnose epilepsy from apoplexy by the fact
that in the first there is continual spasmodic action, while in the
last there is not the slightest motion; in the one there is frothing
at the mouth, in the other it occurs but rarely; in apoplexy the
respiration is slow and stertorous, and the pulse full and slow,
while in epilepsy respiration is of usual frequency without ster-
tor, and the pulse is small and frequent. We diagnose it from
hysteria by the previous history of the case, and by the fact that
we are able to determine that there is not complete loss of consciousness in the latter case.

PROGNOSIS.—So far as regards the cure of the disease, the
prognosis is unfavorable, unless the means here recommended
prove more serviceable than those heretofore used. But, as before remarked, it runs a course of years, and the patient dies
finally of some other affection in a great many cases.

POST-MORTEM EXAMINATION.—In a majority of cases the
scalpel reveals no lesion to account for the severe disturbance of
the system during life, and what lesions are found generally have
no relation to the epileptic affection. In some cases the evidence
of slow inflammatory action is found in the brain or spinal cord,
or, in rare cases, a morbid growth in the nervous substance, or
from the meninges or bones, is observed, and in others a change
of structure, usually softening, has occurred. These, however,
form but a small fraction of the cases. In other instances some
organ, as the stomach, kidneys, uterus, etc., is found diseased,
and as the epilepsy made its first appearance with the symptoms
of these diseases, we have good reasons to believe that they acted
as exciting causes.

TREATMENT.—The treatment in these cases is of two kinds: that for the arrest of the paroxysm, and that for the radical cure of the disease. If called to see a person suffering from an attack of epilepsy, we would place the patient in such a position that he would not be likely to injure himself, and if the convulsive action was severe, get a friend to hold a cork or piece of soft wood between the teeth to prevent biting the tongue. Usually this is all that is necessary, except in cases where the patient has a succession of attacks. In these cases, as soon as the first paroxysm commences passing off, we may administer the compound tincture of Lobelia and Capsicum in half-teaspoonful doses every five or ten minutes, until nausea is induced, which, in a large majority of cases, will prevent a return of the convulsion; or we may use the tincture of Gelseminum for the same purpose, giving it in doses of from ten to twenty drops, or even half a drachm of the common tincture, every ten or fifteen minutes, until the full relaxant influence of the remedy is produced. A combination of sulphuric ether, liquor ammonia, and tincture of asafœtida may be used for the same purpose, but it is not as efficient as the preceding measures. If need be, stimulant applications may be made to the lower extremities and to the spine, but usually this is not necessary.

As regards a radical cure, we may attempt it in all cases in which there is no structural lesion of the spinal cord or brain, or their enclosures, to account for the disease. If there is, the case becomes one for the surgeon rather than the physician, though operations thus far have proven very unsuccessful. If we can detect any lesion of function, especially if it seem to bear a relation to the epileptic seizure, we would employ remedies for its removal. Thus, in rare cases, a cure will result from the removal of worms, and relief of irritation of the intestinal canal; from the relief of menstrual irregularity; by establishing and maintaining free secretion of the kidneys, when functional lesion of these organs has been prominent, etc. In some cases the disease appears to be dependent upon spinal irritation and determination of blood, and occasionally a cure may be effected by the use of the irritating plaster to the spine, the administration of tincture of Gelseminum and the use of those other measures recommended under the head of spinal irritation. Belladonna, ergot, and nux vomica may be used when there seems to be feeble circulation in the nervous substance and tendency to congestion, manifested by symptoms of paralysis, or a feeling of deadness, coldness, or tingling as if the part were asleep.

In a large majority of cases, however, there is no lesion that would seem sufficient to occasion the epileptic seizure; and even where there is, and we have removed it, and restored all the functions of the system, the nervous disease will still continue. Here our treatment will be, to a great extent, empirical; it is true we correct all lesions of function, and try to get the system in as healthy a condition as possible, but after this we give remedies simply because they have proven efficient in other cases. I have employed the bromide of ammonium in my practice, with the most marked success, sometimes using it alone, and at others in combination with other remedies. I prescribe it in the proportion of half an ounce of the salt to four ounces of water, of which the dose is half a teaspoonful four or five times a day. If we are to expect success, the remedy must be persevered with, and if the quantity named is not sufficient, it should be increased to such an extent as to hold the paroxysms in check. After twenty years experience, I can still recommend the bromide of ammonium, especially in early life, having cured scores of cases with it.   Do not substitute bromide of potash for it, for this remedy has a widely different field. If the patient is stout, has a vigorous circulation, and suffers from excitement of the reproductive organs, bromide of potassium is the remedy. Some most persistent cases have yielded to this treatment, and I am in hopes that it will prove curative in many of these distressing cases.

All undue excitement must be avoided in epilepsy, the sufferer leading the most regular life. Some employment should be furnished that would amuse the mind, and keep it normally active, but much mental exertion is injurious; novel reading, or anything in which the mind becomes deeply absorbed, proves hurtful. Above all things else, excessive sexual excitement is most injurious, either as solitary vice or irritation of the organs from disease, and it will become the practitioner's duty to examine into the case with reference to this matter, and give the necessary advice and treatment.

CONVULSIONS.

Convulsions occur far more frequently during childhood than after puberty, though they may be occasionally noticed at all ages. The causes giving rise to them are various. Sometimes they are produced by disease of the brain and spinal cord, as in determination, inflammation, and some obscure structural lesions; at others they arise from an external irritation, it being transmitted to the spinal cord, and giving rise to excited reflex action. According to Dr. Marshall Hall, convulsions are dependent upon irritation of the *true* spinal system, and though this occurs in some cases from causes acting directly upon the nervous system, it more frequently depends upon an irritation of some distant part transmitted to the spinal cord through the nerves. Thus, we find convulsions arising in this way during dentition, from crude or acrid ingesta, from irritation of the stomach or bowels, from the irritation produced by worms, and from inflammation of internal organs, or disease of the surface, attended with great irritation and pain.

SYMPTOMS.—If convulsions occur during disease, they are generally preceded by tolerably well-marked symptoms, by which the close observer may anticipate their approach; and though not always constant it is well to give them due consideration. The most marked of these is a sudden, jerking, involuntary movement of the extremities, and quick grasping movement of the hands. This will be observed as well when the child sleeps as when awake, and is sometimes increased by motion. Usually the child sleeps with its eyes partly open, and we observe that the globe of the eye is drawn upward and rolled about, and this involuntary movement of the eye may be frequently noticed when awake. With these symptoms there may be excitement of the nervous system, manifested by restlessness, fits of crying in children, and sleeplessness; or we may have the reverse, the patient being dull, impassible and somnolent.

The attack is always sudden, the patient losing consciousness, and being to a great extent insensible. The convulsion is usually very marked, but in some cases we find it slight or entirely absent, the patient being rigid and remaining in one position. Respiration is labored, in many cases very markedly so, and in these the countenance is turgid and purple, and the features distorted. The pulse is very frequent and small, or it is soft, feeble and small, and but little increased in frequency. In the severer cases, deglutition is almost impossible, and from the falling backward of the tongue respiration is snoring. These symptoms may continue for a moment or two to fifteen minutes, or half an hour, in the milder cases terminating in a return of consciousness, but the severer in a deep sopor, from which the patient can not be aroused. One convulsion may terminate the attack, but in many cases one succeeds another for from one to twenty-four hours. The interval between the spasms is frequently marked by nothing more than a relaxation of the entire system, and restoration of the power of deglutition, the patient being in a semi-comatose condition, and totally unconscious. Children having convulsions once, are usually more liable to them than others, and they will frequently come on from slight causes.

DIAGNOSIS.—The diagnosis of convulsions is very easy, there being no possible chance of mistaking the symptoms. The sudden loss of consciousness, convulsive movement, difficult respiration, and frequent, small pulse, can not be confounded with any other disease. It is true that we can not distinguish between simple convulsions and epilepsy, except by the lapse of time.

PROGNOSIS.—The prognosis is usually favorable, though it is very difficult in some cases to arrest the convulsive action. Occasionally cases will be seen that will prove fatal in spite of treatment.

POST-MORTEM EXAMINATION.—The scalpel reveals no constant lesion to account for the symptoms. When there has been determination to or inflammation of the brain, we of course will find the evidence of these lesions. But when the disease has arisen from an *extrinsic* irritation, there is not the slightest evidence of disease of the nerves.

TREATMENT.—Our primary object is to arrest the spasmodic movement which is so alarming to the friends, and, no matter how often seen, to some extent so to the practitioner. Calmness and decision are very important requisites in this case, as all around the patient is excitement, and a hundred expedients to benefit the sufferer are proposed.

A quick examination of the patient will enable us to clasify our cases into two groups, and select the appropriate remedy. In the one there is marked vascular excitement, with irritation and determination of blood to the brain and spinal cord; in the other the circulation is feeble, and the condition is one of atony. In the first case the remedies are sedative, in the second they are stimulant.

If the convulsion passes off so that remedies can be given by mouth, we will add to a half glass of water Tinct. of Aconite or Veratrum (as indicated) gtt. v., Tinct. of Gelseminum gtt. x. to gtt. xxx., and give a teaspoonful every ten or fifteen minutes, until the convulsions are arrested, then less frequently.

If there is a markedly pinched expression about the eyes or base of the brain, Rhus will take the place of Gelseminum. It is especially indicated by sudden startings and a shrill cry (*cry encephalique*). If the patient has been suffering from severe pain, and the convulsions have come on during a paroxysm of it, we would give—℞. Hydrate of Chloral ʒij., syrup, water, *aa.* ʒj., a half teaspoonful or teaspoonful sufficiently often to arrest the convulsions. Sometimes the single dose will be sufficient, sometimes the, child will require as much as a half ounce of the mixture. When the patient is full blooded, and the face and neck are full or red, Bromide of Potassium may be added to the mixture.

Asafœtida is a remedy when there has been irritation of the stomach and intestinal canal, with gaseous accumulation. It is a stimulant, and ought to be classed with the second group, but the gastro-intestinal irritation is the best indication.

Of the second group of remedies (stimulants), Lobelia will take the first place. It is indicated by an oppressed circulation and respiration, fullness of tissue, and want of expression, when the convulsion has passed off. Frequently we prescribe from our pocket-case— ℞ Tinct. Lobelia Seed gtt. x. to gtt. xx., water ʒiv.; a teaspoonful every fifteen minutes until the convulsions have ceased. If we have it at hand, the old-fashioned Comp. Tincture of Lobelia and Capsicum may be given in doses of half to one teaspoonful, to nausea and vomiting.

Sulphuric ether is a good remedy in these cases, and may be given in doses of ten drops on sugar or in mucilage, and repeated frequently. It may also be used as an anæsthetic.

Chloroform may occasionally be given internally in doses of

ten drops, but I prefer to use it by inhalation. If the convulsions are severe, with but little interval between them for the administration of medicines, or if, owing to their severity, we can not see the indications for remedies, chloroform should be used by inhalation to the extent of complete arrest of convulsive action, and continued until we are satisfied that we have the case well in hand. The indicated remedies should be given as soon as the patient can swallow them.

One of the best, if not the very best remedy to prevent a recurrence of convulsions when they have been arrested, is the bromide of ammonium. For children I usually prescribe it ʒij. to water ʒiv., and give teaspoonful doses as often as may be required (usually every three or four hours). It may be given for the arrest of convulsions in some cases with most excellent results. When children are inclined to convulsions the preparation of bromide of ammonium should be kept in the house, and its administration directed when the child shows the slightest symptoms of convulsive action.

As convulsions of childhood will sometimes run into confirmed epilepsy, we can not be too much on our guard. I prescribe the remedy under consideration, and insist that there shall never be any neglect to administer it if indicated.

If a hot foot bath can be rightly used, it will be of advantage in many cases; there must be a bucket of water, so that the legs may be immersed to the knees; it must be as hot as the child can bear it, and hot water must be added every ten minutes to keep up the temperature, and both child and bucket must be surrounded by a blanket or shawl. Thirty minutes is the least time for using the hot foot bath.

The sitz bath may sometimes be employed with the same precautions, and even a general bath. When, however, there is need of stimulating the skin, the hot blanket pack may be used, or the child may be rapidly sponged with hot water.

When the temperature is very high, and the head hot, an enema of five or six ounces of cold water, with a teaspoonful of tincture of lobelia may be given. If there is marked prostration, hot water may be used for this injection and Compound Tinct. of Lobelia and Capsicum added.

If we are satisfied that the convulsions are due to irritant or crude ingesta, an emetic may be given at once. Ten grains of Ipecac, any preparation of Lobelia, or even salt water, will answer our purpose. If we have reason to believe that there is irritant material in the intestinal canal, use an enema of warm water to which we have added compound powder of jalap and senna ℨss.

In the treatment of disease we want to notice the symptoms of convulsions when they first appear, that remedies may be selected to remove them. "An ounce of prevention is worth a pound of cure," at least it saves us much trouble. Gelseminum, Belladonna, Rhus, Lobelia, Apis, Veratrum, Bromide of Ammonium, as indicated, will usually relieve the nervous irritation. The rule may be repeated—"the indicated remedy is the remedy to prevent convulsions."

CHOREA

This affection, known commonly as *St. Vitus' Dance*, occurs most generally about the age of puberty, though it sometimes appears as early as the sixth or eighth year, and as late as the thirteenth, and in some cases later than this. It is confined principally to the female sex, but in rare cases it is met with in the male. Most generally it is associated with some derangement of the sexual organs, and it is not unfrequently associated with hysteria. We usually find it in persons of feeble health, and precocious mental development, but in some cases, in persons of the opposite character, in which it may be induced by torpor of the liver and bowels, deranged secretion of the skin and kidneys, and from close confinement or sedentary occupations.

The modern disease received its name, doubtless, from the dancing manias of the middle ages. The "dancing plague" or St. Vitus' dance, commenced in Strasburg, in 1418, and is thus described by Burton : "Chorus Sanctæ Viti, the lascivious dance, as Paracelsus calls it, because they that are taken with it can do nothing but dance till they are dead or cured. It is so called for that the parties were wont to go to St. Vitus for help, and after they had danced there a while they were certainly freed. 'Tis strange to hear how long they will dance, and in what manner, over stools, forms, tables; even great-bellied women sometimes (and yet never hurt their children) will dance so long that they can stir neither hand nor foot, but seem to be quite dead. One in red clothes they can not abide; music above all things they love; and therefore magistrates in Germany will hire musicians to play to them, and some lusty, sturdy companions to dance with them."

Another form of the dancing mania, termed St. John's dance, commenced in 1374, and extended over the greater portion of Europe. "At Cologne the number possessed amounted to more than five hundred, and at Metz the streets are said to have been filled with eleven hundred dancers. Peasants left their plows, mechanics their workshops, housewives their domestic duties, to join the wild revels, and this rich commercial city became the scene of the most ominous disorder; secret desires were excited and two often found opportunities for wild enjoyment; numerous beggars, stimulated by vice and misery, availed themselves of this new complaint to gain a temporary livelihood. Girls and and boys quitted their parents, and servants their masters, to amuse themselves at the dances of those possessed, and greedily imbibed the poison of mental infection. Above a hundred unmarried women were seen roving about in consecrated and unconsecrated places, and the consequences were soon perceived; gangs of idle vagabonds, who understood how to imitate to the life the gestures and convulsions of those really affected, roved from place to place seeking maintenance and adventures, and thus, wherever they went spreading this disgusting spasmodic disease like a plague; for in maladies of this kind, the susceptible are infected as easily by the appearance as the reality." (Hecker.)

This gives the origin of the name of the affection we are now considering, and though there is no similarity between the ancient and modern St. Vitus' dance, the description just given illustrates the ease with which nervous affections of this kind may be propagated. And it is a fact, proven by numerous instances in hospital practice, that attacks of hysteria, epilepsy, and chorea, will be excited by witnessing the malady in another.

As regards the pathology of the affection, we must conclude that there is an irritation of the true spinal cord, arising sometimes from debility, and at others from extrinsic causes of irritation. In either case the excitation of the nervous system is indicative of debility, rather than strength, and in many cases is based upon feeble nutrition of the nerve substance.

SYMPTOMS.—The first evidences of chorea are occasional involuntary movements of the hands and the facial muscles, and an inability to sit quietly in one position. Very frequently the fingers are quickly and involuntarily moved, and when the patient uses the hands it is with a quick unnatural movement. As the disease progresses the involuntary movements become continuous, some part of the body being constantly in motion, and the movements are now very much exaggerated. If the patient attempts to do anything, she seems to have but partial control over her muscles, and while they are being directed to the end intended, they are going through a succession of movements entirely independent. So great is this, sometimes, that the patient can not sit still, nor even keep the hands quiet for a moment, and her walking is irregular from the same cause. The facial muscles are sometimes very much involved, and the attempt to speak, or give expression to the emotions, is followed by various contortions of the countenance, which would be laughable were they not connected with so serious a malady. Sometimes it is almost impossible for the patient to express herself intelligibly, owing to spasmodic action of the muscles of the mouth and of the larynx.

As before remarked, the general health is usually impaired previous to the commencement of the disease, and this becomes more marked as it progresses; symptoms of anæmia are of common occurrence, the skin being blanched, the pulse feeble, the lips and gums pale, variable appetite, imperfect digestion and constipation of the bowels. The mind is more or less affected, the patient being low-spirited, and desiring solitude, the countenance being pale, languid and vacant. In some instances confirmed chlorosis will be developed during the progress of the disease. It will be noticed, that the child has no disposition to play or to take exercise, and does not desire to associate with others, but prefers rather to get where her infirmity will not be noticed; the sensitiveness in this respect being sometimes very great.

DIAGNOSIS.—Chorea is marked by such distinctive symptoms that it is easily recognized, the continual partly voluntary and partly involuntary movements not being observed in any other disease.

PROGNOSIS.—Though in some cases very obstinate, the disease is almost always curable. It may last for two or three weeks,

27

or as many months, and in some rare cases for years. Usually it disappears as the general health is improved.

POST-MORTEM EXAMINATION.—In fatal cases the evidences of anæmia are usually very marked, the tissues being very pale, soft and flaccid. The different organs have been found more or less diseased, but these were complications and bore no relation to the spasmodic action. We would expect to find lesions of the brain and spinal cord; but except in those cases terminating in general convulsions, or in inflammation, no change of structure has been noticed.

TREATMENT.—Various plans of treatment have been adopted, and many remedies used as specifics in this affection, and as is usual, we find that where the means are so abundant they are not very efficient. We had much better adopt a rational plan of treatment, by correcting any dyscrasia, and getting a normal performance of the various functions of the body, rather than depend upon any one remedy, no matter how highly it is praised.

The disease is intimately related to rheumatism in many cases and the remedies will be selected from the class of anti-rheumatics, to which the reader is referred. Macrotys probably stands first on the list, as it has probably benefited more patients than any other remedy. In some cases it is given alone, in others it may be combined with Aconite or Veratrum, in others with Valerian, and in still others with Arsenic. One of the best preparations I have ever used is—℞ Tinct. Macrotys ℥ss., Tinct. Valerian ℥iss., in doses of five to twenty drops every three hours. With Arsenic the prescription will be—℞ Tinct. Macrotys ℥j., Fowler s Solution of Arsenic ℥ss., water ℥iv.; a teaspoonful every four hours.

Sticta, Bryonia, Apocynum, and Colchicum, have all been used with good results, as has the iodide of potassium when the tongue is broad and of a bluish pallor.

If the disease is associated with amenorrhœa, or irregularity of the menstrual function, this must be attended to. In some cases the emmenagogue pill of the Dispensatory will prove useful, both as a cathartic and for its action on the uterus. The wild ginger is another agent that will prove useful in some of these cases. The Macrotys, or Cimicifuga, has already been named as a remedy in cases where the patient complains of wandering pains in various parts of the body, or pain in the back and limbs. We sometimes associate it with Valerian or Scutellaria, and sometimes with the bitter tonics. The extract of Indian hemp has been employed with benefit, in doses of half a grain three times a day, and good results are said to have attended the administration of small doses of Stramonium. The sulphate and oxide of zinc have been prescribed oftener possibly than any other agents, and we must believe, from the favorable reports given, that they have an action in these cases ; these remedies may be given, commencing with half-grain doses four times a day, and gradually increased until five or ten grains are administered.

If there is tenderness on pressure over the spinal cord, counter-irritation will often prove very efficient, and the same will be the case when there is tenderness over the epigastrium.

Electricity has been frequently resorted to in chorea, and the reports of its action differ materially. When passed through the limbs it is not only useless, but sometimes positively injurious; but when applied to the spine it is almost always beneficial. The common electro-magnetic battery may be employed, the negative pole being applied to the sacrum, and the positive passed backward and forward over the spine. The better plan, however, is to insulate the patient, and by the old-fashioned electric machine, charge the patient and withdraw the spark from the back. In one case I have employed the bromide of ammonium in addition to a tonic treatment, and seemingly with marked benefit.

Very much will depend upon the home management of the patient. All causes of irritation must be carefully avoided, and she should be encouraged to take suitable exercise, and try to control the involuntary movements. Out-door exercise, pleasant company, and something to constantly occupy the mind with, exert an important influence, and it will sometimes be found that where the patient is allowed to have her own way, if not decidedly improper, she will get along better. In some cases the disease results, in both male and female, from sexual excitation and onanism; this should be looked into, and if reasonable evidence exists, means should be employed to put a stop to it. The manner of doing this will have to be left to the discretion of the physician, and will vary in different cases.

TRISMUS NASCENTIUM.

Infantile tetanus is of very rare occurrence, and a physician may be in practice a lifetime without seeing a case. It occurs most frequently in the children of the poor, especially in badly ventilated dwellings, and when the mother and child have had insufficient food, clothing, etc., and a want of cleanliness. That it does not always depend upon these conditions, I have evidence in one case that came to my knowledge, in a family in comfortable circumstances.

CAUSES.—The conditions above named may be considered as predisposing causes. The exciting cause is an unhealthy inflammation of the navel; the disease is, therefore, really traumatic tetanus.

PATHOLOGY.—The tetanic convulsion is dependent upon irritation of the spinal cord and excess of reflex action. Post-mortem examination shows an inflammatory condition of the umbilical vessels, and of the peritoneum about the umbilicus and reflected upon the artery and veins. Examination of the spine shows injection of the arachnoid and pia mater, with effusion of serum, lymph, and sometimes of blood.

SYMPTOMS.—The disease may make its appearance in a day or two after birth, but more frequently about the ninth day. The child is observed to be restless and fretful, sleeps badly, and has paroxysms of crying which seem like colic. It moves the lower extremities in a peculiar manner, does not nurse well, and has greenish and slimy discharges from the bowels.

These symptoms continuing for a time, it is attacked with convulsions, which are, however, very irregular. "These convulsive motions recur at uncertain intervals, and produce various effects. Sometimes the agitation is very great; the mouth foams; the thumbs are rivited into the palms of the hands; the jaws are locked from the commencement, so as to prevent the action of sucking and swallowing; any attempt to wet the mouth or fauces, or to administer the medicines, seems to aggravate the spasms; and the face becomes turgid and of a livid hue, as do most other parts of the body. From this latter circumstance nurses speak of this form as the 'black fits.' The conflict lasts from eight to thirty hours, and in some rare instances to about forty hours ; when the powers of nature seem to sink exhausted and overpowered by their own exertions."

TREATMENT.—Put the child upon the use of Veratrum and Gelseminum in full doses, as—℞. Tincture of Veratrum gtt. v., Tincture of Gelseminum ʒss., water ʒij. ; a teaspoonful every hour. As it is not very easy to give the entire teaspoonful at once, it may be given in smaller doses through the hour, or the quantity of water may be lessened. When the disease is very persistent I should not hesitate to double or even quadruple the quantities of the remedies named.

Locally the application of a warm emollient poultice to the abdomen over the umbilicus will be found beneficial. Ulmus fulva, flaxseed, or anything of a similar character may be used ; or in place of these, application of a cloth spread with the mild zinc ointment will answer a good purpose.

NIGHT TERRORS.

We are consulted occasionally with regard to the sudden waking of the child at night, manifesting every evidence of terror. Generally it wakes with a piercing scream, and starts up in bed with affright, and it is some minutes before it can be assured that it is safe, and quieted. There may be but one paroxysm through the night, but occasionally there will be two or three. In some cases it will also occur when the child sleeps in the daytime.

Many persons think that this grows out of frights from playmates or servants, or fearful stories about ghosts, giants, robbers, etc., which are so frequently told to children. While this gives rise to a morbid fear, manifesting itself especially after dark, but also following the person in the day, it has little or nothing to do with this case.

In this there is an irritation of the brain with determination of blood, and the night terrors are similar in their cause to convulsions and epilepsy, the first being of the cerebrum—the intellectual portion, the second of the spinal cord—or of automatic movement.

The use of small doses of Aconite with Rhus, given in the afternoon and evening, will frequently remove the unpleasantness. In some cases the indications are very clear for Gelseminum and we use it in place of Rhus. In other cases there is an irritation of the skin associated with the unpleasant dreams, and Apis or Belladonna will come in play. If the child is nervous and sobs in its sleep, Pulsatilla will be the remedy, and it may be given three or four times daily.

In some cases we will find a torpid condition of the bowels, which will require the administration of a mild cathartic. In some, very small doses of podophyllin, thoroughly triturated and combined with hydrastia, will be found to answer the purpose well.

A tonic and restorative treatment will almost always be necessary. Here I would put the child upon the use of our common prescription: R. Tincture of Muriate of Iron 3ss., Glycerine 3ij.; a teaspoonful three times a day; or the compound syrup of the hypophosphites.

PARALYSIS.

Infantile paralysis is a very different disease from that we are called to treat in the adult. We are accustomed to think of paralysis as a loss of motion or sensation; in this case there is but diminution of one or both. In slight degree it is met with quite frequently, and I have no doubt that it is often overlooked, or attributed to something else.

CAUSES.—It is very difficult to determine the causes of infantile paralysis. In most cases I believe there is a congenital feebleness of the nerve centers, and afterward from imperfect nutrition the disease is developed. In other cases it is the result of injury to the spine, usually from bad nursing or from a fall.

PATHOLOGY.—The pathology of the disease is pretty clearly stated above—enfeeblement of the nerve centers and impaired nutrition. I do not think that I have seen evidence of any other lesion in the cases I have examined.

SYMPTOMS.—In the majority of cases we will first notice the difficulty in feebleness of the lower extremities, and an imperfect command over them. Usually one side is principally affected, and the disease seems to be of one or more articulations. I have noticed it oftenest in the hip, and from the eversion of the toes, and shortening of the limb, have felt confident there was a dislocation. Careful examination, however, could detect no lesion. Iu other cases the whole pelvis seemed to be affected, having no strength to support the body, giving a very feeble point of action for the muscles of the extremities.

More rarely we see cases of paralysis of the upper extremities, yet they are easily recognized by the sameness of the symptoms. The mother thinks the child's arm has been dislocated, or has been injured in some way, and attributes the loss of power to this. On first looking at the arm, we suspect a dislocation, its position is so unnatural and its movements so feeble, but an examination shows the bones in their proper place.

Facial paralysis is still more rare, but cases are on record. Paralysis of the third pair of nerves, causing falling of the upper eyelid, or *ptosis*, is occasionally seen. Again, a single muscle may be affected, as when from paralysis of a sterno-mastoid we have *torticollis*.

TREATMENT.—Of course the treatment will be, to some extent, dependent upon the cause, and a careful examination will be instituted to determine the lesion which has given rise to it. If from irritation of the stomach and intestinal canal, this should be removed, and means employed to restore digestion. If worms are thought to be present, the Podophyllin and Santonine may be given. If from irritation of the urinary apparatus, this must be relieved.

Outside of these wrongs, which may be the cause of paralysis, we look to the nerve centers for the wrong, and here we will be guided by the special indications for remedies. Nux vomica will frequently be indicated, there being evidences of an enfeebled circulation. With flushed face and contracted pupils the patient will have Gelseminum, usually with one of the sedatives. If the patient is dull and inclined to sleep, Belladonna will be given; and with sluggish muscular action, Ergot. Rhus is indicated by the sharp pulse, frontal headache, and red papillæ of tongue; Bryonia by the corded pulse, and flushed right cheek; Macrotys by muscular pain and soreness.

Inunction with quinine has been of marked benefit in these cases; even friction with a fatty matter (lard) is of advantage. Brisk friction to the spine and the part affected, with salt and water, using the open hand, is good. In some cases the use of water is injurious, and I direct dry friction with flannel or silk;

this is especially useful in cases in which there seems a deficiency of electricity.

Benefit is derived from the moderate use of electricity, passing the current downward from the spine through the muscles. The Voltaic electricity, used by the method called " Faradization," I believe to possess the most active curative properties.

The child should be taken into the open air, and have an abundant supply of light, and even sunlight, where it passes most of its time. It should also be instructed to call the paralyzed parts into action as much as possible, and control their movements. Practice in this direction is often attended with the happiest results.

CHAPTER X.

DISEASES OF THE EYES.

The eye is one of the most important organs of the body, and though its diseases do not endanger life, their favorable termination is as anxiously watched for by both patient and friends as that of the more grave maladies. As regards the pathology of these affections, we will find it the same as in other portions of the body, and, as a general rule, the same treatment will be applicable. Inflammation of the structures of the eye is the same disease as inflammation of any other part of the body, differing only as regards the peculiarity of structure and function of the parts. And in the treatment of this affection the same general principles apply in the one case as in the other. So it is in all other diseases, and he who properly understands the pathology and nature of the affection need be at no loss for appropriate treatment.

The organ of sight, it will be recollected, consists of two parts, the eye itself and its appendages, the latter being two palpebra or lids, the conjunctiva or investing membrane, the lachrymal apparatus, the muscles moving the eye, and the cellular and adipose tissues which form its bed. Each of these parts may be diseased, but some of them so rarely that it is hardly worth while to notice them in this place. The globe of the eye is composed of three tunics—the external, composed of the sclerotic and cornea; the middle, of the choroid; and the internal, the retina or expansion of the optic nerve. It has a muscular septum dividing it into two parts—the iris; and it has three humors possessing different degrees of density—the aqueous, the vitreous, and the crystalline lens.

OPHTHALMIA NEONATORUM.

The sore eyes of the newly born child vary very greatly in character, from simple conjunctivitis to that grave inflammation which, involving structure after structure, destroys the organ in a few days.

CAUSES.—In some cases it is the result of awkward washing, either from the irritation of soap in the eye or rubbing; these are always mild and of short duration. In others the conjunctivitis arises from exposure to a bright light, which induces irritation; and in some it is caused by cold.

The severer cases are thought to arise from the introduction of purulent matter into the eyes from the soft parts of the mother while the child's head was passing through the vagina. The severest inflammation of the eyes known, is caused by inoculation with gonorrhœal virus, and we have abundant evidence that other purulent discharges from the vagina are quite as virulent.

PATHOLOGY.—In the milder cases the inflammation is confined to the conjunctiva, and is but little more than determination of · blood. In the severer cases the inflammation is at first of the conjunctiva, but of very active character, impairing the vitality of the tissues, and giving rise to free suppuration. In many cases it gives rise to separation of tissues, producing *phlyctenula*, and extending to the cornea, causes ulceration. In still severer cases it extends to the deep-seated structures of the eye, causing softening, suppuration, and other lesions that destroy the organ.

SYMPTOMS.—In the milder cases the eyes are observed to be injected, and the child dislikes to expose them to a strong light. In a few days a tenacious secretion accumulates at the margin of the lids, and agglutinates them together. The eyes may remain sore in this way for a week or two, occasionally causing the child to be restless and fretful.

In the severer form of the disease, the eyes are noticed to be red, and the lids much swollen, and from the first the child seems to suffer much pain in them, and keeps them tightly closed the most of the time.   By the second day all these symptoms are increased, the swelling being very marked, and when the eyelids are separated, the entire surface is seen to be involved in the inflammation, the conjunctiva being injected and thickened.   By the third or fourth day, there is a free discharge of yellowish purulent matter from the eye, and the intolerance of light is extreme.   When the lids are opened the conjunctiva is seen to be much swollen, and many times the *chemosis* is so great that the cornea can hardly be seen.

Continuing in this way, the disease may run its course to the destruction of the eye in three or four days, or, as in the majority of cases, it will last as many weeks, and the child will slowly recover.

TREATMENT.—I have great faith in the value of internal remedies in disease of the eyes, though the proper local means should not be neglected.   Prepare for the child Aconite in the usual doses, with Gelseminum, Rhus, Belladonna, Phytolacea, or Apis, as may be indicated.   The child may also be medicated through the mother, if her health is impaired.   Not unfrequently the unpleasant coating upon the tongue, the bad breath, and imperfect digestion, will call for sulphite of soda or sulphurous acid.

The eyes should be kept free from the discharge (clean) by washing them every two or three hours with warm water, to which may sometimes be added a small portion of salt.   In some cases the water may be used as hot as it can be borne.   To thoroughly cleanse the eye, the lids may be opened and the water allowed to trickle through, to wash the pus away.

A lotion of Aconite or Veratrum (ℨj. to ℥j.) applied above and around the eyes, and sometimes to the lids, will give relief.

In cases of moderate severity a collyrium may be made by adding Tinct. Belladonna or Gelseminum to water.   Or in a more advanced stage a solution of atropia may be employed (gr. ½ to water ℥j.), or an infusion of Baptisia filtered, or a solution of Hydrastia (Berberin), grs. ij. to water ℥j.

If the case is very serious, and the eye is endangered, nitric acid may be applied to the surface of the lids (everted), using a pine pencil.   I think it preferable to the nitrate of silver which is so commonly used in these cases.

DISEASES OF THE APPENDAGES OF THE EYE.

The *eyelids* may be the subject of phlegmonous inflammation, usually associated with erysipelas. They are swollen and livid, and very painful, and occasionally the inflammation extends to the cellular tissue of the orbit. It may terminate in resolution or suppuration, the pain being severe and throbbing when pus has formed, and the constitutional symptoms tolerably well marked. If the inflammation is dependent upon erysipelas, we may apply equal parts of tincture of muriate of iron and glycerine, every two or three hours, keeping a cloth wet with the same over it ; if from other causes, a poultice of equal parts of Hydrastis and Ulmus, or cloths dipped in a decoction of Cornus, with a small portion of tincture of Aconite. The bowels may be moved with any simple laxative, and if necessary, a diaphoretic and diuretic given. If suppuration occurs, the abscess should be carefully opened as soon as it is detected, as if it remains it increases in size, and sometimes causes great destruction.

Furuncle, or *boils* of the eyelid, are of very frequent occurrence, and sometimes occasion much suffering. Occasionally they pass through their stages rapidly, a week sufficing for their removal, but in other cases they are very chronic. When formed on the edge of the eyelid, they are called styes, and are smaller but not less painful. They require but little attention, except in such cases as would be injured by the continued pain and restlessness produced by them. In such cases they may be incised, and if pus has not yet formed, touched freely with a crystal of sulphate of zinc, and a poultice applied.

In such cases lime water may be given with benefit, to prevent the recurrence of them. The general health is also improved. In place of lime water we may employ sulphide of lime in doses of one eighth of a grain four times a day.

Ptosis, or falling of the upper eyelid, is caused by paralysis of the third pair of nerves, or by disease affecting the eyelid, or the levator muscle. In cases of paralysis, it may be relieved sometimes by the use of electricity or local stimulant applications and the proper internal remedies; failing in this and in the cases not dependent upon paralysis, a surgical operation is demanded. *Entropium* or inversion of the eyelids, and *ectropium* or eversion, are only remediable by surgical operations.

Trichiasis, or inversion of the eyelashes, is popularly known as "wild hairs in the eye," and is often a source of great irritation or inflammation. The trouble is owing to a misdirection of the cilia, a portion of them being turned inward, so as to come in contact with the eye. The cause is usually easily detected by turning the patient's eye to a strong light and slightly raising the lid, the faulty hairs being seen to pass inward to the conjunctiva. They are usually of a light color, smaller and much more flexible than the normal ones, and for these reasons are sometimes detected with difficulty. In cases of partial trichiasis the treatment is easy, and consists simply in removing the offending cilia with a pair of forceps. I can yet feel the mortification I once experienced, in which, after treating a case of "sore eyes" for two weeks, the patient was cured in forty-eight hours by an old woman removing these faulty hairs. In severe cases this will not answer, a surgical operation being necessary.

DISEASES OF THE LACHRYMAL APPARATUS.

The lachrymal gland is so protected within the orbit that it is rarely the seat of disease. Inflammation sometimes occurs, and is marked by pain in the region of the gland and dryness of the eye from arrest of secretion. When the inflammation subsides there is usually too free secretion and epiphora, but this soon subsides. It should be treated as any other inflammation.

Inflammation of the lachrymal sac is of frequent occurrence, and requires care in its management. It makes its appearance usually as a diffused, erysipelatous-like redness and swelling of the parts near the internal canthus, with deep-seated pain, and more or less irritation of the conjunctiva, increased lachrymation, and passage of the tears over the eyelid. The inflammation continuing for some days, the parts become much swollen and very painful, and at last pus having formed, it discharges through the integument. In some cases the pus finds its way through the lachrymal canals by pressure, and the inflammation becomes chronic, but without the formation of a fistula. Usually there is closure of the nasal duct, which remains permanent unless an operation is undertaken for its removal, though sometimes the closure of the nasal duct is the primary affection, the inflammation of the lachrymal sac being caused by it.

The child should have the right internal treatment as for any other disease, in the use of Aconite or Veratrum, with Phytolacca, Rhus, Gelseminum, Belladonna, or whatever remedy may be indicated. If now the part is penciled with Tincture Aconite or Tincture Veratrum in the first stage, or has a wet dressing of ℞ Salicylic Acid, Borax, aa., gr. x., water ℥iv., if suppuration is threatened, we usually get along well. In some cases the deep redness will indicate Tincture Muriate of Iron ℥j., Glycerine ℥ss., applied to the part. If pus forms, an incision should be early made for its removal, thus preventing change of the lachrymal sac, and especially distension, and permanent closure of the nasal duct. An injection of ten or twenty grains of sesqui-carbonate of potash to the ounce of water, will now assist very much in effecting a speedy cure. As soon as the inflammation subsides, if the nasal duct seems closed, a style should be inserted.

Closure of the nasal duct, producing *fistula lachrymalis,* frequently results from the above inflammation, though it may be produced by injuries of the bones or soft parts, or an extension of inflammation to its mucous lining from the nose, or from the conjunctiva. In a majority of cases there is a fistulous opening over the lachrymal sac, or a continuous suppuration and discharge of pus at the internal canthus, through the puncta, with more or less frequent attacks of acute inflammation of the sac, and discharge through the integument. In some of these cases a fungous-looking mass of considerable size is found upon the site of the lachrymal sac, which is constantly discharging pus mixed with tears. This and the constant flowing of the tears over the eyelid, is very unpleasant, and occasionally it keeps up continuous irritation of the eye, and causes imperfect vision. The disease is only cured by an operation, which consists in opening the lachrymal sac and introducing a silver style made for the purpose. The usual means to relieve irritation are then made use of, and the style retained until there is evidence of the free passage of tears and restoration of the mucous membrane lining the duct, when it is removed, and the external opening allowed to heal.

The *puncta or canaliculi* may be obstructed from inflammatory action, and occasionally from other causes. In these cases there is also the overflow of tears and irritation of the lid. If it is produced by inflammation, the means heretofore named may be used to arrest it. If from other causes an Anel's probe may be passed into the puncta and through the canaliculi into the sac, with the result of removing the obstruction.

The *caruncula lachrymalis* is sometimes the seat of inflammation very similar to that in ophthalmia tarsi, and by displacement of the puncta will produce watering of the eye. It sometimes gives rise to considerable uneasines and pain. It may be treated in the same manner as the other inflammations named, but when persistent is best removed by the use of the mild zinc ointment, or ophthalmic ointment. Occasionally they are the subject of chronic enlargement, forming a' red, soft, tuberculated tumor, of considerable size, which bleeds readily on pressure. It may be occasionally removed by the application of a saturated solution of tannic acid, or the solid nitrate of silver, but in many cases will have to be excised, one-half or more being cut away, the remainder disappears.

PTERYGIUM.

This is strictly a disease of the conjunctiva, and consists of a thickening of a circumscribed portion of it extending between the internal canthus and the cornea, though occasionally it is found on the temporal side. It is divided into two kinds, the membranous and fleshy, both kinds being triangular, with the apex toward the cornea. It commences from without, and grows inward, occasioning but little disturbance until it reaches the cornea. If it commences to involve the conjunctiva-cornea, it gives rise to irritation, and may be attended with serious consequences. It may be arrested by cauterizing with nitrate of silver, nitric acid, or other escharotics, but the easiest plan is to dissect off the half next the cornea, when the remainder will generally disappear without trouble.

OPHTHALMIA TARSI.

Inflammation of the edges of the eyelids is noticed more frequently in children than in the adult, and is frequently associated with some depraved habit of body, as scrofula. When primary, it may be the result of cold, smoke, impure air, or filthiness; but it is most usually a sequence of catarrhal ophthalmia or scrofulous conjunctivitis. The disease is located in the edge of the lid and meibomian follicles, and in many cases so affects the roots of the eye-lashes as to cause them to fall out, hence that appearance termed "blear-eyed." The eyes look sore and tumid, and the patient complains of a sensation of roughness, and as if there were sand in the eye, when the lids are moved, and thus there is the constant tendency to keep them partially closed. They are agglutinated together in the morning, sometimes so much so that the patient has to soften them before he can open them, and it is even then attended with pain. Ophthalmia tarsi is essentially a chronic affection, with but little tendency to spontaneous recovery, and is sometimes very difficult to cure; and if the meibomian glands are closed, the edge of the lid has a shining, glistening appearance.

TREATMENT.—As there is almost always a faulty constitution, with marked evidence of some cachexia, we find it important to put the patient upon an alterative and tonic course of treatment. The compound tincture of Corydalis, or compound syrup of Stillingia, with iodide of potassium, may be administered in the usual doses. Some preparation of iron should be given with this,—frequently the tincture of muriate of iron will answer best,—and if necessary, the bitter tonics may be added. Once in a while cod-liver oil is of advantage, and occasionally we give our patient Fowler's solution of arsenic, with Phytolacca.

Very much depends upon keeping the eyes clean, and removing the tenacious secretion without causing pain and irritation. Hence the eyes should be frequently bathed during the day with warm water, or a weak decoction of Cornus or Hydrastis, keeping them as entirely free from the secretion as possible. Glycerine answers a very good purpose in some cases, usually combined with an equal quantity of rose-water, and applied freely.

The parts being perfectly cleansed, we apply once or twice daily, a very small portion of mild zinc or ophthalmic ointment; or, instead of this, we may use a mild collyrium of sulphate of zinc or borax, or one or two drachms of nitrous ether and vinegar in eight ounces of water, and followed by the glycerine lotion. The application I now place most dependence upon is the brown citrine ointment, one part to two or three parts of simple cerate. In very severe cases, the faulty cilia may be removed, the crusts carefully taken off, and the ulcers lightly touched with nitrate of silver. In the application of warm water, or the decoctions named, or to foment the eye, we can accomplish our purpose best by the use of a very soft sponge.

Counter-irritation to the nape of the neck, or behind or before the ears, with the blister or irritating plaster, is often of great advantage.

CATARRHAL CONJUNCTIVITIS.

The conjunctiva covering in the globe of the eye, and lining the lids, is exquisitely sensitive, and though abundantly protected, is frequently exposed to the causes of inflammation. Temporary inflammation is often seen as the result of dirt or sand in the eye, or even exposure, but very soon disappears with rest. The disease we are now describing may arise from cold, sudden changes of temperature, extension of inflammation from the mucous membrane of the nose, or from inoculation with the secretion of a diseased eye. This last cause should be carefully guarded against, as we not unfrequently observe whole families attacked with the disease from the indiscriminate use of towels.

SYMPTOMS.—The disease commences with a sensation of dryness and smarting of the eyelids, with a feeling as if dirt or sand had got into the eye, and it is with difficulty that the patient gives up this idea, the impression is so strong. In a short time the eyes seem tumid and swollen, the unpleasant sensations have increased, and a more or less abundant secretion, sometimes opaque and puriform, is established. If the eyes are now examined, the palpebral conjunctiva will be found red, and swollen, and more or less reticular redness of the ocular conjunctiva. As the inflammation progresses, the last portion of the conjunctiva becomes more completely involved, and we sometimes observe ecchymosis or extravasated blood under it. In a still severer form the conjunctiva is remarkably injected and swollen to the point where it passes into the cornea, so much so occasionally as to partially cover up this part of the eye; this swelling is termed chemosis. Catarrhal ophthalmia is frequently periodic, the exacerbation always occurring in the evening, and sometimes attended with headache; the pain and itchiness cease a short time after going to bed, and the patient sleeps well, but it re-appears in the morning on attempting to use the eyes.

In many cases the disease continues thus for a week or ten days, and then gets well without further change; but in some cases it is more persistent. Sometimes we notice a small blister on the ocular conjunctiva, which rupturing forms an ulcer, constantly throwing off an abundant puriform secretion; it may attain the size of a half-dime, or be even larger than this, and is usually very painful. The cornea is sometimes obscured and hazy from the inflammation, and in that variety of the disease termed phlyctenular has a tendency to ulcerate. This last form of the disease occurs most frequently in children and young persons, and is usually connected with a scrofulous constitution. The symptoms are, marked pain and intolerance of light, free secretion of tears, deep redness of the eyelids, but slight of the ocular conjunctiva, sometimes but three or four vessels being seen to pass across to the cornea. Soon we notice the production of one or more blisters on the cornea, which discharging, forms an ulcer; this may increase in size until it involves a considerable portion of the cornea, or it may rapidly increase in depth until it perforates it, and causes a discharge of the aqueous humor. In some of these cases, the phlyctenula are absorbed, leaving a small, white spot, called *albugo;* or a cicatrix results from the ulceration; called *leucoma.* If the ulcer penetrates the cornea, the iris is almost always thrown forward by the escape of the aqueous humor, and passing into the opening becomes adherent, and is termed *synechia anterior.*

DIAGNOSIS.—Catarrhal conjunctivitis is usually recognized with ease; the inflammatory action commencing in the palpebral conjunctiva, and subsequently extending to the ocular portion, with secretion of muco-pus, are the characteristic symptoms. In phlyctenular ophthalmia, there is inflammation of the conjunctiva, but the disease is principally confined to the cornea; the appearance of the small vesicles or ulcers in the cornea marks the distinction. That form described as *pustular,* is marked by the formation of pustules, terminating in ulcers in the ocular conjunctiva near the cornea.

PROGNOSIS.—In the milder forms of catarrhal ophthalmia we usually succeed in arresting the disease in a week or ten days, but if allowed to progress or badly treated, it may endanger the integrity of the eye and last for months. The phlyctenular form is more difficult to treat and not unfrequently leaves the marks already mentioned. Pustular ophthalmia is usually very perverse, but with care may be managed so as to leave no bad

28

result.   Either of these forms may become chronic, and develop structural change which will impair vision to a greater or less extent.

TREATMENT.—The treatment of inflammation of the eyes will be conducted on the same general principles that govern our therapeutics in all forms of inflammation. Though the part involved is but small, it requires exactly the same general means as if it were an entire lung or other large organ. Remedies, also, will be found just as definite in their action, and as prompt in arresting the inflammatory process.

The patient has the indicated sedatives, Aconite or Veratrum gtt. v. to a half glass of water, a teaspoonful every hour. Other remedies are given according to the indications. Rhus if there is burning pain in the eyes, or sharp pulse with frontal headache ; Gelseminum if there is flushed face and active circulation to the brain ; Belladonna if there is dullness and capillary congestion ; Bryonia if there is tensive pain in the eye and through the orbits ; Apocynum if the lids are markedly swollen (œdematous) ; Phytolacca if the cervical glands are involved or there are phlyctenulæ.

Occasionally a very foul tongue with sense of oppression will call for an emetic, or full veins and full tissues generally will want Podophyllin. The simple pallid tongue wants a salt of soda ; the broad, pallid, dirty tongue, sulphite of soda ; the red dirty tongue, sulphurous acid.

In malarial regions the frequent necessity for antiperiodics will be noted, and many times they will serve quite as good a purpose as in other diseases with periodic complications. I have seen a conjunctivitis arrested in a couple of days with quinine alone. In some cases Alstonia will prove a good remedy. When antiperiodics cannot be given by mouth, quinine can be used by inunction.

The first five days the patient should be kept in a darkened room, and when exercise is taken, the eyes should be well shaded. It is of especial importance that the nervous system be kept quiet, and fretfulness and crying prevented ; this will require considerable care and management upon the part of the mother.

Various collyria are recommended, in fact so many that the young practitioner does not know which to select. ℞ Atropia

gr. ½, water ʒj. ; ℞ Sulphate of Morphia grs. ij., Aromatic Sulphuric Acid gtt. x., distilled water ʒij. ; or ℞. Tincture of Belladonna gtt. x.; Tincture of Gelseminum ʒss., water ʒij.; dropping
it in the eye every three or four hours. Associated with this, if
the case is acute, I direct that the eye be fomented with hot
water, a piece of very soft sponge being used to make the application. In some cases a poultice of equal parts of Hydrastis
and Ulmus, answers a good purpose, but I now prefer the fomentation. In all cases the patient should be kept perfectly still;
in a darkened room, and use a light and easily digested diet.
Mackenzie strongly recommends the use of a solution of nitrate
of silver, gr. iv. to water ʒj. ; a large drop being applied to the
eye two or three times daily. For a few minutes the eye feels
easy, and then for ten minutes there is a sharp pricking pain,
which subsiding leaves the eye almost free from pain for five
or six hours, when the application should be repeated. Dr.
Williams recommends a solution of sulphate of zinc in rose-water
in the proportion of from two to four grains to the ounce.

OPACITY OF THE CORNEA.

Opacities of the cornea are distinguished by different names,
according to their density and the character and situation of the
lesion. *Nebula* is the slightest degree, and is most generally
situated in the superficial layers, though occasionally deep seated ;
sometimes it is general, and is the result of pressure, or of serous
effusion into the substance of the cornea. *Albugo* is that
form of opacity in which the spot has a pearly appearance, and
generally results from effusion of plastic lymph in the anterior
layers of the cornea. It usually results from the phlyctenula
which have receded without bursting. *Leucoma* is an opaque
cicatrix closing an ulceration; it has usually a contracted and
circumscribed appearance, and is depressed in its center.

TREATMENT.—"All the three kinds of speck—*nebula, albugo*
and *leucoma*—have a natural tendency to disperse as soon as the
disease giving rise to them subsides or is removed, and whether
they depend on primary inflammation, spreading to the cornea,
or secondary inflammation of that part arising from the irritation
of inverted eyelashes or granular conjunctiva. We must, then,
in every case, endeavor to remove the ophthalmia or the

mechanical irritation on which the opacity depends, assured that if we succeed in this, nature, by the process of absorption, will accomplish the whole amount of recovery which is possible. In children and young persons many very dense and extensive opacities are removed in the natural process of growth, which would be quite immovable in adult life." (Mackenzie.)

Patience and perseverance are the great elements of success in these cases, and abundant time, from three months to as many years, is necessary to the accomplishment of the purpose. The inflammation should be entirely removed in the manner heretofore named, and if the person is scrofulous this should be counteracted as much as possible and the general health improved. Frequently this is all that is necessary, the opacity disappearing as the inflammation is removed. If after this we deem it necessary, we prescribe a mildly stimulant collyrium, as wine of opium, pure or diluted; Common salt grs. ij. to grs. x., to water ℥j.; or a solution of sulphite of soda grs. ij. to grs. v., to water ℥j. Sulphite or sulphate of soda, very finely powdered, a small portion dropped in the eye once or twice a day, is sometimes an excellent stimulant, and the eye clears up under its use. In other cases, all that is necessary is to give nature sufficient time to remove the deposit; and to prevent injurious meddling with the eyes, we will in these cases prescribe some mild and grateful application simply to occupy the attention of the patient and prevent discouragement.

CHAPTER XI.

DISEASES OF THE EARS.

Disease of the ears is of more frequent occurrence in the child than in the adult; and partial deafness, annoying the person for a lifetime, frequently has its origin in early life. Some of these affections are marked by prominent symptoms, while others are very obscure, and run their course many times without being recognized.

Diseases causing structural change arise mostly from two causes—the scrofulous constitution, or an impairment of the blood following the eruptive fevers. In either case the constitutional disease must be recognized, and remedies directed to it, if we are to expect success.

The *symptoms* of disease of the ears are those of unpleasant or painful sensations, discharges from the ears, and deafness. The infant expresses its suffering in these cases by frequent movement of the head, carrying the hand to the affected part, rubbing the side of the head on the pillow, or the mother's arm, and a peculiar contraction of the facial muscles of that side.

If there is pain, as from neuralgia or inflammation, the child is restless and fretful, crying out suddenly and piercingly, and at an advanced age sobbing during sleep.

EXAMINATION OF THE EAR.—The majority of persons fail in their examination of the ear from defective light, or the want of ability or means to use it. I have seen professed aurists make the most lamentable failures here, though working with fine and costly apparatus.

The means are very simple, as is the use. We require sunlight or a good lamp, an ordinary toilet mirror, and an ear speculum. Of the last there are several kinds, and most persons who use them express a preference for the simple silver cylinders, which come in nests of two or three. I like the old-fashioned bivalve, as being most readily employed, and adapted to all ages and conditions.

If we have sunlight, the patient is placed so that the ear to be examined is from the light, the head at such level that the rays may be readily thrown from the mirror to the ear. The speculum is then introduced, and held with one hand, while with the mirror in the other we catch the rays of light and throw them to the bottom of the meatus. If we use a lamp, it is set on a table by the patient's head, the mirror being employed as before to concentrate the light and throw it into the speculum.

The advantages of this method are two-fold. First, we employ a larger volume of light, and have greater command over it, directing it where and as we choose. In the second place, we are able to see the parts without interposing our head between the ear and the light. Those who employ the ophthalmoscope or the laryngoscope, may use the perforated mirrors of these instruments as reflectors.

There is no difficulty with these means of making a thorough examination of the ear, even in the child, and of determining the condition of the *external auditory meatus*, and membrani tympani.

FOREIGN BODIES IN THE EAR.

Children more frequently than the adult, get foreign bodies in the ear. Often a pea, bean, grain of corn, or other material, will be introduced purposely by the child or some companion in play. Rarely an insect or other living thing accidentally gains entrance, and causes considerable disturbance.

The foreign body does not at first produce any or but little unpleasant feeling, but the efforts at removal, very often rough, excite the sensibility of the organ, and after a time excite an extreme irritability. The history of the case is plain, and a slight examination determines the presence of the body.

TREATMENT.—There are two methods of removal, both good, which we choose to suit the particular case in hand. The one is, to take a wire, and bend it upon itself, forming a loop, introducing this by the side of the object until it has passed clearly beyond. Then separating the strands of wire, they are drawn gently forward, and catching the object in the loop, it is withdrawn. It is the old plan of removing a cork from a bottle.

The other plan is to remove the foreign body by a jet of water thrown from a syringe. Taking an ordinary hard rubber or glass syringe of small size filled with warm water, the jet is thrown in the ear with some force; the rebound causes the detachment and removal of the body. It is better that the child's head be placed in such position that the ear will be dependent.

Under no circumstances should we undertake to remove a foreign body with a scoop, forceps, or like instruments.

EARACHE.

Earache may not strike the reader as being worthy of much study, as the term has little of medical terminology, and smacks of the nursery. The ache, however, is a very troublesome reality, producing a very unpleasant commotion in the household, and demanding speedy relief.

CAUSES.—The cause of this neuralgic pain in the ear is most frequently cold, though it may be, in part, from derangement of the digestive organs and arrest of secretion.

SYMPTOMS.—The child is very restless and uneasy, moves its head from side to side, sleeps but little, and wakes with a start, crying loudly. The pain is undoubtedly paroxysmal in character, as, for a little time, the child is easy, and then breaks out in a piercing cry, as if hurt.

Sometimes there is slight febrile action with suffused eyes, stopped nose, and other evidences of cold. In some cases the child's face is flushed, and the temperature of the affected side increased; but at others the face is pallid, the eyes dull, and the patient evidently depressed.

An examination of the ear determines whether it is neuralgia or inflammation. If the first, there is no swelling, or evidence of structural change, while in inflammation, the swelling, redness, heat, and tenderness on pressure, are prominent symptoms.

TREATMENT.—When there is excitement of the circulation, I put the child upon the use of tincture of Aconite in the usual doses, adding Gelseminum if there is much irritation of the nervous system, or Rhus if there is the contraction about the eyes, and the sharp stroke of the pulse. If, on the contrary, there is dullness, the face pale, the pupils dilated, and evidently a sluggish circulation, Belladonna should be used with the Aconite. In the last case a full dose of quinia once or twice daily, will be an important addition to the treatment.

As a local application I generally prescribe: ℞ Tincture of Aconite, Tincture of Opium, aa. ʒij., Glycerine ʒss.; mix. Warm a spoon and from it drop one or two drops in the ear, and also apply it around the ear with the finger. A warm flannel applied over the ear will complete the treatment.

OTITIS.

Quite a number of different affections have been grouped together under the head of Otitis, and as they are all inflammatory, present similar symptoms, and require nearly the same treatment, it will hardly be worth while to endeavor to make the distinction. Inflammation of the external auditory meatus and cavity of the tympanum are usually produced by sudden changes of temperature, though it may be caused by the introduction of irritants, or even from accumulation of cerumen.

SYMPTOMS.—Inflammation of the external auditory meatus commences with a feeling of stiffness, fullness, and uneasiness about the meatus, which is increased when the ear is pressed upon. In a short time the pain becomes very severe, is tensive, darting, lancinating, and seems to affect the entire side of the head to some extent. Frequently there are marked chilly sensations with the accession of the severe pain, and these are followed by febrile reaction. On examination we will find the lining membrane of the meatus tumid and red, sometimes swollen so as almost entirely to close the opening. The pain continuing for from two to six days, secretion takes place, or pus is formed and discharged, sometimes in considerable quantity. At first it is usually thick, but at last is thin, and in some cases is secreted in very large quantity. The discharge continuing for a short time, the symptoms of inflammation entirely disappear, and the part is restored to its normal condition.

Acute inflammation of the cavity of the tympanum is a far more serious affection, and may result in permanent impairment of the hearing, or even in death by extension to the brain. In children this is usually very severe at night, with comparative ease during the day, though the child is restless and irritable. There is usually considerable fever at night, and even during the day the skin is dry and the pulse hard.

DIAGNOSIS.—Inflammation of the ear presents such marked symptoms that it is not easily mistaken. The severity of the pain and its location, and attendant constitutional disturbance, are sufficiently characteristic. If the external meatus is the seat of the disease, it will be found red and swollen, as is the case if the membrana tympani is affected. If confined to the cavity of the tympanum, all the symptoms are more severe and there is an absence of external signs of inflammation. When the inflammation extends to the mastoid cells, the constitutional disturbance is very marked, and when pus forms, the deep throbbing and marked disturbance of the brain show the character of the lesion.

PROGNOSIS.—Though quite painful, inflammation of the external meatus is not dangerous, nor attended with worse results than otorrhœa in occasional cases. If the tympanum is affected there is some danger of affection of the brain, and considerable of impairment of the hearing. Inflammation of the mastoid cells, if it progresses to suppuration, is always dangerous.

TREATMENT.—Put the patient upon the use of Aconite or Veratrum in the usual doses, aiding the action by the use of the hot foot-bath. If there are indications for other remedies these may be added to the sedative solution, or given in alternation. We will thus find in different cases indications for Gelseminum, Rhus, Bryonia, Belladonna, Apis, Macrotys, Phytolacca, Apocynum, etc. In some cases—a full tongue heavily coated—emesis will give the speediest relief. If the tongue is pallid and dirty, give sulphite of soda; if red and dirty, sulphurous acid.

Locally we may apply the lotion of Aconite and Opium, heretofore named, around the ear, and even drop it into the ear.

This may be followed by the use of the vapor of water, and hot fomentations of Stramonium. Occasionally much relief is obtained from the use of a lotion of equal parts of tincture of Aconite and Belladonna, applied around the ear. In some cases, the fever being very intense, we may employ the vapor of tincture of opium, stramonium, lobelia, tobacco, etc., directly to the external meatus and membrana tympani, by means of a guttapercha tube. Chloroform and ether may be used in the same way, as may also carbonic acid gas.

With throbbing pain in the ear, we are satisfied that suppuration has occurred, and sometimes an inspection of the meatus will show the yellowish color of pus. The little abscess should be at once opened with a sharp-pointed bistoury, when we will have immediate relief.

If the disease is of the middle ear, and the pain becomes tensive and throbbing, we may be satisfied of the presence of pus. An examination of the ear will show the membrana tympani pushed outward (convex) and blanched and yellowish. If it is now incised with a small tenotome, or small, sharp-pointed bistoury, the escape of a drop or two of pus gives relief. In place of destroying the tympanic membrane by incision, we frequently preserve it, for the continuous presence of pus would cause its entire destruction, and sometimes the destruction of the internal membrana, whilst if the pus is discharged, the inflammation ceases, and the wound closes up, the membrane showing but a slight cicatrix.

If the disease seems to extend to the mastoid portion of the bone, I should apply a blister immediately over it, and follow it with the irritating plaster. In some cases, suppuration having undoubtedly taken place, and dangerous symptoms occurring, it becomes necessary to open into the mastoid cells through the bone, in order to permit the escape of pus.

OTORRHŒA.

Purulent discharges from the ear may be occasioned by chronic inflammation of the external meatus, or disease of the bony canal, or it may proceed from chronic inflammation of the tympanum, or disease of adjacent parts, the membrana tympani having been ruptured or destroyed, so as to permit its escape. In either case there is more or less deafness, uneasiness in the ear, and an offensive discharge. The most frequent causes of otorrhœa are inflammation attending the eruptive fevers, injuries, the direct action of cold, and chronic inflammation resulting from acute attack. Some families seem to have a predisposition to this affection, the majority of their children having such discharge. In such cases it is almost always associated with scrofula and feeble vitality.

Otorrhœa from disease of the external auditory meatus is the most frequent form of the affection, and might properly be called chronic catarrh. It is of frequent occurrence after scarlet fever and measles, and is often seen in infancy or up to the age of two or three years, becoming more rare as we advance to adult age, except in the cases named. It is true, that the disease commencing at the age of two or three years may continue through life, but this is not very common when the patient has sufficient vitality to reach adult age. Further than the discharge from the ears of an offensive purulent matter, and some dullness of hearing, there are no prominent symptoms, if we except the almost invariable cachectic appearance of the child. On examining the ear, we will find the bone in a carious condition. When the hearing is much affected, we will find the membrana tympani opaque, and its dermoid layer thick and vascular. In some cases the discharge is produced by a small polypoid formation in the ear, and in others by a hardened cerumen.

Otorrhœa from disease of the middle ear occurs only when the membrana tympani has been destroyed or ruptured, and may arise from chronic inflammation of the lining membrane, disease of the ossicles, or disease of the bony walls. It is most generally the sequence of acute inflammation, which terminating in suppuration, the membrana tympani gives way, and the inflammation gradually assumes the chronic form. There is always deafness, sometimes but slight, at others marked. There may not be pain or unpleasant sensations in the ear, though usually if there is but a slight opening in the membrana tympani it occasionally becomes closed, and dizziness, ringing in the ear, etc., result from the pressure of the retained secretion.

The condition of the tympanum varies greatly; in some cases there is but slight change of structure, in others the ossicles become diseased, and are cast off, the mastoid cells and eustachian tube are affected to some extent, and the hearing is nearly entirely destroyed. It may occasionally terminate fatally by an extension of the inflammation to the membranes of the brain.

TREATMENT.—In all cases it becomes necessary to pay attention to the general health, for as long as the child or adult continues cachectic, or shows evidence of a wrong of the blood, it is almost impossible to arrest the discharge. The use of lime water, sulphite of soda if there is eczema, sulphurous acid, cod liver oil, arsenic, iodide of potash, Donovan's Solution, Phytolaca, and some of the vegetable alteratives, are suggested. It is impossible to name a remedy which will be good in all cases, for here, as in other diseases, the remedy will be selected according to the special indications in the case.

Prominent among external applications in all forms of this affection, except when occasioned by a foreign body lodging in the ear, or a polypoid growth, is counter-irritation over the mastoid process. It should rarely be neglected, but pursued steadily until the cure is complete. The best agent that I have employed is the cantharides, which may be repeated sufficiently often to keep up a continued influence.

In common chronic inflammation of the external meatus, washing the ear out thoroughly with tepid water, and dropping three or four drops of—℞ Tinct of muriate of iron ʒij., glycerine ʒj., into the ear once or twice daily, will effect a cure in one or two weeks. In place of this, nitric acid gtt. x., to water and glycerine aa. ʒss., may be employed with advantage. A weak solution of thymol, of sulphate of zinc, or acetate of lead, from four to ten grains to the ounce of water, may be used in some cases. ℞ Chlorate of potassa, gr. xx., glycerine ʒss., water ʒj., also forms a good application. An infusion of Hamamelis, Hydrastis, Cornus, Geranium, Sage, etc., is found useful in some cases. Occasionally I have employed oxide of zinc gr. x., rubbed up with glycerine ʒss., adding a small portion of morphia if necessary, to relieve irritation.

Failures in these cases will frequently depend upon a faulty use of local means. When injections are used they do not reach the affected surfaces. In place of the syringe, a long camel's hair pencil may be used to clean the ear first, and afterward to carry the local remedy close to the membrana tympani if necessary. This will be readily done by twirling the pencil between the fingers as it is introduced.

In some cases the brown citrine ointment, with one to three parts of cod liver oil or simple cerate (according to the soreness), may be applied to the entire diseased surface by twisting a soft piece of cotton cloth, smearing it with the ointment, and twisting it as it is pressed into the meatus. In other severe cases, nitric acid applied lightly with the pine pencil, will be found a most excellent plan of treatment.

If the tympanic cavity is the seat of disease, we will pay especial attention to the general health, and keep up continuous counter-irritation near the ear. Cleanliness is of prime importance, and hence the ear should be thoroughly washed out, once or twice daily. This may be followed by some of the lotions above named, being careful that they are brought in contact with the diseased surface.

If the discharge is produced by accumulations of hardened cerumen acting as a foreign body, this should be softened and removed with a scoop. If from a polypus, and it is not red and vascular, it may frequently be removed by the application of a saturated solution of tannin, or the careful application of chloride of zinc. The best plan, however, in all cases, is to catch it with a strong pair of ring forceps, and detach and remove it.

DEAFNESS.

Partial loss of hearing depends upon various causes, some of which are remediable; total deafness depends upon disease of the internal ear, and if of any considerable duration is incurable.

We wish, therefore, in this place, to inquire into the causes of partial deafness, and see how far they are amenable to treatment. We may sum them up, as: 1st, from disease of the external meatus; 2d, from disease of the membrana tympani; 3, from disease of the tympanum; and, 4th, from disease of the eustachian tube. Diseases of the internal ear are beyond our powers of diagnosis, though we are able occasionally to determine with considerable certainty that the deafness is dependent upon partial paralysis—we call this nervous deafness. The ear-speculum should always be used, so as to make an accurate diagnosis.

1. The external meatus suffering from chronic inflammation will give rise to hardness of hearing, as we have already seen. In other cases the lining membrane is thickened and dry, and in addition, the ceruminous glands seem to pour out a very inspissated secretion, which desiccating sometimes fills up the bottom of the meatus. In this case we would use injections of tepid water and the scoop, to thoroughly cleanse the ear, and then use the lotion of tincture of muriate of iron and glycerine, heretofore mentioned. The lotion of thymol and glycerine may also be used in these cases. If there is irritation of the structures, much benefit will be derived from counter-irritation over the ear. Polypi obstructing the meatus should be removed, as before mentioned.

2. A condition of chronic inflammation of the membrana tympani, giving rise to a fleshy vascular appearance when examined with the speculum, is sometimes a cause of deafness; quite frequently it is associated with catarrhal inflammation of the meatus, though it may persist afterward. An injection of a decoction of Cornus and Hydrastis, and the local application with a camel's hair pencil of the oxide of zinc, morphia, and glycerine, heretofore named, is usually sufficient. Counter-irritation over the mastoid process is also employed. *Relaxation* of the membrana tympani is not of frequent occurrence, but may occasionally be met with as a cause of deafness. It is readily determined by the use of the speculum, the membrane being remarkably concave on its external face, and is diagnosed from the same appearance resulting from closure of the eustachian tube, by its being thrown outward by swallowing with closed nostrils. Pond's Hamamelis and glycerine, equal parts, make a good local application. Tincture muriate of iron 3j. to glycerine 3j. may be used in some cases, and a solution of Hydrastia,

gr. ij. to rose-water ℥j., will give tone to the structures in others. Perforation of the membrana tympani is a frequent cause of deafness, and is readily detected with the speculum. It seems, however, that the deafness depends in part upon thickening of the mucous membrane of the tympanic cavity, for when this is marked, the patient can hardly hear at all, while in other cases the deafness is but slight. We should therefore endeavor to remove all irritation by the use of counter-irritants and appropriate local applications, and we will then have placed the patient in the best condition for the use of the artificial membrana tympani; this is formed out of vulcanized rubber, and has been very successfully employed.

3. Various changes in the tympanic cavity, resulting from inflammation, may be the cause of deafness, but there is only one, so far as we know, that can be reached by remedial measures. We have already noticed that a chronic inflammation of these structures might continue for years, attended with secretion; and examination shows us in some cases, a thickening of the lining membrane, with increased vascularity. In these cases, the persistent use of counter-irritation and the local means heretofore named will do much toward the relief of the deafness.

4. Obstruction of the eustachian tube always gives rise to partial deafness, though, as the causes are usually temporary, the deafness is not of long duration. Dr. Toynbee notices three points of obstruction: 1, at its *faucial orifice*, a thickening or relaxation of the mucous membrane; 2, at its *tympanic orifice*, from thickening of the mucous membrane, or a deposit of fibrin; 3, in the *middle part* of the tube, from a collection of mucus, a stricture of the osseous or cartilaginous portions, or membranous bands connecting the walls. If the eustachian tube is impervious, we will find the membrana tympani sunken in, of a dull, leaden hue, and its surface unnaturally glossy, and swallowing with the nose closed or forcible expiration will not have any effect on it.

If the patient has had disease of the tonsils, fauces, or posterior nares, we may reasonably suppose that the disease has been caused by this, and is at the faucial extremity of the tube. The inflammation sometimes extends to the mucous membrane lining the tube, and its tumefaction causes the disease. In other cases the swelling of the mucous membrane at the termination of the tube, is the cause of it, and in another class it results from relaxation. In these cases, appropriate measures to relieve inflammatory engorgement in the one instance, and to remove the atony and relaxation in the other, should be adopted. The orifice of the eustachian tube may be reached through the mouth or inferior meatus of the nose, and local applications may be made with a probang or syringe. In some cases it is proposed to remove obstructions by means of a probe passed into the eustachian tube, but no permanent benefit results from it. We may, however, introduce a catheter, for the purpose of using an injection into the tube, using the same remedies that would be indicated in other situations, as, for instance, those recommended in otorrhœa.

NERVOUS DEAFNESS.—Toynbee remarks that "As some cases of deafness dependent upon the derangement of the nervous apparatus connected with the organs of hearing appear to be caused by the condition of the brain generally, or of that part in intimate relation with the acoustic nerve, it has seemed desirable to divide the nervous diseases of the ear into two classes; to the first of which belong those cases where the special nervous apparatus of the organ is alone affected; to the second, those where the brain, conjointly with the ear, seems to be injured. The first class may be divided into diseases arising from: 1, concussion; 2, the application of cold; 3, various poisons, as that of typhus, scarlet, or rheumatic fevers, of measles and mumps, of gout, of an accumulation of bile in the blood, and of quinine in large doses. And the second into diseases arising from: 1, excess of mental excitement; 2, physical debility.

In the first class of cases there is not unfrequently ringing and singing in the ears, with other morbid sounds, and sometimes a feeling of giddiness and unsteadiness extremely unpleasant. From its commencement there is frequently a continuous increase in the deafness; but in other cases it remains the same, and in still others there is gradual improvement. It is generally conceded that in very many cases there is congestion of the nervous apparatus of the internal ear, though if it continues for a considerable time it will very likely terminate in structural change.

The treatment will vary in these cases, in some the remedies being selected from those which influence the nervous system directly, in others from those which influence the blood, and in

still others from those which influence waste and excretion. Of
the first class we find marked benefit from Rhus, Bryonia, Sticta,
Belladonna, Apocynum, Bromide of Potash, Bromide of Ammonium, etc. Of the second, Fowler's Solution of Arsenic, Cod
liver Oil, Donovan's Solution, Phytolacca, Alnus. The third
class will embrace the alkaline diureties, and remedies that
increase secretion from skin and bowels.

Persistent counter-irritation over the mastoid portion of the
temporal bone, with cantharides, or the irritating plaster, is one
of the most important parts of the treatment. These measures,
followed up for months, will occasionally produce the most
marked benefit, the hearing being sometimes completely restored; but in other cases no benefit results.

In the second class of cases we will have more or less evidence
of cerebral disturbance, though frequently the symptoms are imperfectly marked. No treatment can be laid down for these
cases, as the symptoms are so variable and changing. They
should be treated on general principles, and we will sometimes
be agreeably surprised at a favorable termination in cases which
had seemed hopeless; and not unfrequently we will fail when we
seemed to have the best chance of success.

CHAPTER XII.

DISEASES OF THE SKIN.

Difficult of classification in all cases, we find diseases of the
skin doubly so in the child. True, some cases are so well defined that the merest tyro will assign them their proper place;
but others are obscure from the first, and more become so as
they progress.

Probably the simplest classification is that adopted by
Cazenave, into eight orders. The following brief description of
each will enable the reader to classify many cases:

Exanthemata.—This term is applied to patches of a reddish
color, varying in intensity, size and form, disappearing under
pressure of the finger, and terminating in delitescence, resolution or desquamation.

Vesiculæ.—A vesicle is a slight elevation of the epidermis, containing a serous and transparent fluid, which, however, is occasionally opaque or sero-purulent. The vesicle may terminate in absorption of the fluid, slight desquamation, excoriation, or the formation of small, thin incrustations.

Bullæ.—Generally speaking bullæ differ from vesiculæ merely in size; they are small superficial tumors caused by effusion of serum underneath the epidermis.

Pustulæ.—This term should be strictly confined to circumscribed collections of pus on the surface of the inflamed mucous layer. The contents of the pustules in drying produce scales, and they may be followed by chronic induration, or by red inflamed surfaces, or sometimes by slight excoriation.

Papulæ.—These are small elevations, which are solid, resisting, and never contain any trace of fluid; they may likewise give rise to ulceration, but generally terminate in resolution and furfuraceous desquamations.

Squamæ.—The term squamæ is applied to the scales of thickened, dry, whitish, friable, and degenerated epidermis, which cover minute papular elevations of the skin; they are easily detached, and may be reproduced for an indefinite length of time by successive desquamations.

Tuberculæ.—These are small hard tumors more or less prominent, circumscribed in form, and persistent; they may become ulcerated at the summit, or suppurate partially. In this definition we consider tubercles as elementary lesions, and not those which appear after abscesses.

Maculæ—Are permanent changes in color, in certain points of the skin, or of the whole cutaneous envelope, but unattended with any general derangement of the health.

ORDER I.—EXANTHEMATA.

The general characteristics of this order are well marked at first, though in the progress of the disease they may so change that they will approximate some of the others. They always commence with redness of the skin, which is effaced for the moment by pressure, returning as soon as this is removed. Some of them, as erysipelas, rubeola and scarlatina, are attended with marked constitutional disturbance, and in the last two, as we have already seen, the cutaneous disease is associated with disease of the throat and respiratory apparatus, and in all three of the diseases named there is in some cases marked lesion of the blood.

ERYTHEMA.

Erythema is one of the mildest of the exanthemata, and usually is not accompanied with febrile action, though in the severer cases there is arrest of secretion and some constitutional disturbance. It may be produced from mechanical irritation of the skin, but the most frequent causes are cold and arrest of cutaneous secretion, or gastric and intestinal derangements.

SYMPTOMS.—The disease appears in the form of patches of variable size, of a light, superficial red color, readily effaced by pressure, and most frequently on the face, chest and limbs. In some cases they spread so as to cover a considerable portion of the body, but this is not frequent. One form, termed *erythema nodosum*, is preceded by slight constitutional disturbance, and comes out in oval, red patches, from half an inch to an inch in diameter, most generally on the lower extremities. When more fully developed they are slightly elevated above the adjacent skin, and in a few days form small, red, painful tumors, which seem inclined to suppurate, and in severer cases give a suspicious sense of fluctuation, but at last disappear without any change of structure. The first form may last but a few hours, or in rare cases it may continue two or three weeks; the second ususually continues for from three to six days.

Before the eruption makes its appearance the child is sometimes quite sick, and the febrile action is marked, and if there is a sudden retrocession, we sometimes find the child restless and uneasy, with nausea, pain in the head, and frequent pulse.

TREATMENT.—Usually we put the patient upon the use of small doses of Aconite, with Belladonna if the eruption is not out freely; Rhus if there is frontal pain, startings in sleep, and red papillæ on tip of tongue; Apis if there is itching of the surface; Gelseminum if the face is flushed and the patient restless; Apocynum if the eyelids are swollen.

The surface should be bathed with a week solution of carbonate of potash, (alkaline bath,) and in some cases we would use the warm foot-bath. In the second form of the disease, I have usually prescribed a gentle laxative, with a solution of acetate of potash, and very small doses of Aconite. The use of the alkaline bath gives great relief, and it may sometimes be repeated several times a day. In some rare cases there seems to be a tendency to excoriation, and in such cases I would advise a lotion of—℞ Glycerine, ℨj., Chlorate of Potash gr. xx., rose water ℨij.; mix. Or the part may be dusted with powdered subnitrate of bismuth.

ROSEOLA.

Roseola, or *rose rash*, is a mild exanthematous eruption, continuing from one to six or seven days, and attended by more or less febrile action. The causes are obscure, though arrest of secretion and gastro-intestinal irritation, are the most frequent. It sometimes occurs as an epidemic, especially in warm seasons, and sporadically, from over-heating the body, severe exercise, etc. Four varieties have been distinguished: R. infantilis, R. œstiva, R. autumnalis, and R. annulata.

SYMPTOMS.—*Roseola infantilis*, as its name indicates, is usually met with in young children, and arises form gastro-intestinal irritation, or from dentition. It comes out in the form of deep rosy-red patches about one-fourth of an inch in diameter, and circular in form. When severe, they are very much crowded together so as to give a general red appearance to the surface, but yet each one is well defined. They may continue for several days, or vanish and reappear for several days. Usually the fever is but slight, but the child shows symptoms of irritation, being cross and fretful.

Roseola œstiva is usually ushered in by marked febrile action, and in children delirium or convulsions sometimes supervene. The eruption usually appears about the third or fourth day on the face and neck, and in a few hours involves the greater part of the body. "The spots are of a deep red color, more irregular in shape than those of measles, and their original color soon passes into a light rosy hue. There is also present a considerable degree of itching and pain, and often difficulty in swallowing." The disease runs a very variable course, but the eruption usually disappears in three or four days without desquamation. A bastard measles, which is contagious, resembles this

form of roseola.   It is known as *ratheln*, or German measles,
and as it occurs where children have had true measles, many
have supposed that persons could have the disease twice.

Roseola annulata comes out in the form of rose-red rings, in
the center of which the skin retains its natural color ; it is said
to be principally observed on the abdomen and buttocks.   It is
not usually accompanied with much fever, but is occasionally
very persistent, and is usually associated with gastro-intestinal
irritation.

DIAGNOSIS.—Roseola may be distinguished from measles by
the spots being larger, circular, circumscribed, and of a deep
rose color, while the patches of measles are small, irregular, and
of a common red color.   The eruption of scarlet fever consists
of a great number of small red points of a scarlet or raspberry
color, and grouped together so as to form irregular patches.

TREATMENT.—The patient has the proper sedative in small
doses, to which may be added such remedy influencing the skin
as may be indicated.   If there is dullness or inclination to sleep,
Belladonna is given until the eruption comes freely to the surface, and the nervous system is relieved.   If the pulse is sharp,
the tongue shows the red papillæ, and there is frontal pain, or
burning of the surface, Rhus is the remedy.   If the itching is
severe, add Apis to the solution of Aconite.   Phytolacca is given
if there is soreness of the mouth and throat; the carbonate of
soda, if there is a pallid tongue ; sulphite of soda, if the tongue
is broad, pallid and dirty ; and sulphurous acid, if the tongue is
red and dirty.

When there are frequent recurrences of roseola, we may suspect some wrong of the gastro-intestinal apparatus, or a failure
in waste and excretion.   Sulphite of soda and sulphurous acid
are frequently indicated in these cases, and their administration
effects a cure.   Lime-water or minute doses of sulphite of lime,
will frequently prevent a recurrence of the attacks.   In some
cases the patient will be benefited by Nux, in others by arsenic,
and in still others by cod-liver oil.   Occasionally tincture of muriate of iron in small doses exerts a good influence, the disease
resembling a mild form of erysipelas.   In other cases Pulsatilla
will prove a good remedy, and in still others a second trituration
of Podophyllin may be given.

URTICARIA.

Urticaria, or *nettle-rash*, occurs most frequently in childhood, though we occasionally see cases of it in the adult. The most common cause is doubtless gastro-intestinal irritation, though the milder forms may be caused by sudden changes of temperature, or excessive mental emotion. Sometimes it is an acute affection, but more frequently it assumes a chronic form, and may last for months or years, reappearing on the slightest imprudence of diet or change of habits.

The reader will know urticaria by its more common name, *hives*, and if he has had much experience with it, he will recall many unpleasant cases. It not only varies in intensity and duration, but also in the appearance of the eruption, in size, form, and color; but in all there is the common symptoms, intense persistent itching, and sometimes burning of the skin.

SYMPTOMS.—Though divided into several varieties, it will suit our purpose to consider it as *febrile* and *non-febrile*. In the first case the eruption is preceded for a day or two by slight febrile symptoms, irritation of the stomach, and pain in the epigastrium. The eruption then comes out in the form of red or pale-red blotches, irregular in shape, elevated above the adjacent skin, hard around their edges, and surrounded by a bright red or scarlet border. An intolerable pruritus and burning accompany the eruption, aggravated by warmth, and usually by scratching or rubbing the part, and is sometimes so severe as to prevent the patient's sleeping. The eruption is not constant, but goes away and re-appears sometimes every few hours. The disease usually continues for seven or eight days, with some constitutional disturbance during the entire period, and at last disappears, leaving but slight itching; in severe cases there may be some desquamation.

In some cases the eruption fails to come out, and the patient suffers from extreme nausea, nervous irritation (sometimes going on to convulsions), and occasionally marked febrile reaction. The swollen face and slightly mottled skin, will frequently suggest hives, and we relieve the patient by bringing the eruption to the surface. When there is a retrocession of the eruption, the patient is sometimes very sick, so that between the intense itching and burning when the hives are out, and the nausea and great prostration or febrile action when they go in, the patient passes a very unpleasant week. In some cases marked croupal symptoms are observed, and this becomes an important feature.

The non-febrile form is usually chronic, and has been divided into two varieties, U. evanida and U. tuberosa. In the first, the eruption appears at irregular intervals sometimes for months or years, is not attended by febrile action, and has not the red border just noticed; the spots look more like those produced by whipping, and are only accompanied by itching.

The last form is very rare, and instead of the slightly elevated blotches, there are broad, hard, deep-seated and painful tuberosities which impede motion. It passes off and reappears like the preceding variety, but almost always leaves the patient fatigued and depressed.

DIAGNOSIS.—There is but one disease (*lichen urticatus*) with which this can be mistaken, and from that it may be distinguished by the large irregular blotches, while in litchen, the papulæ are rounder, less prominent, smaller, harder, and of a deeper color. Urticaria may be complicated, however, with erythema, roseola, impetigo and litchen.

TREATMENT.—In the majority of cases the prescription will be—℞ Aconite gtt. iij. to gtt. v., Tincture Belladonna gtt. v.; a teaspoonful every hour until the eruption comes out freely and the patient is relieved. If now we substitute—℞ Tincture Apis gtt. v., Tincture Phytolacca gtt. x., water ʒiv., a teaspoonful every two hours to relieve the itching, our patient will make a rapid recovery.

When there is severe burning as well as itching the patient will have Rhus with the Aconite. In some of these cases the irritation of the nervous centers would call for Rhus if there was not the prominent symptom, "burning with itching."

In rare cases, the evidences of accumulations in the stomach—heavily coated tongue and fullness in the epigastrium—will demand an emetic. Usually the nasty tongue and breath will be met with sulphurous acid, or sulphite of soda.

If there is marked oppression of the respiratory apparatus, with cough, and increased secretion of mucus, we may use Lobelia, either with Veratrum, in the usual way, or as named heretofore. ℞ Tincture Lobelia ʒj., Comp. Tincture Lavender

℥iij., simple syrup ℥iss., Croupal symptoms are best met by an application of stillingia liniment over the larynx, and sometimes its internal administration in half drop or drop doses. If convulsions are threatened the patient should have bromide of ammonium.

When the skin is feeble, and the eruption does not appear, the patient may be rapidly sponged with hot water; before a fire, being careful to prevent chill. A hot foot bath (a full half hour and plenty of water *hot*) will sometimes give great relief. When the eruption is out, and the patient complains of excessive itching, the old-fashioned alkaline bath may be employed, or a salt water bath will sometimes give relief.

ERYSIPELAS.

Erysipelas would have been more properly classed with febrile diseases, as in many respects it resembles them. The constitutional affection is the primary disease, and of which the local inflammation is but the outgrowth.

It is not so frequently met with in children as in the adult, but occasionally is quite a serious affection. Thus we will see it during the first week of life in erysipelatous inflammation about the umbilicus; at a later period affecting the genitalia and nates; and from two years upward attacking the extremities. It is a singular fact that children rarely have erysipelas of the face, which is the most common seat of the disease in the adult.

CAUSES.—Erysipelas undoubtedly belongs to the class of zymotic diseases, in which there is a blood-poison. What this peculiar poison is, we do not know, or how it is generated, farther than that it seems to be produced in certain depressed conditions of the system, where there is rapid waste of tissue and imperfect excretion. When once generated, it has the power of propagating itself, and at times will become eminently contagious.

PATHOLOGY.—As above remarked, the disease is undoubtedly produced by a blood poison, and the first lesion is of the fluids, manifested by the usual symptoms of fever. Following this, we have an inflammation of the skin, and occasionally of the deeper tissues. The inflammation does not at first seem to differ from that produced by other causes, further than that the circulation

is more impaired. But as it advances the tissues seem to lose
vitality, and the exudation is of a depraved serum rather than
coagulable lymph.

SYMPTOMS.—The symptoms vary in different cases, not only
in intensity, but in their order. Usually, the child seems depressed for a day or two, the appetite being impaired, the breath
bad, and the sleep broken. Following this there is febrile
action, frequently coming up slowly, but sometimes rising
quickly and running an active course.

With the appearance of the fever the cutaneous disease shows
itself, usually in the form of a red spot, where the skin is evidently swollen, hot and painful. As the disease progresses, we
find the redness extending, with the same induration of the skin,
and other evidences of inflammation.

Occasionally in the course of two or three days the epidermis
is loosened and distended with a yellowish serum, forming
bullæ of larger or smaller size, and these rupturing pour out
their secretion, and sometimes become covered with thin incrustations. The redness usually fades, and the inflammation
commences to disappear by the fifth or sixth day, leaving the
epidermis wrinkled and yellowish, and at last it desquamates
over the entire surface.

The fever is in some degree dependent upon the local lesion.
When this is acute it runs a very active course, and gradually
subsides as the inflammation passes off. In other cases the fever
is of an asthenic type, and sometimes presents marked typhoid
symptoms.

We observe another form of the disease in children of from
six to twelve years. They complain of soreness of the legs and
feet and lameness, and when examined the parts are slightly
swollen, the skin smooth and glistening, and presenting a
peculiar mottled red and purplish discoloration. The child's
health has not been good for some days; its appetite is impaired, its bowels irregular, breath bad, and it is evidently losing
flesh.

DIAGNOSIS.—The peculiar redness of the skin, with swelling,
and the burning heat that attends it, are characteristic. Associated with this is the marked constitutional disturbance, rendering
the diagnosis certain.

PROGNOSIS.—Usually we will have but little difficulty in controlling the local and general disease, and remedies will give much satisfaction. Some cases will be quite intractable, but upon the whole the mortality will be but small.

TREATMENT.—The treatment of erysipelas illustrates the truth of specific medication, both as regards the action of the small dose, and the advantage of following special indications for the selection of the remedy. If the reader will consult the article "Anti-erysipelatous Remedies," in the first part of this work, he will be able to see this better. The remedies in common use are Veratrum, Rhus, Tinct. Muriate of Iron and Sulphite of Soda.

Veratrum is indicated by a full pulse, and the ordinary redness of inflammation, all the inflammatory symptoms being marked. It is given internally in the proportion of—℞ Tinct. Veratrum gtt. v., water ℥iv., a teaspoonful every hour, and applied locally of full strength, to one part to ten of water, or water and glycerine.

Rhus is indicated by the brighter redness, burning pain, vesication, sharp pulse, and irritation of the nerve centers. We prescribe it as follows—℞ Tinct. Rhus gtt. iij. to gtt. v., Tinct. Aconite gtt. iij., water ℥iv., a teaspoonful every hour. When indicated its action is very prompt and decided.

Tincture of muriate of iron is given in small doses, where the mucous membranes are red, either dusky or livid. The remedy is best given with glycerine as heretofore named. This is also the remedy for the last form named, and its continuance for a few days will entirely remove the unpleasant symptoms, without local applications.

Where the tongue is pallid, broad, and covered with a pasty white coat, sometimes pitting where it comes in contact with the teeth, sulphite of soda is the remedy. To a child two years old it may be administered in doses of three to five grains every two hours.

In young children the best local application is soft cloths spread with fresh lard. In other cases we will find an excellent remedy in one part of the tincture of muriate of iron to five parts of glycerine. When the erysipelatous inflammation is deep seated, and likely to progress to sloughing, a solution of Permanganate of Potash, ℥j., to water, Oj., used as a wet dressing, will be found an admirable remedy.

ORDER II.—VESICULÆ.

The distinguishing characteristic of this order is, the formation of small vesicles by an elevation of the epidermis, which are filled with a serous fluid. This fluid, at first transparent, in severe cases becomes yellowish and opaque, and is finally either absorbed, or dries and forms scales or incrustations. The vesicle is always round, and may or may not stand upon an inflamed base. One variety of this order, *varicella*, has already been described with the eruptive fevers,

MILARIA.

Milaria, or *sudamina*, most generally appears as an attendant upon other diseases, more especially typhoid, and the advanced stages of other fevers and inflammations. There are exceptional cases in "which it assumes an idiopathic form, as for example, when it appears in healthy subjects after violent exercise in warm weather; in these instances it is generally accompanied with copious perspiration. The eruption is then attended with a disagreeable sensation of heat and itching. The number of vesicles is sometimes very considerable, but they are ephemeral, and disappear in the course of twenty-four hours."—(Cazenave.) The miliary vesicle is small, not larger than a pin's head, and the contents being clear and transparent, it can not be seen well unless we look across the surface. They are usually grouped together in patches, upon the thorax and neck, and in rare cases become confluent, forming bullæ. They demand no treatment, being simply symptomatic of other diseases.

ECZEMA.

Eczema, *humid tetter*, or *running scall*, is characterized by an eruption of small vesicles grouped and crowded together, and forming more or less well defined patches. It may be divided into the acute and chronic form, and these have to be still further divided into several varieties. The causes of eczema are very obscure, and it is non-contagious, except in rare cases where the disease affects the genital organs.

SYMPTOMS.—*Eczema infantilis* is the *milk-scall* that is sometimes found so troublesome during the first year, and occasionally continuing after this. It makes its appearance in the form of transparent vesicles about the ears, sometimes upon the face. These increase in number, gradually extending from part to part, until a large surface is diseased. The vesicle continues but a day or two, then rupturing, a crust is formed, beneath which pus in found, and the crust is renewed as often as detached. Thus running together, large surfaces are covered with single crusts, from which there is an offensive discharge.

In many cases the disease manifests but little tendency to recovery, sometimes continuing for months.

Eczema simplex commences with a sensation of itching, which is soon followed by the appearance of numerous small transparent vesicles, flattened, and set close together; after a time the fluid they contain becomes opaque, and they finally rupture, forming a small thin scab which is soon detached. They appear more frequently upon the fore-arm, and where the skin is thin and delicate, and frequently between the fingers, somewhat resembling the itch.

Eczema rubrum is accompanied with considerable heat and tension of the skin, and at first the vesicles may be observed as small solid points, but they soon become true vesicles, which attain the size of a pin's head, and finally disappear about the sixth or eighth day. In some cases the vesicles coalesce and rupture, a disagreeable excoriation producing repeated incrustations being left.

In *eczema impetignodes* the inflammation of the skin is very marked and it is swollen, the vesicles are larger, and the contained fluid loses its transparency and becomes purulent, and finally they rupture, forming a scab, which is thrown off and re-formed sometimes for two or three weeks. Acute eczema of the two last forms is usually attended with well marked febrile action, which continues for two or three days, and sometimes for a longer period. The eruption is always accompanied by itching, which is sometimes very severe and troublesome.

Chronic eczema most generally results from an acute attack, and may continue for months, or even years. In these cases the skin becomes deeply inflamed and excoriated, and fissures form about the joints; a continuous ichorous discharge is kept up, which increases the irritation, and forms thin crusts, or coming in contact with the clothing agglutinates it to the part, and when removed there is much pain and smarting, and sometimes a considerable flow of blood. When the crusts are detached, the surface is found reddened, soft, and swollen. In other cases there is less exudation, the skin being dry, inflamed, and fissured, and covered by slight crusts. "Chronic eczema is invariably attended with intense itching, more distressing than the severest pain. The patient in vain struggles against it, but he can not, however, resist the urgent desire to scratch himself, and thus increases his suffering. After a certain period the itching begins to subside, the serous exudation gradually ceases, the scaly incrustations dry up, and the skin is less inflamed. Finally, the disease becomes reduced to a small, dry, red surface, which is covered with extremely thin, laminated crusts. The surrounding skin is smooth, tense, and firm, and only slowly resumes its natural state." (Cazenave.)

DIAGNOSIS.—It may be distinguished from itch by the flatness of the vesicles, and their being grouped together, while in itch. they are pointed and isolated. The diagnosis of chronic eczema from lichen is sometimes difficult, but usually the presence of papulæ near the red inflamed surface is sufficient.

TREATMENT.—In acute eczema, if there is some febrile action, we prescribe Aconite with Rhus; or if there is no frequency of pulse or increase of temperature, Rhus with Phytolacca. Rhus exerts a specific influence upon the skin, and if there is no indication for other remedies, it will serve a good purpose in all these cases. In some it can be given in full doses, but in many it must be used in small doses, for in the usual quantity it increases the eruption. Phytolacca is especially a good remedy when there is fullness of the lymphatic glands.

If I were called upon to select a specific remedy for eczema, I should take sulphite of soda, especially in the chronic form. In many cases the indications for it will be found at first, in the moderately full and pallid tongue, dirty. In other cases sulphurous acid will be the remedy, in doses of gtt. x. to ʒss. three or four times a day.

In chronic eczema with a pallid and atonic skin, Fowler's Solution gtt. x., Tinct. Phytolacca gtt. x., water ʒiv., a teaspoonful four times a day, will prove beneficial. In rare cases Donovan's Solution may be substituted for Fowler's.

Cod-liver oil, the compound syrup of the hypophosphites, lime-water, hypophosphite of lime, and occasionally some of the vegetable alteratives, especially the Alnus, Rumex, and Scrophularia, will come in place. Sulphur in the old-fashioned dose may sometimes be used, if the disease is very persistent.

In milk-scall I have obtained marked advantage from the use of an ointment from the inner bark of the common elder (Sambucus). Occasionally a glycerole of starch answers the purpose well, or a small portion of oxide of zinc may be added to it. In very severe cases of chronic eczema, I employ brown citrine ointment one part with two parts of simple cerate. It is to be employed carefully, and not on too large a surface at once. In other cases an infusion of Alnus, Rumex and Quercus Rubra, is an excellent local application.

In some cases a general bath, rendered emollient by the addition of mucilage or gelatin, will be beneficial; it should be about ninety degrees Fahrenheit, and continued for an hour or longer. In place of this we may use the vapor bath, repeating it two or three times weekly.

HERPES.

Herpes is most generally an acute disease, and is characterized by an eruption of vesicles grouped together on an inflamed base. The causes are unknown. Five varieties are distinguished : H. phlyctenoides, H. labialis, H. præputialis, H. zoster, and H. circinatus.

SYMPTOMS.—*Herpes phlyctenoides* is usually attended by slight indisposition, loss of appetite, and constipation. The patient feels a smarting, burning sensation of some part, and upon examination finds a number of slightly red spots, upon which in a short time is developed six or eight firm and prominent vesicles from the size of a millet seed to that of a small pea. At first they are transparent, but in the course of a day become opaque and milky; there is frequently a sensation of itching, and sometimes the part feels quite painful. They commence to decline about the fourth or fifth day, drying up and leaving larger or smaller incrustations, and by the eighth or tenth day they have entirely disappeared, nothing but redness of the surface remaining.

Herpes labialis is usually preceded by slight indisposition and fever, and hence the vesicles are often termed *fever blisters*. It usually comes out at the junction of the skin and mucous mem-

brane, but may appear in the mouth, or as far back as the
pharynx. It is usually preceded for a few hours by redness,
and sometimes the part is swollen and painful. The vesicles are
of various sizes, the largest about the size of a small pea; at first
they are transparent, but in two or three days become opaque
and yellow, and in two or three days more desiccate, forming
brownish crusts.

Herpes præputialis appears on the external surface of the
prepuce, small inflamed spots being first noticed, which, in the
course of a few hours, are covered with groups of small globose
vesicles. It runs a similar course to that just noticed, but in
some cases continues to reappear for years, causing great annoyance to the patient.

Herpes zoster, or *shingles,* is usually the severest form of the
disease, being attended in many cases with marked febrile action.
It usually makes its appearance on the trunk in irregular
patches of a red-color, which are soon covered with vesicles; new
patches coming up, the disease may pass entirely round the body,
though Cazenave states that it never appears but upon one side
at a time. The vesicles resemble those already described, but
are sometimes larger; they usually disappear in four or five days,
leaving at some points thin, brown incrustations, which are soon
detached. The disease usually lasts for ten or fourteen days,
and sometimes longer.

Herpes circinatus, or *ringworm,* appears most frequently upon
the face, neck and arms, though it may come out on any portion
of the body. It comes out at first as a red spot about the size
of a dime on which shortly appear numerous small vesicles ar-
ranged in rings, hence the common name of ringworm; it is not
attended with constitutional disturbance, and generally disappears in ten or twelve days.

DIAGNOSIS.—The diagnosis of herpes is generally easy, the
vesicles being round, prominent, and grouped together on one
inflamed or red base; the symptoms of the different forms are
usually sufficiently marked for their easy distinction, as above
described.

It is generally thought that but little if any treatment is required in herpes, but many times this is a mistake, for the eruption is but a symption of a deeper disease, and if not looked after
it may leave permanent impairment of the health. In the first

form of the disease, the patient may have the usual doses of Aconite with Rhus, or Phytolacca, until the eruption has ceased and the crusts are thrown off. Sometimes a day's treatment will thus remove the disease when otherwise it might have lasted for ten days.

The same treatment may be adopted for herpes labialis, or Aconite with Phytolacca. But occasionally in both these forms there will be indications for sulphurous acid or sulphite of soda, and the general health will be much improved by their use. Herpes labialis is sometimes an unpleasant symptom. If persistent and associated with some pain in the chest, and slight cough, it indicates disturbance of all the vegetative functions, and eventually tuberculosis. Treatment cannot be commenced too early in such cases, and the patient should be carefully watched for some time.

Herpes præputialis is frequently cured with soap and water, a good castile soap, or Colgate's glycerine soap; sometimes we order a lotion of borax and rose water, salicylic acid with borax, or in some cases the application of the citrine ointment.

Shingles is treated with the special sedatives and such additional remedies as may be indicated. If there is nausea and prostration from non appearance or retrocession of the eruption we add Belladonna; if there is severe burning of the surface with nervous disturbance it is Rhus; with intense puritus Apis; with pallid tongue and soreness of the mouth and throat it will be Phytolacca; when the tongue is pallid and dirty, sulphite of soda is given; when it is red and dirty sulphurous acid.

Herpes circinatus is usually cured by the local application of a saturated solution of iodide of ammonium, or sulphurous acid. In some very persistent cases we will find internal remedies needed, and sulphurous acid, sulphite of soda, minute doses of arsenic, and cod liver oil may be suggested, as indicated.

SCABIES.

Scabies, or itch, though a vesicular disease, is produced by an animal parasite—the acarus scabiei—and hence, as this insect possesses a very tenacious vitality, the disease is rendered contagious by its transmission from one to another. The acarus is usually found a short distance from the vesicle, in a small furrow leading from it. With good sight or a magnifying glass, it can be seen as a small, round, grayish body, sometimes moving, sometimes at rest. Under the microscope, its body is seen to be oval, the back convex, and marked with curved lines, its head covered with fine hairs, and eight legs passing from its abdomen. The insect passes from one part to another, by burrowing under the epidermis, but is only conveyed to distant parts by the fingers, after scratching, and by the clothing.

SYMPTOMS.—Scabies almost always makes its first appearance between the fingers and front part of the wrist, in the form of small pointed vesicles, containing a clear, limpid fluid, and a a very fine line leading from it, and marking the situation of the acarus. An intense but pleasurable sensation of itching attends their appearance, and the patient can not resist the inclination to scratch or rub the part, though this sometimes gives rise to a sensation of smarting if too severe. As the disease progresses, the irritation of the skin by the nails usually produces suppuration in the vesicles, the result being the formation of larger or smaller scabs, and some inflammation and stiffness of the skin. In severe cases we occasionally see in the interspace between the fingers a large festering surface covered with thick scabs, and the hands so stiff and painful that they can hardly be used. Sometimes the itch is confined to the hands, but in others it is conveyed to the flexures of the joints, to the perineum, around the anus, and in fact wherever the skin is thin and delicate. In all these situations we may have the suppurative action above named, so that occasionally instead of a mild vesicular disease, the patient will be covered with foul, painful, ulcerating sores.

Itch never terminates spontaneously, but may last for years. In some cases it never passes the vesicular form first named, but in a majority, especially where cleanliness is neglected, it goes on to the formation of hard scales, and induration of the skin.

DIAGNOSIS.—The diagnosis of itch is generally not difficult, as the vesicles are pointed and solitary, while in eczema they are flattened, and in prurigo the eruption is first papular, as it is also in litchen, and in neither case does it appear between the fingers, the frequent seat of scabies. The sulcus passing from the visicle in itch is a good diagnostic feature, though not usually very well marked. In the severer stages of the disease, there would be difficulty in the diagnosis were it not for the constant reappearance of the disease in its original form.

TREATMENT.—The object of treatment is to destroy the itch insect, and whatever will accomplish this with the greatest certainty, and in the least time, will prove the best remedy. Sulphur has formed the basis of most applications, and is I believe the best remedy. We may use it in the form of ointment mixed with lard, or with an alkali, as—℞ Sulphur Sub. ℨij., Subcarbonate of Potash ℨj., Lard ℥viij; mix. Or, ℞ Prepared Chalk ℨiv., Sulphur, Tar, aa. ℥vj., Soft Soap, Lard, aa. ℥xvj.; mix. These ointments should be thoroughly applied to the parts affected. It is hardly sufficient to say, cleansed with soap and water, for it requires a thorough saturation of the affected part with soap, and then its removal with soft warm water (soft soap is best). The parts should be thoroughly dried before the ointment is applied, and it may be toasted in before a fire. If now the patient's clothing is entirely changed, and the old clothes boiled, we may expect an immediate cure; if not the process should be repeated.

I have used a combination of—℞ Sulphuret of Potassium, ℥ss., Oils of Rosemary and Lavender, aa. ℨj., Lard ℨvj., mix, and apply as before. Cazenave states that after repeated trials they determined that the two following formulæ yielded the most satisfactory results—℞ Essence of Peppermint, Rosemary, Lavender and Lemon, aa. gtt. iv. to gtt. vj., Alcohol, ℨjss., weak infusion of Thyme, Ovj.; it was freely used, and the cure resulted in eight days. ℞ Iodide of Sulphur, Iodide of Potassium, aa. ℨjss., water Oij.; the mean duration being six days. They say, whatever the lotion employed, it is necessary not only to wet the affected parts, but to prolong its application, so as to produce that kind of maceration which is required to destroy the insect. A solution of sulphuret of lime, ℨij. to the pint of water is very efficient, the cure being effected sometimes with three or four applications.

In the milder forms of the diseases no internal treatment is necessary, but the patient should be guarded against cold, dampness, and sudden changes of temperature, and change his entire under-clothing every day. In the more persistent cases, we may give equal parts of sulphur and cream of tartar, to the extent of keeping the bowels open, and in some cases where the

30

patient is cachectic, the bitter tonics and iron. I have cured the itch with a local application of the Phytolacca and Podophyllum, but I prefer the remedies first named.

ORDER III.—BULLÆ.

This order, it will be recollected, is characterized by the formation of large *blebs*, or blisters, from the size of a pea to a hen's egg, sometimes with and sometimes without redness of the skin. Properly speaking, there is but one variety, *pemphigus*, but some authors class *rupia* under this order. Both affections are usually chronic, and may appear in succession, on any part of the body.

We have no knowledge of their causes further than they are usually associated with a cachectic condition of the system.

PEMPHIGUS.

Pemphigus is almost always associated with general debility, and imperfect performance of the various functions of digestion, assimilation and secretion, though the person may seem to enjoy tolerably good health. It makes its appearance in the form of blebs, or blisters, from the size of a split pea to an inch or more in diameter, containing a thin transparent serum. They frequently increase in size for two or three days, the fluid becoming straw-colored when they are ruptured, and a thin brownish crust forms. Sometimes the surface heals at once, but at others these crusts are reproduced for several days or even weeks.

DIAGNOSIS.—The diagnosis is always easy when they first appear, as in no other skin disease do we see such a large elevation of the epidermis. When they have ruptured the diagnosis is more difficult, but it may usually be distinguished from other affections by the brown thin scab, and by the dark-red irregular spot when it is removed.

TREATMENT.—In the milder form of the disease, Tinct. Rhus gtt. v., Tinct. Phytolacca gtt. x., water ℥iv., in teaspoonful doses every two or three hours, will be sufficient to effect a cure. In cases where the face is full and purplish, Baptisia may replace the Rhus.

If the tongue is pallid and dirty, give sulphite of soda, grs. ij. to grs. v. every two hours; if red and dirty, sulphurous acid will be the remedy. If the disease persists, we may prescribe ℞ Fowler's Solution gtt. x., Tinct. Phytolacca, gtt. x., water ℥iv.; a teaspoonful every four hours. In some cases cod-liver oil will be of service; in others the hypophosphite of lime, compound syrup of the hypophosphites, or some preparation of malt after meals to aid digestion.

If the bullæ are large, and the surface painful when they rupture, it may be dressed with equal parts of lime-water and linseed oil, or powdered elm, flour or hydrastis may be sprinkled on it to absorb the discharges.

RUPIA.

This, like the preceding disease, is almost always associated with a cachectic condition of the system, and enfeebled vitality, and appears most frequently among the poor, destitute, and ill-fed, though occasionally when the patients have all the comforts and luxuries of life. Its only relation to the preceding disease, or to this order, is in its first appearance, and it soon loses this resemblance. It is always a chronic affection, lasting from two or three weeks to many months. Three varieties are distinguished: R. simplex, R. prominens, and R. escharotica.

SYMPTOMS.—*Rupia simplex* appears in the form of bullæ, about the size of a dime, round and flattened, and without evidence of inflammation. The contained fluid is at first a limpid serum, but it soon becomes opaque and purulent, and finally concretes, forming thick flat crusts, of a brownish color. These fall off in a few days, leaving a superficial ulcer of the skin, which soon cicatrizes, but a livid-red color remains for some time afterward.

Rupia prominens makes its appearance in a smilar manner, but the bullæ are frequently larger, and the ulceration deeper, and the scales thicker. Usually the skin is reddened, and sometimes there is a burning sensation and pain. The scab seems to grow, in many cases, by continued additions at the base, and becomes one-fourth or even half an inch in thickness, and conical, and resembles to some extent a snail's shell. When the scab is removed, a new one frequently takes its place, and they may be thus reformed for months. In some cases the ulcer is healed with difficulty, the edges being livid and tumefied, the center pale, and bleeding on slight pressure.

Rupia escharotica occurs most frequently in children up to two years of age. It commences with the appearance of slightly prominent livid patches, upon which irregular and flattened bullæ are soon formed ; when the bullæ break, ulcerated surfaces are left which secrete a disagreeable unhealthy pus. "The infant suffers from acute pain, and much fever and insomnolency. When the disease assumes an intense form, death may ensue in one or two weeks. When it does terminate favorably, the ulcerations are very long in healing." (Cazenave.)

DIAGNOSIS.—Rupia is diagnosed with ease, in most cases, by the prominent, conical, brown scabs, those of pemphigus being flat. Ecthyma resembles it most in some cases, and it will be difficult to distinguish between them in its later stages, but the hard and inflamed base, irregular scabs, and superficial excoriations, are usually sufficiently diagnostic.

TREATMENT.—The treatment in this disease, as in the preceding, should be strictly tonic, arrest of the skin disease depending to a great extent upon the restoration of the general health. I have obtained the best results from the administration of Fowler's Solution of Arsenic with Phytolacca as heretofore named. Cases will be found requiring Rhus, Apis, Corydalis and Alnus, and either of these may be associated or alternated with Arsenic. When the tongue is pallid and dirty, Sulphite of Soda may be given in the usual doses, and continued for some weeks. Occasionally Sulphurous Acid will be indicated. In rare cases Donovan's Solution of Arsenic will prove better than Fowler's. If there is much derangement of the stomach, we frequently derive benefit from an emetic, and in many cases, excretion needs to be stimulated with small doses of the alkaline diuretics.

Associated with these means, the patient should have a daily bath of salt and water, or in some cases of a decoction of Cornus or Hydrastis. The diet should be nutritious, and exercise should be taken in the open air.

When the local affection is very persistent, we may dress the ulcer with three parts of glycerine, and one of tincture of muriate of iron; or with the mild zinc ointment, black salve, or an ointment made of the inner bark of the elder. Sometimes a decoction of equal parts of Cornus, Alnus and Rumex, answers an excellent purpose, or the tinctures of the same agents may be used. Salicylic Acid with Borax (ʒj. of each to water Oj.) is a good dressing; or Thymol grs. ij, to grs. iv., to water ʒj., may be employed with good results. When the ulcers are very persistent, they may be cauterized with a saturated solution of chloride of zinc, or a paste made with this and Hydrastis; and after the slough is cast off, the part usually heals kindly with any simple dressing.

ORDER IV.—PUSTULÆ.

This order is distinguished by the formation of small elevations containing pus, and hence termed pustule They are almost invariably situated on an inflamed base, which usually precedes the eruption, though in some cases the inflammation comes on after the appearance of the eruption, and is more or less diffused. The diseases included under this order are both acute and chronic, two of them, *variola* and *vaccinia*, heretofore described, being eminently contagious, and one, *porrigo*, being propagated by contact. The others seem to depend upon some unknown internal cause.

ECTHYMA.

Ecthyma may be divided into the two forms, acute and chronic, the first occurring most frequently in children and young persons, the second in the adult, though sometimes in children.

SYMPTOMS.—In the acute form it is usually preceded by lassitude and indisposition, and its appearance is frequently marked with slight chills and febrile action. It makes its appearance in the shape of red, circumscribed, inflamed spots, which soon suppurate at their apices. In some cases the eruption is attended with pain, the inflammation being quite severe, but in others it is simply a sense of stiffness. Some of the pustules terminate by resolution, while others are succeeded by a thick, adherent scab, which, in falling off, leaves a deep-red mark, and in some cases a cicatrix. It usually lasts for one or two weeks.

In *chronic ecthyma* there is a successive appearance of the eruption, sometimes for months, the general health being much depressed. It may present the same character as that just described, or it may become confluent in large suppurating surfaces. A variety termed ecthyma cachecticum, occurs in old persons, and those who have broken their systems down by intemperance. "The skin is inflamed and more swollen than in the common forms of the disease. It assumes a deepened color, and in about six or eight days the cuticle is raised over the pustules, is blackish and infiltrated with blood. It soon bursts and forms a thick, dark scab, raised at the center; the edges are hard, callous, and more or less inflamed. The scabs are very adherent and do not become detached for several weeks, sometimes for months. If they fall accidentally, an unhealthy ulceration ensues, and the scab is with difficulty removed. Sometimes febrile symptoms precede or accompany the eruption, but they generally disappear with the disease."—(Cazenave.)

DIAGNOSIS.—Ecthyma is usually recognized with ease by the hard and inflamed base, suppuration commencing on the surface, and not deep as in furunculi, acne, and sycosis, which are most frequently mistaken for it, but in these the base is hard, not inflamed, and the pustules are small and slowly developed.

TREATMENT.—In the acute form of the disease we give the special sedative, Aconite or Veratrum, using the warm bath with glycerine soap. In some cases Rhus will be indicated, in others Apis, and in still others Phytolacca. If the tongue is broad and pallid, sulphite of soda will be found an admirable remedy, and will sometimes effect a speedy cure. If the tongue is red, with brownish coat down the center, sulphurous acid will be given.

In the chronic form of the disease, the patient may have Fowler's Solution, with Phytolacca, or cod-liver oil; or, if congenital syphilis is suspected, Donovan's Solution, or iodide of potassium is indicated. In some cases the old combination of aa. Alnus, Rumex, and Quercus, in infusion, may be given internally, and used as a local application.

If the eruption is very perverse, a local application of brown citrine ointment one part, with cod oil or simple cerate two or three parts, may be used. It needs to be used with care, but it is very certain in its action.

In some cases a dry dressing is preferable, and the part is dusted with subnitrate of bismuth.

PORRIGO.

Porrigo, or *tinea*, is a disease of the scalp, and is generally known by the name of *scald-head*. It is undoubtedly contagious, and is propagated from one to another by contact; hence the necessity for care in the use of articles of clothing, combs, baushes, towels, etc. Two varieties are distinguished: P. favosa, or tinea capitis, and P. scutulata, or tineà annularis.

Porrigo favosa commences with an eruption of minute, round, yellow pustules, which seem to be imbedded in the skin. At first they are distinct, and situated on a hard base, but as the disease progresses they become confluent, the entire scalp being inflamed or indurated. In a short time after their formation, the yellowish fluid begins to concrete, and when they are distinct forms a scab with a marked depression in the center, but when close together they form one large scab. If this is allowed to remain, it becomes thick, whitish, and brittle; if removed, slight erosions are seen under it, and it is not re-formed, except by the appearance of a new crop of pustules.

"This affection is never accompanied with febrile symptoms, but a troublesome and annoying itching is often present during its progress, which is aggravated by want of cleanliness. A number of lice are often seen under the scabs, causing the patients to scratch themselves, and by this means increase the inflammation. In these cases there is a strong, disagreeable odor, similar to that of cat's urine, given off from the head. After the head is cleansed from the scabs the odor becomes sickening. The excoriations on the surface, which often reach to the hair-bulbs, and thus produce baldness, are not covered with the regular cup-shaped favus pustules, but a reddish and fetid sanies oozes out, which concretes into irregular-shaped scabs. Fresh pustules, however, soon appear, which give rise to fresh favus scabs. Small subcutaneous abscesses may sometimes appear, accompanied with sympathetic engorgement of the lymphatic glands of the neck. It has been remarked that the growth of those persons who have been affected with porrigo is often arrested, and the development of the mental as well as the physical powers is slow and imperfect. The duration of the disease is very variable and uncertain; and the hair, when reproduced, is rarely the same as the original either in color or consistence." (Cazenave.)

Porrigo scutulata commences with the appearance of red circular patches, upon which small yellow pustules are soon developed. Each pustule has a hair passing through it, and has the same cupped appearance as in the preceding variety; and they appear more frequently upon the circumference of the spot than at its center. The scabs increase in thickness for some time, and when removed a large furfuraceous patch with an uneven surface is left, from which the hair frequently falls off. It spreads by spontaneous development, or by inoculation of other parts by scratching; marked and sometimes intense itching attending the eruption. Like the preceding affection, its duration is variable, but if allowed to run its course it would probably continue for years, resulting in permanent baldness.

DIAGNOSIS.—The presence of the small, rounded, yellow pustule, depressed in its center, is the diagnostic feature of both forms. Porrigo scutulata is determined by the appearance of the eruption in circular patches, though when these are numerous, they are so crowded together as to cover the entire surface, and the distinction then between this and porrigo favosa can not be made out.

TREATMENT.—Cleanliness is of great importance in this affection, and to secure it we would have the hair cut close, and the head frequently washed with castile soap and water. It may be necessary at first to soften the incrustations by continuous emollient applications, or in some cases with poultices, using soap and water freely in the meantime. Having thus exposed the scalp we would apply—℞ Oxalic Acid, grs. x. to grs. xx., Creosote gr. x. , water ℥ij.; M., and follow it in half an hour with free inunction of mild zinc ointment. The ointment of iodide of sulphur is a very efficient remedy, and should when used be gently rubbed over the parts night and morning, the scalp being kept perfectly clean by the use of soap and water. A nicer application and a more effectual one will be found in brown citrine ointment one part, simple cerate two parts. The head may be thoroughly cleansed with a solution of salicylic acid and borax (*aa.* ℥j. to water Oj.,) and the ointment then carefully applied.

As regards internal remedies, we will find it necessary to give, the vegetable alteratives heretofore named, associated with some preparation of potash, as the iodide, acetate, carbonate, etc. Sulphite of soda is an admirable remedy in some cases, and sulphurous acid is occasionally indicated. Usually the bitter tonics and iron will be required to some extent, and occasionally cod-liver oil will prove beneficial.

ORDER V.—PAPULÆ.

This order is characterized by small, firm, solid elevations of the skin, always attended with more or less itching, and never contain pus or serum, though occasionally from irritation these surfaces become ulcerated and covered with incrustations. They are developed without any appreciable cause, are rarely attended with febrile symptoms, and are not contagious. They are most generally chronic, but sometimes acute. Two diseases are in cluded under this order—lichen and prurigo.

LICHEN.

Lichen appears as small, hard elevations, but slightly red, or of the color of the skin, and attended with severe pruritus. We may distinguish three forms : L. simplex, L. agrius, L. urticatus.

Lichen simplex comes out in the form of small and aggregated papulæ, being attended with severe itching, and sometimes burning. It most frequently appears on the face and arms, and the neck and breast, though it may extend to all parts of the body. They remain stationary for three or four days ; when the redness gradually declines, there is slight furfuraceous desquamation, and the disease terminates in seven or eight days, unless there is a new eruption. In many cases it continues for weeks or months by the appearance of successive crops of papulæ.

Lichen urticatus usually appears suddenly in the form of large and numerous papulæ, attended with a burning, distressing pruritus. It appears most frequently on the face, neck, and extremities, and is irregular and transitory, subsiding and reappearing with great rapidity. "The papulæ are clustered, and they are either white or surrounded by a faint-red areola ; sometimes they are prominent, and considerably inflamed, and at first bear considerable resemblance to flea-bites." When scratched or otherwise irritated, they frequently bleed, and dark scabs form on their surface. The eruption may disappear with one crop of papulæ, but it is occasionally very obstinate, lasting for months, by their successive reproduction.

Lichen agrius may appear spontaneously, or it may succeed lichen simplex. When it appears spontaneously, the papulæ are very small, red, accuminated, inflamed and developed on an erythematous surface of limited extent, which is generally attended with heat and painful tension. Instead of subsiding on the fourth or fifth day, they continue increasing; slight ulcerations form on their apices, whence issues a sero-purulent fluid, which concretes and forms yellowish, prominent crusts, soft and slightly adherent. These incrustations fall off, and are then replaced by thin scaly scabs. Sometimes the redness diminishes, the inflammation disappears, slight desquamation ensues, and the disease terminates about the twelfth or fifteenth day. But frequently the discharge continues, and new crusts are formed, by which the disease is prolonged considerably. The itching which accompanies it is often so intense that the patient seeks the hardest substances to rub himself with, and thus invariably aggravates the pruritus. It may continue in this manner for several weeks, or it may pass into the chronic state, when the scaly incrustations disappear, and are succeeded by slight exfoliation, and the skin is often considerably hypertrophied. This form may last for months.—(Cazenave.)

A peculiar form of disease has prevailed extensively in the Western country, for the last twenty years, known as *Illinois itch, soldiers' itch*, etc., presenting many of the characteristics above named. Its symptoms seem so variable, that it is difficult to classify it, as it sometimes resembles eczema, and at others impetigo, and in others, again, it presents to some extent the characteristics of all three. It appears most generally upon the wrists and hands first, and then extends to various parts of the body, and is remarkably persistent and annoying.

Diagnosis.—The diagnosis of lichen is very difficult, as it may be mistaken for eczema, porrigo, scabies, or impetigo, but it may usually be determined by the presence of some of the characteristic papulæ.

Treatment.—*Lichen agrius* is more difficult to manage, and no remedy seems to answer in all cases. In some, I have had very good success with glycerine and tincture of muriate of iron in the proportion three parts of the first to one of the last, given internally in teaspoonful doses four times a day. A lotion of muriate of ammonia has been frequently employed, composed of—℞ Hydrochlorate of Ammonia ℥j., vinegar ℥iv., water Oj., and applied freely to the affected parts. In some cases the internal administration of the compound tincture of Corydalis, with iodide of potassium, and a wash of a decoction of equal parts of Cornus, Alnus, and Rumex, has answered a good purpose. In other cases a lotion of—℞ Glycerine ℥ij.; Oxide of Zinc, ℥ss., Morphia grs. v.; Rose-water ℥iv., has answered an excellent purpose, as has the ointment of elder and the mild zinc ointment. In other cases, good results will be obtained by the internal use of sulphur, and its local employment as a bath, wash, or ointment.

PRURIGO.

Several varieties of this disease are described, but many of them are named, not from any prominent difference of symptoms but more on account of their location. The disease is characterized by the appearance of papulæ, usually larger than those of lichen, and without discoloration of the skin, which are attended by very severe pruritus, and sometimes burning. Three varieties may be named, P. mitis, P. formicans, P. senilis.

Prurigo mitis is the mildest form of the disease, and is usually acute. The papulæ are slightly prominent, but very small, and accompanied with intense itching. In prurigo formicans the papulæ are much larger, and flattened, and distinct, and accompanied with an intolerable pruritus, which increases at night, and by the warmth of the bed. If not irritated by scratching, they frequently disappear in the course of one or two weeks, but frequently the skin is torn in the efforts for relief, and the part bleeds, and a dark thin scab is formed on its surface. It may continue for a considerable time by continued development of the eruption. In old people, or in weakly children, the papulæ are frequently large and prominent, and the skin becomes thickened and inflamed; vesicles, pustules, and boils form, and being opened by scratching give rise to unpleasant excoriations and superficial ulcers, and a most intense burning and itching. It may thus last for months, or even years. Prurigo may attack any part of the body, but is most severe when it attacks the genitial organs, or is situated around the anus.

DIAGNOSIS.—Prurigo may be distinguished from *lichen* by its larger papulæ, and the dark incrustations which are sometimes formed on them; from *scabies* by the accuminated vesicles of the latter, and their rose-colored base. It may be associated with lichen, scabies, eczema, impetigo and ecthyma, and in such cases the diagnosis will of course be difficult.

TREATMENT.—In the milder forms of the disease, the removal of any internal irritation, and soothing local applications, are all that is required. Frequently it is desirable to keep the bowels open with a saline purgative, and give an alkaline diuretic, with some gentle diaphoretic. As a local application, the glycerine lotion will answer a very good purpose; or we may use it with chloroform, adding ten or fifteen drops of it to each ounce of the lotion. A solution of borax answers a very good purpose, as—℞ Borax ℨij., Morphia gr. v., Rose-water ℨvj. A decoction of Hydrastis or Cornus, with borax and morphia, is frequently beneficial.

NÆVUS MATERNI.

It is generally supposed, and I think with truth, that certain impressions made upon the mother's mind during pregnancy will affect the growth and structure of the child, and in some manner deform it. It is true that we can not account in a rational manner for any such occurrence, but the instances are so numerous that we can not dispute the connection between the impression and the mark.

Very frequently we find women making anxious inquiry with regard to the matter, as they are strongly impressed that *they* have received some shock or sudden impression which will mark or deform the child. As it is so rarely that nævi result from these impressions, we are safe in giving such assurance to the prospective mother as will quiet her fears.

Numerous instances are related to prove the relation between the impression on the mother's mind and the deformity of the child. The severest case of the kind that ever came under my observation was a child born with a hand so completely deformed as to be useless, and which was attributed to the mother's witnessing the dressing of a hand that had been crushed in a threshing machine. In another case the child had a vascular nævus on the cheek, immediately below the eye. The mother attributed the mark to her husband's throwing a cherry at her, which, striking upon the same part of the cheek, burst, and not only startled her at the time, but firmly impressed her mind that the child would be marked.

The deformities and distortions, that are rarely met with, may frequently be remedied by the surgeon, as in talipes, hair-lip, cleft-palate, etc., but we will not attempt to describe them here.

The most common form of mother's mark is a discoloration of the skin, from an increase in the size of the blood-vessels. It may be located upon any part, and usually increases in size as the child grows. In some cases it is so large at birth that it is useless to interfere, as where it involves one-fourth, or sometimes one-half of the face. In other cases it is but the size of a five or ten cent piece, and frequently not larger than a grain of wheat.

There is much difference in the rapidity of growth. In some it is very slow, so that the spot will not have doubled in size to adult years; in others it grows rapidly, and will have doubled its size the first month. In the first operative interference is not so necessary, and may be postponed; in the second the nævus should be removed as early as possible.

The vascularity varies in different cases. In some it is a true *anastomotic aneurism*, occasionally distinctly pulsating. In these much care should be used to prevent hemorrhage in any operation that may be undertaken.

In others the disease is wholly confined to the capillaries, and the operation outside of the spot will not cause greater hemorrhage than for any other purpose.

The bright vivid coloration is met with in those cases where the arterial capillaries are involved; the purplish discoloration where the veins are dilated.

We rarely meet with a case upon the skin where the growth is such as to form a red tumor, projecting from the skin. But when it is at the junction of the skin and mucous membrane, as on the lip, or wholly of the mucous membrane, as sometimes of the tongue or mouth, and occasionally about the reproductive organs, we frequently find it assuming the form of an erectile tumor. Occasionally the structure gives way in such cases, and there is profuse hemorrhage.

TREATMENT.—A great many plans of treatment have been recommended for nævus. The most commonly adopted are by excision, strangulation by ligature, and removal with escharotics.

Where the growth is not markedly vascular, the smaller capillaries being alone involved, the best plan is to excise the growth. Where the tissues are loose, or when the growth is oblong or oval in form, the edges may be drawn together with silver sutures, and union obtained without any scar.

If the surgeon fears hemorrhage, he inserts his ligatures before operating. The needle is carried through the sound skin, and beneath the nævus, drawing the wire through and clipping it at proper length. Enough are inserted in this way to insure an arrest of bleeding.

Strangulation by ligature is performed by transfixing the base of the growth one or more times, and tying it in separate parts so tightly as to cut off the circulation.

Probably nitric acid is the best caustic for the removal of nævi. Sir B. Brodie remarked : " Caustics may be used with advantage in congenital tumors, nævi, etc. Little vascular spots on children's faces are an object of anxiety. If you look at these, you will see one large vessel and several branches supplying them. You may destroy them in the following manner : Take a glass pen [a pine stick pointed answers the same purpose] which will hold nitric acid, and apply it to the principal vessel, puncture it; and insert into the puncture a fine point of potassa fusa ; a moment's touch will be sufficient to destroy the vessel; if the potassa extends further than you intended, apply vinegar. You may thus obliterate the vessel without leaving a scar. There are some congenital nævi abounding in the skin, formed by an intricate mesh of vessels; the skin is elevated and of a mulberry color. If these are of a large size, they must be destroyed by ligature or the knife; if of smaller size you may use caustics not unprofitably. The nitric acid is the best application; this makes a slough, the blood coagulates and the parts become indurated. This is only applied when nævi are of small size. In subcutaneous nævi, which are not of the same color but purple, caustics may be applied to effect their destruction, whether of a large or small size; the great object is to destroy them with caustics rather than the ligature. These nævi have been cured by application of vaccine matter, which acts by producing a slough. You may cure these subcutaneous nævi upon the same principle; puncture them with a finely-pointed lancet, then having a probe armed with nitrate of silver, introduce it into the puncture—the caustic presently causes sloughing, and the vessels are obliterated."

Lately the removal of nævi by the use of hypodermic injections into the growth, has been strongly recommended. The solution of per-sulphate of iron has been employed in this way, and it is claimed with great success. One fatal case, however, has been reported, in which the death was evidently due to the injection.

When the nævus is quite superficial, its removal may be attempted by the use of a strong tincture of iodine. A very good formula is—℞ Tincture of Iodine (to saturation), Aqua Ammonia, *aa.* ℥j.; let it stand in a well-corked bottle for a week, and it is ready for use. Apply this daily with a camel's-hair brush, and the nævus will gradually yield.

INDEX.

www.ingramcontent.com/pod-product-compliance
Lightning Source LLC
Chambersburg PA
CBHW020900210326
41598CB00018B/1733